AF782848

European Studies in Philosophy of Science

Volume 4

Series editors
Dennis Dieks, Institute for History & Foundations of Science, Utrecht University, The Netherlands
Maria Carla Galavotti, Università di Bologna, Italy
Wenceslao J. Gonzalez, University of A Coruña, Spain

Editorial Board
Daniel Andler, University of Paris-Sorbonne, France
Theodore Arabatzis, University of Athens, Greece
Diderik Batens, Ghent University, Belgium
Michael Esfeld, University of Lausanne, Switzerland
Jan Faye, University of Copenhagen, Denmark
Olav Gjelsvik, University of Oslo, Norway
Stephan Hartmann, University of Munich, Germany
Gurol Irzik, Sabancı University, Turkey
Ladislav Kvasz, Charles University, Czech Republic
Adrian Miroiu, National School of Political Science and Public Administration, Romania
Elizabeth Nemeth, University of Vienna, Austria
Ilkka Niiniluoto, University of Helsinki, Finland
Samir Okasha, University of Bristol, UK
Katarzyna Paprzycka, University of Warsaw, Poland
Tomasz Placek, Jagiellonian University, Poland
Demetris Portides, University of Cyprus, Cyprus
Wlodek Rabinowicz, Lund University, Sweden
Miklos Redei, London School of Economics, UK
Friedrich Stadler, University of Vienna, Austria
Gereon Wolters, University of Konstanz, Germany

This new series results from the synergy of EPSA - European Philosophy of Science Association - and PSE - Philosophy of Science in a European Perspective: ESF Networking Programme (2008–2013). It continues the aims of the Springer series "The Philosophy of Science in a European Perspective" and is meant to give a new impetus to European research in the philosophy of science. The main purpose of the series is to provide a publication platform to young researchers working in Europe, who will thus be encouraged to publish in English and make their work internationally known and available. In addition, the series will host the EPSA conference proceedings, selected papers coming from workshops, edited volumes on specific issues in the philosophy of science, monographs and outstanding Ph.D. dissertations. There will be a special emphasis on philosophy of science originating from Europe. In all cases there will be a commitment to high standards of quality. The Editors will be assisted by an Editorial Board of renowned scholars, who will advise on the selection of manuscripts to be considered for publication.

More information about this series at http://www.springer.com/series/13909

Carlo Cellucci

Rethinking Knowledge

The Heuristic View

Springer

Carlo Cellucci
Department of Philosophy
Sapienza University of Rome
Rome, Italy

ISSN 2365-4228 ISSN 2365-4236 (electronic)
European Studies in Philosophy of Science
ISBN 978-3-319-53236-3 ISBN 978-3-319-53237-0 (eBook)
DOI 10.1007/978-3-319-53237-0

Library of Congress Control Number: 2017934078

© Springer International Publishing AG 2017
This work is subject to copyright. All rights are reserved by the Publisher, whether the whole or part of the material is concerned, specifically the rights of translation, reprinting, reuse of illustrations, recitation, broadcasting, reproduction on microfilms or in any other physical way, and transmission or information storage and retrieval, electronic adaptation, computer software, or by similar or dissimilar methodology now known or hereafter developed.
The use of general descriptive names, registered names, trademarks, service marks, etc. in this publication does not imply, even in the absence of a specific statement, that such names are exempt from the relevant protective laws and regulations and therefore free for general use.
The publisher, the authors and the editors are safe to assume that the advice and information in this book are believed to be true and accurate at the date of publication. Neither the publisher nor the authors or the editors give a warranty, express or implied, with respect to the material contained herein or for any errors or omissions that may have been made. The publisher remains neutral with regard to jurisdictional claims in published maps and institutional affiliations.

Printed on acid-free paper

This Springer imprint is published by Springer Nature
The registered company is Springer International Publishing AG
The registered company address is: Gewerbestrasse 11, 6330 Cham, Switzerland

Preface

From its very beginning, philosophy has been viewed as aimed at knowledge and methods to acquire knowledge. In the past century, however, this view of philosophy has been generally abandoned, with the argument that, unlike the sciences, philosophy does not rely on experiments or observation but only on thought.

The abandonment of the view that philosophy aims at knowledge and methods to acquire knowledge has contributed to the increasing irrelevance of the subject, so much so that several scientists, and even some philosophers, have concluded that philosophy is dead and has dissolved into the sciences. The question then arises whether philosophy can still be fruitful, and what kind of philosophy can be such.

In order to answer this question, this book attempts to revive the view that philosophy aims at knowledge and methods to acquire knowledge. Reviving it requires a rethinking of knowledge. The importance of such rethinking depends on the central role knowledge plays in human life. In particular, a rethinking of knowledge requires a rethinking of mathematical knowledge, which raises special problems.

Rome, Italy

Carlo Cellucci

Acknowledgments

I first presented some of the ideas on knowledge developed in this book in "Rethinking Knowledge," *Metaphilosophy* 46 (2015), pp. 213–234, and some of the ideas on mathematical knowledge developed in this book in *Filosofia e Matematica*, Rome, Laterza, 2002, and in "Philosophy of Mathematics: Making a Fresh Start," *Studies in History and Philosophy of Science* 44 (2013): 32–42.

In writing this book, I have drawn from a number of earlier essays, although in most cases I have substantially rewritten the material in question. I thank Springer for the permission to draw from this material. In Chap. 2, I use material from "Rethinking Philosophy," *Philosophia* 42 (2014), pp. 271–288; in Chap. 5, material from "Is Philosophy a Humanistic Discipline?," *Philosophia* 43 (2015), pp. 259–269; in Chaps. 8 and 9, material from "Knowledge, Truth and Plausibility," *Axiomathes* 24 (2014), pp. 517–532; in Chap. 13, material from "Models of Science and Models in Science," in *Models and Inferences in Science*, ed. Emiliano Ippoliti, Fabio Sterpetti, and Thomas Nickles, Springer, Cham 2016, pp. 95–112; in Chap. 20, material from "Is Mathematics Problem Solving or Theorem Proving?," *Foundations of Science* 22 (2017), pp. 183–199; and in Chap. 23, material from "Mathematical Beauty, Understanding, and Discovery," *Foundations of Science* 20 (2015), pp. 339–355.

I am grateful to many people for the help they gave me, either reading a chapter and making remarks, raising questions in correspondence, or making comments at seminars or conferences. For their help in whatever form, I am especially indebted to Atocha Aliseda, Arthur Bierman, Angela Breitenbach, Mirella Capozzi, Anjan Chakravartty, Riccardo Chiaradonna, Tom Clark, Pascal Engel, Michèle Friend, Roger Frye, Maria Carla Galavotti, Donald Gillies, Norma Goethe, Emily Grosholz, Robert Hanna, Reuben Hersh, Hansmichael Hohenegger, Danielle Macbeth, David B. Martens, Nicholas Maxwell, Thaddeus Metz, Dan Nesher, Thomas Nickles, Aaron Preston, Nicholas Rescher, Howard Sankey, Stephen P. Schwartz, Hourya Benis Sinaceur, Nathalie Sinclair, Fabio Sterpetti, Robert Thomas, Francesco Verde, Alan White, Jan Woleński, and Semir Zeki. This does not mean that they share the views expressed in this book or are in any way responsible for any remaining inaccuracies.

I am also grateful to an anonymous referee for useful comments and suggestions and to Arlette Dupuis for reading the manuscript and suggesting several linguistic improvements.

Contents

Part II The Nature of Knowledge

Chapter 1
Introduction

Abstract By invading many areas traditionally covered by philosophy, modern science has made philosophy increasingly irrelevant, and has presented it with the challenge to legitimate itself. This challenge has not been successfully met, so much so that several scientists, and even some philosophers, have concluded that philosophy is dead and has dissolved into the sciences. The question then arises whether philosophy can still be fruitful, and what kind of philosophy can be such. This book attempts to give an answer to this question, reviving the ancient view that philosophy is the acquiring of knowledge. This involves rethinking knowledge and methods to acquire knowledge, including mathematical knowledge, which raises special problems. In addition to describing the aim of the book, the introduction briefly describes the parts and the individual chapters of the book, and outlines some conventions adopted in the book.

1.1 Philosophy and the Birth of Modern Science

"One ought either to philosophize or say goodbye to life and depart hence, because," as compared with philosophy, "all other things seem to be a lot of nonsense and foolishness" (Aristotle, *Protrepticus* Düring, 110).

Is this praise of philosophy even conceivable today? The question is justified, because in the seventeenth century philosophy suffered a trauma from which it has not recovered yet, the birth of modern science. The latter has invaded many areas traditionally covered by philosophy. As a result, the role of philosophy has become problematic, and philosophy has come to need legitimation in the face of science.

Thus Gadamer states that, "since the seventeenth century," what "we today call philosophy is found to be in a changed situation. It has come to need legitimation in the face of science in a way that had never been true before" (Gadamer 1998, 6).

In fact, since the seventeenth century, a great deal of philosophy has been an attempt to provide an answer to the trauma caused by the birth of modern science. There have been both radical answers and moderate answers.

© Springer International Publishing AG 2017

C. Cellucci, *Rethinking Knowledge*, European Studies in Philosophy of Science 4,
DOI 10.1007/978-3-319-53237-0_1

1.2 Radical Answers

A radical answer is that, with the birth of modern science, philosophy has nothing left to speak of, because all questions about which one can say anything sensible belong to the sciences.

Thus, Wittgenstein states that "the correct method in philosophy would really be the following: to say nothing except what can be said, i.e. propositions of natural science – i.e. something that has nothing to do with philosophy" (Wittgenstein 2002, 6.53). For, "philosophy is not one of the natural sciences" (ibid., 4.111).

This answer, however, is unsatisfactory, because it assumes that the whole field of knowledge is exhausted by the present sciences. Such assumption is unjustified because, as research proceeds, new questions arise which are essentially beyond the present sciences, and may even give rise to new sciences.

Another radical answer, but opposite to the previous one, is that the sciences do not give knowledge, only philosophy provides access to knowledge.

Thus, Heidegger states that "science does not think" (Heidegger 1968, 8). It "is the disavowal of all knowledge of truth" (Heidegger 1994, 5). Therefore, "no one who knows will envy scientists – the most miserable slaves of modern times" (ibid., 6). Science aims at exactness and security but, with its "insistence on what is demonstrable," it blocks "the way to what-is" (Heidegger 1972, 72). Only philosophy opens the way to what-is, because "philosophy is the knowledge of the essence of things" (Heidegger 1994, 29).

This answer, however, is unsatisfactory, because the assumption that only philosophy opens the way to what-is has no foundation. There is no special source of knowledge which is available to philosophy but not to science.

1.3 Moderate Answers

A moderate answer is that we must admit that philosophy is not yet a science. While no one would doubt the objective character of mathematics and natural sciences, the same cannot be said of philosophy. A revolution in philosophy is necessary if philosophy is to acquire the character of a rigorous science.

Thus, Husserl states that philosophy is still "incapable of giving itself the form of actual science" (Husserl 2002, 250). It "does not merely have a doctrinal system at its disposal that is incomplete and imperfect in one respect or another, but has none whatsoever. Anything and everything is controversial here, every position-taking is a matter of individual conviction'" (ibid., 251). But, "the highest interests of human culture demand the elaboration of a rigorously scientific philosophy" (ibid., 253). So, "if a philosophical revolution is to prove itself in our time, it must always be animated by the intention to found philosophy anew in the sense of rigorous science" (ibid.).

This answer, however, is unsatisfactory, because philosophy has no specific field of its own, therefore all attempts to give philosophy the character of a rigorous science have been unsuccessful. Husserl wants to develop a philosophy that is a "universal science of the world, universal, definitive knowledge, the universe of truths in themselves about the world, the world in itself" (Husserl 1970, 335). He aims at "the discovery of the necessary concrete manner of being of absolute subjectivity" in a "life of constant 'world-constitution'" (ibid., 340). He also aims at "the new discovery, correlative to this, of the 'existing world'," which "results in a new meaning for what, in the earlier stages, was called world" (ibid.). His attempt, however, fails because, starting from absolute subjectivity, Husserl arrives at the discovery, not of the existing world, but only of a world as a correlative of subjectivity.

Another moderate answer, but opposite to the previous one, is that we must abandon the idea that science is the paradigmatic human activity, and hence philosophy should try to become a science. Science has no privileged position with respect to philosophy, both science and philosophy must be evaluated in terms of their capacity to achieve the aims we would like to achieve through them.

Thus, Rorty criticizes the logical empiricists who maintained that "science was the paradigmatic human activity," and "what little there was to say about other areas of culture amounted to a wistful hope that some of them (e.g., philosophy) might themselves become more 'scientific'" (Rorty 1991–2007, I, 46). Science has no privileged position with respect to philosophy. Both science and philosophy must be evaluated in terms of their capacity "to be reliable guides to getting what we want" (Rorty 1999, 33).

This answer, however, is unsatisfactory, because Rorty states that we philosophers "are not here to provide principles of foundations or deep theoretical diagnoses, or a synoptic vision" (ibid., 19). When we are asked what we "take contemporary philosophy's 'mission' or 'task' to be," the best we "can do is to stammer that we philosophy professors are people who have a certain familiarity with a certain intellectual tradition," and "can offer some advice about what will happen when you try to combine or to separate certain ideas, on the basis of our knowledge of the results of past experiments" (ibid., 19–20). But "we are not the people to come to if you want confirmation that the things you love with all your heart are central to the structure of the universe" (ibid., 20). That is, we philosophers are not the people to come to if you want to have an answer to the questions you are most interested in. This makes philosophy into a marginal activity with little utility.

1.4 Death of Philosophy?

The inadequacy of these answers to the trauma caused by the birth of modern science, makes one doubt that trying to legitimate philosophy is a feasible enterprise. This doubt is strengthened by the fact that, in the past century, some philosophers have affirmed that philosophy is dead and has dissolved into the sciences.

Thus, Quine states that philosophy, and specifically epistemology, or theory of knowledge, has dissolved into psychology, because "the stimulation of his sensory receptors is all the evidence anybody has had to go on, ultimately, in arriving at his picture of the world. Why not just see how this construction really proceeds? Why not settle for psychology?" (Quine 1969, 75). The "old epistemology aspired to contain, in a sense, natural science; it would construct it from sense data. Epistemology in its new setting, conversely, is contained in natural science, as a chapter of psychology" (ibid., 83). Epistemology "simply falls into place as a chapter of psychology and hence of natural science" (ibid., 82).

Heidegger states that "the sciences are now taking over as their own task what philosophy in the course of its history tried to present in part, and even there only inadequately, that is, the ontologies of the various regions of beings (nature, history, law, art)" (Heidegger 1972, 58). The "development of philosophy into the independent sciences" is "the legitimate completion of philosophy. Philosophy is ending in the present age. It has found its place in the scientific attitude of socially active humanity" (ibid.). The "end of philosophy means the completion" of philosophy, where "however, completion does not mean perfection as a consequence of which philosophy would have to have attained the highest perfection at its end" (ibid., 56). On the contrary, completion means "the end of philosophy in the sense of its complete dissolution into the sciences" (Heidegger 1998, 259, footnote).

But, if philosophy is dead and has dissolved into the sciences, then it is impossible to try to legitimate it in the face of science.

1.5 Criticisms by Scientists

That philosophy is dead and has dissolved into the sciences is also the opinion of many scientists.

Thus Hawking states that questions such as "What is the nature of reality? Where did all this come from? Did the universe need a creator?" are traditionally "questions for philosophy, but philosophy is dead. Philosophy has not kept up with modern developments in science, particularly physics. Scientists have become the bearers of the torch of discovery in our quest for knowledge" (Hawking and Mlodinow 2010, 5).

Dyson states that, "compared with the giants of the past," the present philosophers "are a sorry bunch of dwarfs," which compels us to ask: "When and why did philosophy lose its bite? How did it become a toothless relic of past glories?" (Dyson 2012).

Krauss states that "science progresses and philosophy doesn't," and "the worst part of philosophy is the philosophy of science; the only people" who "read work by philosophers of science are other philosophers of science. It has no impact on physics what so ever," so "it's really hard to understand what justifies it" (Andersen 2012).

Therefore, Pinker asserts that "philosophy today gets no respect. Many scientists use the term as a synonym for effete speculation" (Pinker 2002, 11). What a difference between this spiteful attitude toward philosophy, and the appreciative attitude of Galileo, who stated that he had "studied for a greater number of years in philosophy than months in pure mathematics" (Galilei 1968, X, 353).

1.6 Why Still Philosophy?

Since some philosophers and many scientists have affirmed that philosophy is dead and has dissolved into the sciences, we must ask: Why still philosophy? Can philosophy still be fruitful, and what kind of philosophy can be such? In particular, what kind of philosophy can be legitimized in the face of science? Asking these questions is nothing really new, because philosophy has always called into question everything, including itself. But, with the birth of modern science, such questions have become more pressing, as well as more difficult and embarrassing.

Some philosophers, however, scorn these questions. Thus, Popper states that a philosopher "should try to solve philosophical problems, rather than talk about philosophy" (Popper 1974, 68).

Rorty states that questions about "the nature of philosophical problems" are "likely to prove unprofitable" (Rorty 1992, 374).

Williams states that "philosophy is not at its most interesting when it is talking about itself" (Williams 2006, 169).

But this amounts to taking for granted that philosophy can still be fruitful. This is unjustified because, as we have seen, it contrasts with the opinion of some philosophers and many scientists. This makes it necessary to call philosophy into question.

1.7 Aim of the Book

This book aims to give an answer to the question whether philosophy can still be fruitful, and what kind of philosophy can be such. Briefly, its answer is that philosophy can still be fruitful only if it aims at knowledge and methods to acquire knowledge, because knowledge plays a central role in human life. To a large extent, we are what we know, we reflect reality, and reality is for us what we have access to and we know. Generally our aspirations, desires and hopes essentially depend on what we know.

The view that philosophy aims at knowledge and methods to acquire knowledge may be called the heuristic view, because 'heuristic' is said of methods which guide to acquire knowledge. Developing the heuristic view of philosophy requires a rethinking of logic and a rethinking of knowledge. A rethinking of logic has been carried out in Cellucci 2013a. A rethinking of knowledge is the aim of this book. There are, however, some minor overlappings with Cellucci 2013a, which are motivated by the desire to make the book self-contained.

The questions about knowledge discussed in this book do not exhaust all questions about knowledge. Every investigation is a potentially infinite task, and this book is no exception. However, the questions discussed in this book are essential for the development of a fruitful philosophy.

1.8 Organization of the Book

In order to highlight the organization of the book, the text is divided into five parts, after the present Introduction which occurs as Chapter 1.

Part I examines the nature of philosophy. Chapter 2 presents the heuristic view, according to which philosophy aims at knowledge and methods to acquire knowledge. Chapter 3 discusses the foundationalist view, according to which philosophy aims to justify already acquired knowledge. Chapter 4 argues that the main motivation for the foundationalist view, namely, to save knowledge from sceptical doubt, is unfounded, because absolute scepticism is not logically irrefutable. Chapter 5 argues against the view that philosophy is a humanistic discipline, opposed to the sciences.

Part II examines the nature of knowledge. Chapter 6 explains the central role that knowledge plays in human life. Chapter 7 examines several views about the relation of knowledge to reality. Chapter 8 discusses the view that the aim of science is truth, and considers several concepts of truth. Chapter 9 maintains that the aim of science is plausibility, it distinguishes plausibility from truth, probability, and warranted assertibility, and discusses the relation of science to common sense. Chapter 10 considers the relations of knowledge to certainty, objectivity, intuition, and deduction.

Part III examines the methods to acquire knowledge. Chapter 11 maintains that it is unjustified to say that there is no method to acquire knowledge. Chapter 12 considers various methods to acquire knowledge. Chapter 13 discusses various models of science, and to what extent they are capable of accounting for models in science. Chapter 14 maintains that knowledge is problem solving by the analytic method. Chapter 15 maintains that perceptual knowledge is also problem solving by the analytic method. Chapter 16 discusses the relation of knowledge to error. Chapter 17 considers the relation of knowledge to mind.

Part IV examines the nature of mathematical knowledge. Chapter 18 maintains that mathematics is problem solving by the analytic method. Chapter 19 discusses the nature of mathematical objects, mathematical definitions, and mathematical diagrams. Chapter 20 argues that mathematics is not theorem proving. Chapter 21 examines various notions of demonstration. Chapter 22 considers the question of mathematical explanation of mathematical facts, and the question of mathematical explanation of empirical facts. Chapter 23 discusses the nature of mathematical beauty, and its role in mathematical discovery. Chapter 24 considers the relation of mathematics to the world.

Part V ends and completes the book. Chapter 25 examines the connection between knowledge and the purpose and meaning of human life. Chapter 26 summarizes some of the main theses of the book.

1.9 Conventions

Constant use of 'he or she' may be clumsy, while constant use of 'she' may give rise to misunderstandings. Therefore, I use the generic 'he', while stipulating here that I mean it to refer to persons of both genders.

When I quote Greek expressions, I use the so-called scientific transliteration from the Greek to the Latin alphabet.

When I quote from ancient Greek philosophers, and even from some modern philosophers, translations are mine unless stated otherwise. This is motivated by the fact that every translation is an interpretation, and the interpretations given in this book are often different from those on which current translations are based. Moreover, current translations of different works of the same author by different translators may be inconsistent with each other, so quoting from them would lead to misunderstandings.

References

Andersen, Ross. 2012. Has physics made philosophy and religion obsolete? Interview to Lawrence Krauss. *The Atlantic*, April 23.

Cellucci, Carlo. 2013a. *Rethinking logic: Logic in relation to mathematics, evolution, and method.* Dordrecht: Springer.

Dyson, Freeman. 2012. What can you really know? *The New York Review of Books* 59(17, November 8): 18–20.

Gadamer, Hans Georg. 1998. *Reason in the age of science*. Cambridge: The MIT Press.

Galilei, Galileo. 1968. *Opere*. Florence: Barbera.

Hawking, Stephen, and Leonard Mlodinow. 2010. *The grand design*. New York: Bantam Books.

Heidegger, Martin. 1968. *What is called thinking?* New York: Harper & Row.

———. 1972. *On time and being*. Chicago: The University of Chicago Press.

———. 1994. *Basic questions of philosophy: Selected 'problems' of 'logic'*. Bloomington: Indiana University Pres.

———. 1998. *Pathmarks*. Cambridge: Cambridge University Press.

Husserl, Edmund. 1970. *The crisis of European sciences and transcendental phenomenology*. Evanston: Northwestern University Press.

———. 2002. Philosophy as rigorous science. *The New Yearbook for Phenomenology and Phenomenological Philosophy* 2: 249–295.

Pinker, Steven. 2002. *The blank slate. The modern denial of human nature*. London: Penguin Books.

Popper, Karl Raimund. 1974. *Conjectures and refutations: The growth of scientific knowledge*. London: Routledge.

Quine, Willard Van Orman. 1969. *Ontological relativity and other essays*. New York: Columbia University Press.

Rorty, Richard. 1991–2007. *Philosophical papers*. Cambridge: Cambridge University Press.
———. 1992. Twenty-five years after. In *The linguistic turn: Essays in philosophical method*, ed. Richard Rorty, 371–374. Chicago: The University of Chicago Press.
———. 1999. *Philosophy and social hope*. London: Penguin Books.
Williams, Bernard. 2006. *Philosophy as a humanistic discipline*. Princeton: Princeton University Press.
Wittgenstein, Ludwig. 2002. *Tractatus Logico-Philosophicus*. London: Routledge.

Part I
The Nature of Philosophy

Chapter 2
The Heuristic View

Abstract As a response to the increasing irrelevance of philosophy, this chapter lists the characteristics philosophy should have in order to be fruitful. This results in a view of philosophy that may be called the heuristic view, according to which philosophy aims at knowledge and methods to acquire knowledge. In listing the characteristics philosophy should have according to the heuristic view, the chapter systematically compares them with those of classical analytic philosophy. This is motivated by the fact that, in the past century, classical analytic philosophy has been the prevailing philosophical tradition. The characteristics of philosophy listed in the chapter are not intended to suggest that philosophies without such characteristics are bad, but only that they cannot be expected to be legitimized in the face of science.

2.1 The Characteristics of Philosophy

This part of the book examines the nature of philosophy. In the Introduction, the question has been raised as to whether philosophy can still be fruitful and what kind of philosophy can be such. This chapter answers the question, listing the characteristics philosophy should have in order to be fruitful and legitimized in the face of science. This results in the formulation of a view of philosophy that, as already stated in the Introduction, may be called the heuristic view of philosophy.

In listing the characteristics philosophy should have according to the heuristic view, this chapter systematically compares them with those of classical analytic philosophy. This is motivated by the fact that, in the past century, classical analytic philosophy has been the prevailing philosophical tradition.

Here, 'classical analytic philosophy' refers to the philosophical tradition started by Russell and Moore, which developed through Wittgenstein and logical positivism, knowing a moment of particular fortune with the Oxford ordinary language school of philosophy, and had several followers in the second half of the twentieth century. Dummett includes Frege in this tradition, saying that Frege "was the true father of analytical philosophy" (Dummett 2007, 27). This, however, is controversial. For example, Carl states that "Frege is not an analytic philosopher" (Carl 1994, 24).

In this chapter, special reference is made to Wittgenstein, since he is widely recognized as the single most significant figure in classical analytic philosophy.

© Springer International Publishing AG 2017
C. Cellucci, *Rethinking Knowledge*, European Studies in Philosophy of Science 4,
DOI 10.1007/978-3-319-53237-0_2

The reference to classical analytic philosophy, rather than to contemporary analytic philosophy, is motivated by the fact that, as Glock points out, while classical analytic philosophy is "a historical tradition held together by ties of influence on the one hand, family resemblances on the other," this "tradition is currently losing its distinct identity" (Glock 2008, 231).

The characteristics listed below are not intended to suggest that philosophies without such characteristics are bad, but only that they cannot be expected to be legitimized in the face of science.

2.2 Philosophy and the World

According to the heuristic view, philosophy is an inquiry which primarily aims at acquiring knowledge about the world, including ourselves, since we are part of the world. Like mathematics and science, philosophy is a means by which we make the world understandable to ourselves.

This contrasts with classical analytic philosophy. Thus, Wittgenstein states that "philosophy gives no pictures of reality" (Wittgenstein 1979, 106). It does not give any elucidation of reality, and in fact "one of the greatest impediments for philosophy is the expectation of new, deep, unheard of, elucidations" (Wittgenstein 2005, 309).

Dummett states that "philosophy is concerned with reality, but not to discover new facts about it: it seeks to improve our understanding of what we already know" (Dummett 2010, 10).

But, if philosophy is not concerned to discover new facts about reality, then it becomes self-referential. Thus, Moore states: "I do not think that the world or the sciences would ever have suggested to me any philosophical problem. What has suggested philosophical problems to me is things which other philosophers have said about the world or the sciences" (Moore 1942, 14). In fact, most classical analytic philosophy is self-referential. When it occasionally gets involved with issues outside its own limited domain, its only aim is to cleanse its terrain of alien regrettable intrusions.

Dummett admits that "the layman or non-professional expects philosophers to answer deep questions of great import for an understanding of the world" (Dummett 1991, 1). And "the layman is quite right: if philosophy does not aim at answering such questions, it is worth nothing. Yet he finds most writing by philosophers of the analytical school disconcertingly remote from these concerns," and this complaint "is understandable" (ibid.). According to Dummett, however, the complaint is "unjustified," because "philosophy can take us no further than enabling us to command a clear view of the concepts by means of which we think about the world" (ibid.). Philosophy "concerns our view of reality" only "by seeking to clarify the concepts in terms of which we conceive of it" (ibid., 11).

On the contrary, as the layman expects, philosophy should actually aim at answering deep questions of great import for an understanding of the world. To say that philosophy concerns our view of reality only by seeking to clarify the concepts in terms of which we conceive of it, does not relieve philosophy from the obligation to give pictures of reality. For, in order to enable us to command a clear view of the concepts in terms of which we conceive of reality, philosophy ought to put those concepts in relation with reality. But if philosophy gives no pictures of reality, how can it do that?

2.3 Philosophy and Globality

According to the heuristic view, philosophy is not limited to sectorial questions, it gives a global view. So, there cannot be a philosophy of mathematics alone, or of physics alone, or of biology alone, etc. For, the question of the nature of mathematics, or physics, or biology, etc., cannot be adequately approached locally, that is, in isolation from all other knowledge, but only globally, as part of a general approach to knowledge.

This contrasts with classical analytic philosophy. Thus, Carnap states that philosophers must limit themselves to sectorial questions because, "if we allot to the individual in philosophical work as in the special sciences only a partial task," then "stone will be carefully added to stone and a safe building will be erected at which each following generation can continue to work" (Carnap 2003, xvii).

But if philosophers confine themselves to sectorial questions, then they have no overall plan. This leads them to focus on smaller and smaller questions, thus confirming the motto: Some people know more and more about less and less, until they know everything about nothing, and these are the philosophers.

Classical analytic philosophy adopts the Socratic method of questions and answers, but it retains only the outward form of that method, not the substance, namely, the serious search for answers to general questions. There is no evidence that a minute work on sectorial questions may lead to what is essential. With such a minute work, one risks being merely "a maker of words, incapable of doing any" worthwhile "work" (Plato, *Epistulae*, VII 328 c 5–6). Therefore, philosophy must not be limited to sectorial questions, it must give a global view. As Plato says, "anyone who can have a global view is a philosopher, and anyone who can't isn't" (Plato, *Respublica*, VII 537 c 7).

Carnap's argument for confining oneself to sectorial questions implies a view of philosophy which assimilates philosophy to what Kuhn calls 'normal science', whose aim "is not major substantive novelties" (Kuhn 1996, 35). Normal science accumulates details within a settled paradigm and theory, without questioning or challenging the underlying assumptions of the theory. Only within normal science the individual adds stone to stone, working on a small piece of a broader project. But, as confining science to normal science would lead to a narrow view of science, so confining philosophy to sectorial questions would lead to a narrow view of philosophy.

2.4 Philosophy and Essential Problems

According to the heuristic view, philosophy must not deal with peripheral questions, but with great essential problems in the sense of science. Aiming at acquiring knowledge about the world, including ourselves, philosophy seeks the big picture: to understand the world and our place in it. To this purpose, philosophy necessarily must concern itself with great essential problems.

This contrasts with classical analytic philosophy. Thus, Wittgenstein states that in philosophy "there are no great essential problems in the sense of science" (Wittgenstein 2005, 301). Philosophy "is a tool which is useful only against the philosophers" (Wittgenstein 1932–1933, 11).

But, if in philosophy there are no great essential problems in the sense of science, if philosophy is a tool which is useful only against the philosophers, why should we continue to practice it? How could we avoid concluding that philosophy is only a crossroads of many routes leading nowhere?

Russell himself, though being one of the fathers of classical analytic philosophy, declares: "The new philosophy seems to me to have abandoned, without necessity, that grave and important task which philosophy throughout the ages has hitherto pursued," namely, "to understand the world" (Russell 1995b, 170). It "cares only about language, and not about the world" (Russell 1960, 15). According to the new philosophy, "the desire to understand the world" is "an outdated folly" (Russell 1995b, 162). Wittgenstein makes philosophy, "at best, a slight help to lexicographers, and at worst, an idle tea-table amusement" (ibid., 161). But, "if this is all that philosophy has to offer, I cannot think that it is a worthy subject of study" (ibid., 170).

In fact, the assumption that in philosophy there are no great essential problems in the sense of science, has produced a new kind of scholasticism, characterized by an argumentative style made of dreary distinctions concerning minute, inconsequential questions, incapable of making significant contributions to an understanding of the world, including ourselves. It is no wonder then that, as Fodor states, today "nobody reads philosophy," in particular "it's mainly the laity that seems to have lost interest. And it's mostly Anglophone analytic philosophy that it has lost interest in" (Fodor 2004, 17).

2.5 Philosophy and Knowledge

According to the heuristic view, since philosophy is an inquiry which primarily aims at acquiring knowledge about the world, questions about knowledge are central to philosophy.

This contrasts with classical analytic philosophy. Thus, Searle states that since, in the seventeenth century, "the possibility of certain, objective, universal knowledge seemed problematic" and "the very existence of knowledge was in question," Descartes "took epistemology as the central element of the entire philosophical enterprise" (Searle 2008, 4). As a result, "we had three and a half centuries in which epistemology was at the centre of philosophy" (ibid., 5). But today, "because of the sheer growth of certain, objective, and universal knowledge, the possibility of knowledge is no longer a central question in philosophy," so questions about knowledge no longer "lie at the heart of the philosophical enterprise" (ibid.).

But this is unjustified. If philosophy is an inquiry which primarily aims at acquiring knowledge about the world, then necessarily epistemological questions have a central place in it. The purpose of epistemology is not to inquire into the possibility of certain, objective, universal knowledge, because there is no such knowledge. By Gödel's second incompleteness theorem, even mathematical knowledge cannot be proved to be certain by absolutely reliable means. The purpose of epistemology, or theory of knowledge, is rather to inquire into methods to acquire knowledge, fallible knowledge and yet knowledge. As Kitcher says: "What might we want from a theory of knowledge? There's an obvious answer. A theory of knowledge should enable us to get more of it" (Kitcher 2011, 508). This transfers epistemology from the context of justification to the context of discovery, which includes the context of justification since, in the process of discovery, hypotheses are accepted only when they are shown to be plausible, namely such that the arguments for them are stronger than the arguments against them, on the basis of the existing knowledge.

Dummett replaces epistemology with the theory of meaning as the centre of philosophy. He states that Descartes made epistemology "the foundation of philosophy because he had conceived the task of philosophy as being that of introducing rigour into science" (Dummett 1973, 676). But Descartes was wrong, because "the fundamental part of philosophy which underlies all others" is "the theory of meaning" (ibid., 669). Epistemology deals with questions of justification, but "until we have first achieved a satisfactory analysis of the meanings of the relevant expressions, we cannot so much as raise questions of justification," since "we remain unclear about what we are attempting to justify" (ibid., 667). For this reason, Frege made "the theory of meaning" the "foundation of all philosophy, and not epistemology, as Descartes misled us into believing" (ibid.). Thus Frege "effected a revolution in philosophy," and so we can "date a whole epoch in philosophy as beginning with the work of Frege" (ibid., 669).

But the theory of meaning cannot be the foundation of philosophy, since the main philosophical questions are not questions about the use of language, but questions about the world. Moreover, Frege did not replace epistemology with the theory of meaning as the centre of philosophy. His main purpose was to give a secure foundation for mathematics, which was a question of justification within Descartes' epistemological tradition (see Cellucci 1995). As Descartes conceived the task of philosophy as being that of introducing rigour into science, Frege conceived the task of philosophy as being that of introducing rigour into mathematics. Indeed, in order to achieve a secure foundation for mathematics, Frege asked that "the fundamental propositions of arithmetic should be proved, if in any way possible, with the utmost rigour" (Frege 1960, 4).

2.6 Philosophy and the Armchair

According to the heuristic view, philosophy is not a product of thought alone, hence it is not an armchair subject, it essentially needs inputs from experience.

This contrasts with classical analytic philosophy. Thus, Dummett states that philosophy "is a discipline that makes no observations, conducts no experiments, and needs no input from experience: an armchair subject, requiring only thought" (Dummett 2010, 4). Philosophy is like "another armchair discipline: mathematics. Mathematics likewise needs no input from experience: it is the product of thought alone" (ibid.).

Nagel states that "philosophy is different from science" because, "unlike science it doesn't rely on experiments or observation, but only on thought," since philosophy "is done just by asking questions, arguing, trying out ideas and thinking of possible arguments against them" (Nagel 1987, 4).

But this is based on the assumption that philosophy is not an inquiry which primarily aims at acquiring knowledge about the world, and hence is not concerned to discover new facts about reality. As argued above, this assumption is unjustified. Since philosophy is an inquiry which primarily aims at acquiring knowledge about the world, it necessarily needs inputs from experience, and hence cannot rely only on thought. By thought alone, we can at most reformulate what we already know in other terms.

In particular, saying that philosophy is like another armchair discipline, namely mathematics, is unjustified because mathematics is not an armchair subject, it essentially involves interactions with the world beyond the armchair. As Atiyah points out, "almost all mathematics originally arose from external reality" (Atiyah 2005, 226). In fact, several mathematical problems have an extra-mathematical origin, and the solutions of mathematical problems are only plausible, so they are evaluated in terms of their compatibility with the existing knowledge, and hence with experience.

2.7 Philosophy and the Sciences

According to the heuristic view, philosophy is continuous with the sciences, in the sense that it aims at a kind of knowledge which differs from scientific knowledge in no essential respect, and is not restricted to any area. Thus, the objectives of philosophy are not essentially different from those of the sciences, and philosophy is an activity which is not essentially different from the sciences. The only difference between philosophy and the sciences, is that philosophy deals with questions which are beyond the present sciences. The latter are what we already know, philosophy is about what we do not yet know.

This contrasts with classical analytic philosophy. Thus Wittgenstein states that, in philosophy, "we are not doing natural science" (Wittgenstein 1958, II, xii.230).

Indeed, "natural science" is "something that has nothing to do with philosophy" (Wittgenstein 2002, 6.53). The "word 'philosophy' must mean something whose place is above or below the natural sciences, not beside them" (ibid., 4.111).

Dummett states that "philosophy stands in complete contrast with sciences: its methods wholly diverge from those of science, and its objective differs to an equal extent," moreover "the results of philosophy differ fundamentally in character from those of the sciences" (Dummett 2010, 7).

But if philosophy stands in complete contrast with sciences, if its objectives, methods, and results differ fundamentally in character from those of the sciences, how could philosophy possibly contribute to our knowledge of reality?

Dummett also states that saying that philosophy is continuous with the sciences is a form of scientism, where "scientism is the disposition to regard the natural sciences as the only true channel of knowledge" (ibid., 35). Scientism implies that "the idea that philosophy has a subject matter or a method of its own must be discarded: if it is to contribute to knowledge at all, it must be continuous with the natural sciences," and its task reduces "to that of adding ornamentation to the theories of the scientists" (ibid.).

But saying that philosophy is continuous with the sciences does not amount to regarding the natural sciences as the only true channel of knowledge, or to reducing the task of philosophy to that of adding ornamentation to the theories of the scientists. Scientism is the disposition to regard, not the natural sciences, but the present natural sciences, as the only true channel of knowledge. There are areas of experience which the present natural sciences are incapable to deal with. Dealing with them requires new ideas, not devised by any of the present sciences, and it is the task of philosophy to devise them. It is in this sense that the present sciences are what we already know, and philosophy is about what we do not yet know.

Dummett's view that philosophy stands in complete contrast with sciences is shared, for opposite reasons, by several scientists, who believe that philosophy is, at best, an ornamentation or a parasitic commentary on the achievements of the sciences, and that the sciences are philosophy-free. But, as Dennett says, "there is no such thing as philosophy-free science; there is only science whose philosophical baggage is taken on board without examination" (Dennett 1996, 21).

2.8 Philosophy and the Results of the Sciences

According to the heuristic view, philosophy makes use of the results of the sciences, and this is essential to its progress. Indeed, since philosophy is continuous with the sciences, like the sciences, in order to obtain new knowledge, philosophy must start from the existing knowledge.

This contrasts with classical analytic philosophy. Thus, Wittgenstein states that "one might also give the name 'philosophy' to what is possible before all new discoveries" (Wittgenstein 1958, I, § 126). Therefore, philosophy is independent of any scientific discovery, in particular "no mathematical discovery can advance it" (ibid., I, § 124).

But if, in dealing with philosophical problems, no use is made of the results of the sciences, then philosophy ends up with repeating old idioms, neglecting that often they are based on obsolete views of the world. This is not only acknowledged but even theorized by Wittgenstein, who states that "no new words have to be used in philosophy – the old, ordinary words of language suffice" (Wittgenstein 2005, 309).

Contrary to Wittgenstein's claims, in order to deal with new philosophical problems, philosophy must be able to use whatever is known, starting from the results of the sciences, introducing new idioms adequate to the questions dealt with. Old idioms are often based on obsolete scientific theories, or simply on prejudices.

2.9 Philosophy and Method

According to the heuristic view, the method of philosophy is the same as that of the sciences. Indeed, since philosophy is continuous with the sciences, its method cannot be essentially different.

This contrasts with classical analytic philosophy. Thus, Wittgenstein states that "philosophers constantly see the method of science before their eyes, and are irresistibly tempted to ask and answer questions in the way science does," but this "leads the philosopher into complete darkness" (Wittgenstein 1969a, 18). Actually, the method of philosophy is completely different from that of science, because "the task of philosophy" is "to clarify the use of our language, the existing language. Its aim is to remove particular misunderstandings" (Wittgenstein 1974, 115). Therefore, the method of philosophy is the analysis of language, which removes "misunderstandings concerning the use of words" by "substituting one form of expression for another; this may be called an 'analysis' of our forms of expression, for the process is sometimes like one of taking a thing apart" (Wittgenstein 1958, I, § 90).

Dummett states that the aim of "philosophy is the analysis of the structure of thought," and "the only proper method for analyzing thought consists in the analysis of language" (Dummett 1978, 458). For, "there can be no account of what thought is, independently of its means of expression," namely language, and "the structure of the sentence reflects the structure of thought" (Dummett 1991, 3). Therefore, "the philosophy of language is the foundation of all other philosophy" (Dummett 1978, 442).

But the method of philosophy cannot be the analysis of language, because philosophy is an inquiry which primarily aims at acquiring knowledge about the world, and the analysis of language is inadequate to that purpose. For, questions about the world are not questions of words but questions of things. As Kant states, "in matters over which one has quarreled over a long period of time, especially in philosophy, there has never been at the basis a quarrel of words but always a true quarrel over things" (Kant 2007, 179).

Moreover, the assumption that the only proper method for analyzing thought consists in the analysis of language contrasts with studies on non-linguistic thought. They are important not only with regard to animal thought, whose existence was already acknowledged by Chrysippus, who stated that a hunting "dog makes use of the fifth undemonstrable syllogism when he come to a crossroads" (Sextus Empiricus, *Pyrrhoniae Hypotyposes*, I, 69). They are also important with regard to human thought, for example because several scientists and mathematicians affirm that their most creative work does not involve language.

Thus, Einstein states that "the words or the language, as they are written or spoken, do not seem to play any role in my mechanism of thought. The psychical entities which seem to serve as elements in thought" are, "in my case, of visual and some of muscular type" (Einstein 1954, 142–143).

Hadamard states: "Words are totally absent from my mind when I really think," and "do not reappear in my consciousness before I have accomplished or given up the research" (Hadamard 1954, 75).

Identifying the method of philosophy with the analysis of language is a result of the assumption of classical analytic philosophy, that questions about knowledge no longer lie at the heart of the philosophical enterprise. With respect to this assumption, Hawking states that, while in the eighteenth century philosophers "considered the whole of human knowledge, including science, to be their field," in the twentieth century they have "reduced the scope of their inquiries so much that Wittgenstein, the most famous philosopher of this century, said, 'The sole remaining task for philosophy is the analysis of language'. What a comedown from the great tradition of philosophy from Aristotle to Kant!" (Hawking 1988, 185). Similarly, Rota states that, in the twentieth century, "the classical problems of philosophy have become forbidden topics," and this has led to "the shrinking of philosophical activity to an impoverished *problématique*, mainly dealing with language" (Rota 1997, 98).

Rather than with the analysis of language, the method of philosophy must be identified with that of the sciences. The latter is the analytic method, which is both a method of discovery and justification. The analytic method is the method according to which, to solve a problem, one looks for some hypothesis that is a sufficient condition for solving the problem, namely, such that a solution to the problem can be deduced from it. The hypothesis is obtained from the problem, and possibly other data already available, by some non-deductive rule – such as induction, analogy, metaphor, and so on – and must be plausible, namely such that the arguments for it are stronger than the arguments against it, on the basis of the existing knowledge. But the hypothesis is in its turn a problem that must be solved, and is solved in the same way. That is, one looks for another hypothesis that is a sufficient condition for solving the problem posed by the previous hypothesis, it is obtained from the latter, and possibly other data already available, by some non-deductive rule, and must be plausible. And so on, *ad infinitum*. (For more on the analytic method, see Chap. 12, where it is also argued that Plato first formulated the analytic method as the method of philosophy).

2.10 Philosophy and the Aim to Acquire Knowledge

According to the heuristic view, since philosophy is continuous with the sciences, aiming at acquiring knowledge is part of the deepest nature of philosophy.

This contrasts with classical analytic philosophy. Thus, Wittgenstein states that philosophy "arises neither from an interest in the facts of nature, nor from a need to grasp causal connections;" on the contrary, it is "essential to our investigation that we do not seek to learn anything new by it" but only "to understand something that is already in plain view" (Wittgenstein 1958, I, § 89). In philosophy, "it is not that a new building has to be erected, or that a new bridge has to be built, but that the geography, as it now is, has to be judged" (Wittgenstein 1978, V, § 52). Philosophy has no impact on the growth of knowledge, it "leaves everything as it is" (Wittgenstein 1958, I, § 124). It "simply puts everything before us, and neither explains nor deduces anything" (ibid., I, § 126). It "only states what everyone admits" (ibid., I, § 599).

Dummett states that "philosophy does not advance knowledge: it clarifies what we already know" (Dummett 2010, 21). It "does not seek to observe more, but to clarify our vision of what we see" (ibid., 10).

But a philosophy thus meant has little reason to exist. Therefore, Dummett states that today "it is by no means obvious that universities, and thus ultimately the state, should support philosophy" but for the historical precedent that "the history of Western universities goes back 900 years" and "philosophy has always been one of the subjects taught and studied in them" (ibid., 2). When the first Western universities came into being, philosophy "was not sharply differentiated from what we call 'natural science'" (ibid.). It was then easy to find a justification for philosophy. But, in the twentieth century, "the distinction between philosophy and the natural sciences came to be generally admitted" (ibid., 3). Therefore, finding a justification for philosophy became difficult. Indeed, "if universities had been an invention of the second half of the twentieth century, would anyone have thought to include philosophy among the subjects that they taught and studied? It seems very doubtful" (ibid., 2). It "would be easy to conclude that this is an anachronism" (ibid.).

This is the conclusion to which the rejection of the view that philosophy primarily aims at acquiring knowledge leads. While, in the philosophical tradition, philosophy was inspired by the hope that it would contribute to the advancement of knowledge, classical analytic philosophy conceives of philosophy in such a way that it becomes incapable of making any contribution to our knowledge of the world. Then, as Dummett says, it would be easy to conclude that it is an anachronism.

2.11 Philosophy and the Aim to Obtain Rules of Discovery

According to the heuristic view, philosophy aims not only at acquiring knowledge, but also at obtaining rules of discovery, since nothing guarantees that new knowledge can be acquired by the existing rules of discovery.

This contrasts with classical analytic philosophy. Thus, Hempel states that philosophy cannot aim at obtaining rules of discovery, because hypotheses cannot be obtained by "any process of systematic inference" (Hempel 1966, 15). They "are not derived from observed facts, but invented in order to account for them" (ibid., 15). Their discovery "requires inventive ingenuity; it calls for imaginative, insightful guessing" (ibid., 17). That is, it calls for intuition. To support this view, Hempel mentions the case of Kekulé who one evening, while sitting by a fire, sank into half-sleep and had a dream. He saw atoms fluttering before his eyes, long chains often combined in a denser fashion, all in motion, twisting and turning like snakes, until one of the snakes seized its own tail. This suggested to Kekulé the hypothesis that benzene forms a closed ring of six carbon atoms.

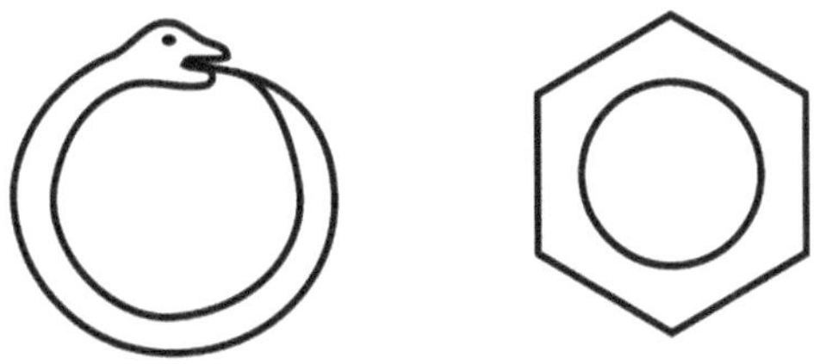

According to Hempel, Kekulé's case shows that, to solve a problem, scientists essentially depend on intuition, only thanks to which they may arrive at "the discovery of important, fruitful theories in empirical science" (ibid.). But, contrary to Hempel's claim, Kekulé's case does not show that. Kekulé himself does not say he had an intuition, but only a dream. Moreover, at Kekulé's time it was well known that the behaviour of a molecule depended on its structure, and the structures already tried for benzene were inadequate. Therefore, Kekulé was well aware of the need to find a new structure, and had already considered various possibilities on that regard. Seeing that a snake biting its tail formed a stable structure suggested to him, by an analogical inference, that a structure for benzene could be a closed ring. Furthermore, Kekulé formulated the hypothesis that the structure of benzene is a closed ring only after comparing the hypothesis with the existing data. Therefore, he formed his hypothesis through a rational process. There is also the question whether Kekulé's report of the event is reliable. It has been argued that Kekulé made up the whole episode. This has been subject of some controversy (see Rocke 2011, Chap. 10).

2.12 Philosophy and the Birth of New Sciences

According to the heuristic view, since philosophy deals with questions which are beyond the present sciences, it must try unexplored routes. By so doing, when successful, philosophy may even give birth to new sciences. This is one of its greatest values.

This contrasts with classical analytic philosophy. Thus, Dummett states that "no practicing philosopher would explain the value of the subject merely as a matrix out of which new disciplines could develop" (Dummett 2010, 4). Philosophy is "what is left when the disciplines to which it gave birth have left the parental home" (ibid.). So, "it was not until the nineteenth century that it made sense to ask for an example of a philosophical problem, as opposed to a problem of some other kind." (ibid., 8). In particular, as opposed to a scientific problem.

But trying unexplored routes, possibly giving birth to new sciences, is one of the greatest values of philosophy. So, in the seventeenth century, philosophy gave birth to modern science, which originated from Galileo's philosophical revolution: the decision to renounce Aristotle's aim to penetrate the essence of natural substances, dealing only with some of their phenomenal properties mathematical in kind, such as location, motion, shape, or size (see Chap. 8).

Since the seventeenth century, philosophy has given birth to several other sciences. For example, in the twentieth century, Turing's philosophical analysis of the computational behaviour of human beings gave birth to computer science. The philosophical character of Turing's analysis is apparent from "the enthusiastic philosophical reception of Turing's approach," which "stands in stark contrast to the very limited attention given to it in print in the following decade" (Mosconi 2014, 38).

There is no reason to think that philosophy will not give birth to new sciences also in the future. For example, knowledge has an important role in evolution. Even simple organisms such as the prokaryotes cannot survive if they do not acquire knowledge about the environment. But the current theories of evolution disregard the role of knowledge in evolution, they do not take into account knowledge processes. Therefore, there is need for a new science which completes the current theories of evolution with a theory of knowledge. As another example, although it is evident that the mind is intimately linked to brain processes, and that the study of such processes will give us much insight into operations of the mind, it should also be understood that the mind involves processes that are external to the mind (see Chap. 17). The present psychology and cognitive science do not take into account such processes. Therefore, there is need for a new science which considers not only the processes internal to the mind but also processes external to it.

These are just two examples of possible new sciences, but other examples could be devised. There is much space for philosophy, because the things which we do not yet know, even on basic questions, are numerous, and philosophy is about what we do not yet know. In this connection, it is worth recalling Seneca's prediction: "Veniet tempus quo posteri nostri tam aperta nos nescisse mirentur [A time will come when our posterity will marvel that we were ignorant of such obvious things]" (Seneca, *Naturales Quaestiones* 7.25). Seneca's prediction should act as a warning also for us.

Of course, trying unexplored routes, philosophy moves on a magmatic ground, so it can offer no theories but only viewpoints. The proper place for theories are the new sciences to which philosophy may possibly give birth. But this does not make philosophy less continuous with the sciences. Obtaining knowledge is a profoundly unitary enterprise.

2.13 Philosophy and the History of Philosophy

According to the heuristic view, philosophy makes use of the experience of the philosophers of the past. Without it, philosophers would keep reinventing the wheel, or hunting down trails that are known to be dead ends. Therefore, the history of philosophy is relevant to philosophers.

This contrasts with classical analytic philosophy. Thus, Wittgenstein states: "What has history to do with me? Mine is the first and only world! I want to report how I found the world. What others in the world have told me about the world is a very small and incidental part of my experience of the world" (Wittgenstein 1979, 82, 2.9.16). The "reason why I give no sources is that it is a matter of indifference to me whether the thoughts that I have had have been anticipated by someone else" (Wittgenstein 2002, 4). A philosopher is like "a king" who has been "brought up in the belief that the world began with him" (Wittgenstein 1969b, § 92).

In fact, in formulating problems and dealing with them, several classical analytic philosophers behave like that king, not only with respect to the philosophical tradition, but also with respect to their own history. They tend to consider only the problems and the solutions their generation proposes, ignoring those of the previous generations. By so doing, they think to behave like those scientists who, while knowing only the most recent literature in their field, give contributions to their discipline. But it is not so. Those scientists deal with questions posed by the world, though very limited ones, and, being very limited, in some cases knowledge of the most recent literature alone may be enough to deal with them. Conversely, philosophers who only know the most recent literature do not deal with question posed by the world, but only with puzzles posed by their colleagues, which are generally irrelevant to an inquiry into the world.

Of course, the history of philosophy is not a substitute for philosophy itself. In this regard, it is worth recalling Kant's complaint that "there are scholars for whom the history of philosophy" is "itself their philosophy" (Kant 2002, 53). According to them, "nothing can be said that has not already been said before" (ibid.). But they are wrong, because the world poses ever newer problems and challenges. Such scholars "must wait until those who endeavor to draw from the wellspring of reason itself have finished their business, and then it will be their turn to bring news of these events to the world" (ibid.).

Nevertheless, being aware of the philosophy of the past is important, on the one hand, to acquire awareness of the implications of certain philosophical assumptions, and, on the other hand, to avoid retracing routes already covered. As Santayana says, "those who cannot remember the past are condemned to repeat it" (Santayana 1948, 248).

2.14 Philosophy and Intuition

According to the heuristic view, philosophy makes no use of intuition, because the method of philosophy is the same as that of the sciences, and intuition plays no role in the method of the sciences.

This contrasts with classical analytic philosophy. Thus, Wittgenstein says that philosophy requires intuition because philosophy is about what lies in front of everyone's eyes, and the philosopher may grasp it because "God grant the philosopher insight into what lies in front of everyone's eyes" (Wittgenstein 1998, 72). Even knowledge of logical relations requires intuition, because "I can't come to this insight through a logical inference, I must see it" (Wittgenstein 1975, 336).

Philosophy requires intuition like mathematics, in which "no investigation of concepts, only insight into the number-calculus can tell us that 3 + 2 = 5" (Wittgenstein 1974, 347). Even "in order to follow the rule 'Add 1' correctly a new insight, intuition, is needed at every step" (Wittgenstein 1969a, 141).

But, being subjective and arbitrary, intuition is unreliable. Bealer states that "denying that intuitions are evidence leads to epistemic self-defeat; it is impossible to have a coherent epistemology without admitting intuitions as evidence" (Bealer 1996, 32, footnote 26). But this is unjustified. It is possible to have a coherent epistemology without admitting intuitions as evidence, if one assumes that the method of philosophy is the same as that of the sciences, namely, the analytic method, and, as it will be argued in Chap. 12, intuition plays no role in the analytic method.

2.15 Philosophy and Emotion

According to the heuristic view, philosophy is concerned with emotions, not only because emotions are important to the quality and meaning of our life, but also because they are a help to reason and play an important role in knowledge, including mathematical and scientific knowledge.

This contrasts with classical analytic philosophy. Thus, Russell states that, although "emotions are what makes life interesting, and what makes us feel it important," when "we are trying to understand the world, they appear rather as a hindrance. They generate irrational opinions" which "cause us to view the universe in the mirror of our moods" (Russell 1995a, 175–176).

But it is not so, emotions are a help to reason, because reason is the capacity to choose appropriate means to a given end (see Cellucci 2013a, Chap. 14). And emotions help us to choose appropriate means to the end of survival, because only individuals who perceive a favorable occasion as a positive emotion, and danger as a negative one, can survive.

Generally, emotions play an important role in knowledge. They play an important role in the choice of problems and hypotheses to solve them. We are able to seek solutions only for a small part of the problems that arise. Emotions may help us to choose which problems to consider and which to disregard. Only if we feel strongly involved in a problem we may have the drive to do the hard work that finding a solution may require. Besides, in solving a problem we may be faced with so many alternative hypotheses, that it would be unfeasible to consider all of them. Emotions may help us to choose which hypothesis to consider and which to disregard.

Even Carnap admits that, also in "the most rational of sciences, namely physics and mathematics," the "basic orientation and the direction of interests are not the result of deliberation, but are determined by emotions" (Carnap 2003, xvii). (For more on the role of emotions in knowledge, see Cellucci 2013a, Chap. 15).

2.16 Philosophy and the Solvability of Problems

According to the heuristic view, philosophy cannot demand and expect conclusive solutions to the questions belonging within it. Solutions to philosophical problems are always temporary and are bound to be replaced sooner or later by others.

This contrasts with classical analytic philosophy. Thus, Wittgenstein states that "philosophical problems must be solvable really completely, in contrast to all others" (Wittgenstein 2005, 310).

But conclusive solutions to philosophical problems are impossible. Philosophical problems, being problems about the world, are like scientific problems, and the latter cannot be solved conclusively. A philosopher is like a scientist who, as Poincaré says, "is always searching and is never satisfied" (Poincaré 1910, VII). Indeed, every solution is based on hypotheses which are only plausible, so it is always provisional and bound to be replaced with another one as new data emerge. Even Wittgenstein acknowledges that "in this work more than any other it is rewarding to keep on looking at questions, which one considered solved, from another quarter, as if they were unsolved" (Wittgenstein 1979, 30, 13.11.14). Therefore, it is arbitrary to insist that philosophical treatments of problems are valuable only if they provide conclusive solutions.

Dennett states that, confronted with the alternative of either (A) solving a major philosophical problem "so conclusively that there is nothing left to say," or (B) writing "a book of such tantalizing perplexity and controversy that it stays on the required reading list for centuries to come," some philosophers "admit that they would have to go for option (B)" (Dennett 2013, 429). Conversely, scientists "tend to opt for (A) without any hesitation," and "shake their heads in wonder (or disgust?) when they learn that this is a hard choice for many philosophers" (ibid.). But, for once, philosophers are right and scientists are wrong. For, not only no major philosophical problem, but also no major scientific problem, can be solved so conclusively that there is nothing left to say.

2.17 Philosophy and Progress

Although solutions to philosophical problems are always temporary and are bound to be replaced sooner or later by others, they advance knowledge. There is progress everywhere, even in philosophy.

This contrasts with classical analytic philosophy. Thus, Wittgenstein states that the solution of philosophical problems produces no progress, because in philosophy we are simply destroying "houses of cards and we are clearing up the ground of language on which they stand" (Wittgenstein 1958, I, § 118). People complain that "philosophy really doesn't make any progress, that the same philosophical problems that occupied the Greeks keep occupying us. But those who say that don't understand" that it must be so because "our language has remained constant and keeps seducing us into asking the same questions" (Wittgenstein 2005, 312). As long as language will remain constant, human beings "will continue to bump up against the same mysterious difficulties, and stare at something that no explanation seems able to remove" (ibid.).

But, if the solution of philosophical problems produces no progress, then philosophy is a futile activity which gets us nowhere. Even Dummett admits that, "if philosophy makes no progress, it is not worth wasting any time on" (Dummett 2010, 148).

As to language, even supposing that our language has remained constant since the Greeks and keeps seducing us into asking the same questions, this does not mean that the solution of philosophical problems produces no progress. The world changes all the time, there are no two instants in which it is exactly the same, so the world poses ever newer questions and challenges. Once again, this shows that it is arbitrary to restrict philosophy to the method of the analysis of language.

2.18 Philosophy and Professionalization

According to the heuristic view, philosophy cannot be a professional activity, because it has no special field to investigate, or special techniques of its own to use.

This contrasts with classical analytic philosophy. Thus, Carnap states that, "while the attitude of the traditional philosopher is more like that of a poet," the analytic philosophers "have taken the strict and responsible orientation of the scientific investigator as their guideline for philosophical work" (Carnap 2003, xvi). The "individual no longer undertakes to erect in one bold stroke an entire system of philosophy. Rather, each works at his special place within the one unified science" (ibid.).

As Rescher points out, this has led to "the turning of philosophy from global general, large-scale issues to more narrowly focused investigations of matters of microscopically fine-grained detail" (Rescher 1993, 731). As a result, today "philosophical investigations make increasingly extensive use of the formal machinery of semantics, modal logic, compilation theory, learning theory, and so forth. Ever heavier theoretical armaments are brought to bear on ever smaller problem-targets" (ibid.).

Austin even claims that, in order to become a professional philosopher, "first we may use the dictionary – quite a concise one will do, but the use must be thorough" (Austin 1970, 186). There are two ways of using it, "one is to read the book through, listing all the words that seem relevant," the other one "is to start with a widish selection of obviously relevant terms, and to consult the dictionary under each: it will be found that, in the explanations of the various meanings of each, a surprising number of other terms occur" (ibid., 186–187). We will then "look up each of these, bringing in more for our bag from the 'definitions' given in each case" (ibid., 187). If we will continue like that, after a while "it will generally be found that the family circle begins to close, until ultimately it is complete" (ibid.). So we will "arrive at the meanings of large numbers of expressions" and we "shall comprehend clearly much that, before, we only made use of ad hoc. Definition, I would add, explanatory definition, should stand high among our aims" (ibid., 189).

But so one does not get at the nature of things, only at the opinion of the authors of the dictionary. Admittedly, the Socratic endeavour was the pursuit of definition. For, Socrates thought that, until one knows the essence of a thing, one cannot answer any other questions about it, and a definition states what a thing is in itself, its essence. But, by stating the essence of things, Socratic definitions are of things, not of words. For example, Socrates asks the question: "What do you think knowledge is?" (Plato, *Theaetetus*, 146 c 3). By this question, he does not want to know what the word 'knowledge' means, but rather "to know what knowledge is in itself" (ibid, 146 e 9–10).

Actually, a professional philosophy is impossible. A philosopher cannot be a professional in the same sense as a mathematician, or a physicist, or a biologist, because philosophy has no special field of its own. As sciences claim special fields, philosophy's field of inquiry changes. It remains the unexplored ground, but what ground is unexplored changes with time. On the other hand, a philosopher cannot be a professional in the same sense as a doctor, or a lawyer, or an engineer, because philosophy has no special techniques of its own. Although the method of philosophy is the same as that of the sciences, namely the analytic method, the latter is only a general framework, and its application requires experience specific to the field. But a philosopher moves on an unexplored ground, on which there is still very limited experience.

Therefore, a philosopher is, and always will be, a great amateur. And yet, just because a philosopher moves on an unexplored ground, on which there is still very limited experience, philosophy is always exposed to the risk of failure but is also capable of surprising developments. Just like those thanks to which, trying unexplored routes, through hazardous though sometimes fortunate moves, philosophy has given birth to new sciences.

2.19 The Heuristic View vs. Philosophy as Criticism of Principles

The heuristic view of philosophy, outlined above, should not be confused with Russell's view of philosophy as criticism of principles.

Admittedly, Russell states that philosophy "aims primarily at knowledge" (Russell 1997, 154). It aims at a knowledge that "does not differ essentially from scientific knowledge," because "there is no special source of wisdom which is open to philosophy but not to science, and the results obtained by philosophy are not radically different from those obtained from science" (ibid., 149). The only difference between philosophy and science is that "science is what we know, and philosophy is what we don't know," but "philosophical speculation as to what we do not yet know has shown itself a valuable preliminary to exact scientific knowledge" (Russell 1950, 24).

So far so good. But then Russell states that philosophy aims at the kind of knowledge "which results from a critical examination of the grounds of our convictions, prejudices, and beliefs" (Russell 1997, 154). It "examines critically the principles employed in science" (ibid., 149). It wants "to see whether they are mutually consistent and whether the inferences employed are such as seem valid to a careful scrutiny" (Russell 1995a, 239).

This view seems unrealistic. Scientists do not have to wait for philosophers to examine critically the principles employed in their sciences, in order to see whether they are mutually consistent, or whether the inferences employed are valid. This is an integral part of their work, and they are much more competent to the task than philosophers, who do not have the necessary qualifications. Moreover, a philosophy in accordance with Russell's view could not be expected to give birth to new sciences.

Conversely, a philosophy in accordance with the heuristic view would permit to give an affirmative answer to the question raised at the beginning of this chapter – whether philosophy can still be fruitful, and what kind of philosophy can be such. A philosophy with such characteristics would be fruitful because it aims at knowledge and methods to acquire knowledge, it pursues this aim by trying unexplored routes, and may even give birth to new sciences.

A remark about the use of the term 'sciences'. Since, according to the heuristic view, philosophy is an inquiry which primarily aims at acquiring knowledge about the world, including ourselves, throughout this book 'sciences' means 'natural and human sciences', unless stated otherwise.

References

Atiyah, Michael. 2005. Joint interview with Isadore Singer, by Martin Raussen and Christian Skau. *Notices of the American Mathematical Society* 52: 223–231.

Austin, John Langshaw. 1970. *Philosophical papers*. Oxford: Oxford University Press.

Bealer, George. 1996. On the possibility of philosophical knowledge. *Philosophical Perspectives* 10: 1–34.

Carl, Wolfgang. 1994. *Frege's theory of sense and reference: Its origin and scope*. Cambridge: Cambridge University Press.

Carnap, Rudolf. 2003. *The logical structure of the world – Pseudoproblems in philosophy*. Chicago: Open Court.

Cellucci, Carlo. 1995. Frege e le origini della logica matematica. Introduction to Gottlob Frege, *Leggi fondamentali dell'aritmetica*.ix-lxvi Rome: Teknos.

———. 2013a. *Rethinking logic: Logic in relation to mathematics, evolution, and method*. Dordrecht: Springer.

Dennett, Daniel. 1996. *Darwin's dangerous idea: Evolution and the meaning of life*. London: Penguin Books.

———. 2013. *Intuition pumps and other tools for thinking*. New York: Norton.

Dummett, Michael. 1973. *Frege: Philosophy of language*. New York: Harper & Row.

———. 1978. *Truth and other enigmas*. Cambridge: Harvard University Press.

———. 1991. *The logical basis of metaphysics*. Cambridge: Harvard University Press.

———. 2007. The place of philosophy in European culture. *European Journal of Analytic Philosophy* 3: 21–30.

———. 2010. *The nature and future of philosophy*. New York: Columbia University Press.
Einstein, Albert. 1954. A testimonial. In Jacques Hadamard, *The psychology of invention in the mathematical field*, 142–143. Mineola: Dover.
Fodor, Jerry. 2004. Water's water everywhere. *London Review of Books* 26(20): 17–19.
Frege, Gottlob. 1960. *The foundations of arithmetic: A logico-mathematical enquiry into the concept of number*. New York: Harper.
Glock, Hans-Johann. 2008. *What is analytic philosophy?* Cambridge: Cambridge University Press.
Hadamard, Jacques. 1954. *The psychology of invention in the mathematical field*. Mineola: Dover.
Hawking, Stephen. 1988. *A brief history of time: From the big bang to black holes*. New York: Bantam Books.
Hempel, Carl Gustav. 1966. *Philosophy of natural science*. Englewood Cliffs: Prentice-Hall.
Kant, Immanuel. 2002. *Theoretical philosophy after 1781*. Cambridge: Cambridge University Press.
———. 2007. *Anthropology, history, and education*. Cambridge: Cambridge University Press.
Kitcher, Philip. 2011. Epistemology without history is blind. *Erkenntnis* 75: 505–524.
Kuhn, Thomas Samuel. 1996. *The structure of scientific revolutions*. Chicago: The University of Chicago Press.
Moore, George Edward. 1942. An Autobiography. In *The Philosophy of G. E. Moore*, ed. Paul Arthur Schilpp, 3–39. La Salle: Open Court.
Mosconi, Jean. 2014. The development of the concept of machine computability from 1936 to the 1960s. In *Constructivity and computability in historical and philosophical perspective*, ed. Jacques Dubucs and Michel Bourdeau, 37–56. Dordrecht: Springer.
Nagel, Thomas. 1987. *What does it all mean? A very short introduction to philosophy*. Oxford: Oxford University Press.
Poincaré, Henri. 1910. *Savants et écrivains*. Paris: Flammarion.
Rescher, Nicholas. 1993. American philosophy today. *The Review of Metaphysics* 46: 717–745.
Rocke, Alan J. 2011. *Image and reality: Kekulé, Kopp, and the scientific imagination*. Chicago: The University of Chicago Press.
Rota, Gian-Carlo. 1997. *Indiscrete thoughts*. Boston: Birkhäuser.
Russell, Bertrand. 1950. *Unpopular essays*. New York: Simon & Schuster.
———. 1960. Introduction. In Ernest Gellner, *Words and things*, 13–15. Boston: Beacon Press.
———. 1995a. *An outline of philosophy*. London: Routledge.
———. 1995b. *My philosophical development*. London: Routledge.
———. 1997. *The problems of philosophy*. Oxford: Oxford University Press.
Santayana, George. 1948. *The life of reason: Reason in common sense*. New York: Scribner.
Searle, John Rogers. 2008. *Philosophy in a new century*. Cambridge: Cambridge University Press.
Wittgenstein, Ludwig. 1932–1933. *Muss sich denn nicht...* . In *Wittgenstein's Nachlass*, TS 219.
———. 1958. *Philosophical investigations*. Oxford: Blackwell.
———. 1969a. *The blue and brown books*. Oxford: Blackwell.
———. 1969b. *On certainty*. Oxford: Blackwell.
———. 1974. *Philosophical grammar*. Oxford: Blackwell.
———. 1975. *Philosophical remarks*. Oxford: Blackwell.
———. 1978. *Remarks on the foundations of mathematics*. Oxford: Blackwell.
———. 1979. *Notebooks 1914–1916*. Oxford: Blackwell.
———. 1998. *Culture and value: A selection from the posthumous remains*. Oxford: Blackwell.
———. 2002. *Tractatus Logico-Philosophicus*. London: Routledge.
———. 2005. *The big typescript TS 213*. Oxford: Blackwell.

Chapter 3
The Foundationalist View

Abstract The heuristic view of philosophy is opposed to the foundationalist view, according to which philosophy does not aim at knowledge and methods to acquire knowledge, but only at justifying already obtained knowledge, by providing a foundation for it. The foundationalist view assumes, first, that there is immediately justified knowledge and all other knowledge is deduced from it, and, secondly, that immediately justified knowledge is absolutely certain, being based on intuition. This chapter argues that both these assumptions are unjustified. One might think of replacing the foundationalist view with a milder version, the weak foundationalist view, according to which immediately justified knowledge, while not being absolutely certain, has at least some intrinsic credibility. But the chapter argues that the weak foundationalist view also is based on some unjustified assumptions.

3.1 The Foundationalist View and the Architectural Metaphor

The heuristic view of philosophy, described in Chap. 2, contrasts not only with classical analytic philosophy, but also with the foundationalist view of philosophy. While, according to the heuristic view, philosophy aims at knowledge and methods to acquire knowledge, according to the foundationalist view philosophy merely aims at justifying already obtained knowledge, by providing a foundation for it.

The foundationalist view is based on two assumptions. The first assumption is that there is immediately justified knowledge, namely knowledge which is justified independently of inference from other knowledge, and all other knowledge is deduced from it. The second assumption is that immediately justified knowledge is absolutely certain, being based on intuition.

The foundationalist view owes its name to the architectural metaphor, according to which knowledge is an edifice built up according to a given plan. The foundation of the edifice consists of immediately justified knowledge, and the body consists of further knowledge anchored to the foundation via deductive inference. The foundation is unshakable, being based on intuition.

The architectural metaphor goes back to Aristotle, who states that scientific knowledge is deduction from principles, where a principle is "that from which a

© Springer International Publishing AG 2017
C. Cellucci, *Rethinking Knowledge*, European Studies in Philosophy of Science 4,
DOI 10.1007/978-3-319-53237-0_3

thing first arises," like "the foundation of a house" (Aristotle, *Metaphysica*, Δ 1, 1013 a 4–5). In particular, a principle is like the foundation of a house which is unshakable, being based on intuition, since it is "intuition that apprehends the principles" (Aristotle, *Analytica Posteriora*, B 19, 100 b 12). And "intuition" is "always true" (ibid., B 19, 100 b 7–8).

From antiquity to the present day, the foundationalist view has had a large following.

For example, Russell states that "what we firmly believe, if it is true, is called knowledge, provided it is either intuitive or inferred" from "intuitive knowledge from which it follows logically" (Russell 1997, 139). A body of knowledge must be arranged "in deductive chains, in which a certain number of initial propositions form a logical guarantee for all the rest. These initial propositions are premisses for the body of knowledge in question" (Russell 1993, 214). They "must be knowledge which is independent of inference" (Russell 1992b, 157). Moreover, they must be knowledge which is absolutely certain, namely knowledge "of whose truth there can be no doubt" (ibid., 178). In order to be such, they must be "intuitive knowledge," namely "immediate knowledge of truths," and hence "self-evident truths" (Russell 1997, 109). Such truths must be "self-evident in a sense which ensures infallibility" (ibid., 135). All other knowledge "consists of everything that we can deduce from self-evident truths by the use of self-evident principles of deduction," so "all our knowledge of truths depends upon our intuitive knowledge" (ibid., 109).

From the first assumption of the foundationalist view – that there is immediately justified knowledge, and all other knowledge is deduced from it – it is clear that, according to this view, knowledge is based on the axiomatic method.

The second assumption of the foundationalist view – that immediately justified knowledge is absolutely certain, being based on intuition – derives from the fact that the main motivation for this view is to save knowledge from sceptical doubt.

Thus, Russell states that he "was troubled by scepticism and unwillingly forced to the conclusion that most of what passes for knowledge is open to reasonable doubt" (Russell 1971, III, 220). The question which worried him was: "Is there any knowledge in the world which is so certain that no reasonable man could doubt it?" (Russell 1997, 7). This kind of knowledge was essential to stop scepticism, by which "the whole attempt to get behind to something more solid, and worthy to be called knowledge, is futile" (Russell 1992b, 159).

3.2 The First Assumption of the Foundationalist View

Is the foundationalist view adequate? In order to answer this question, let us examine the first assumption of the foundationalist view, that there is immediately justified knowledge, and all other knowledge is deduced from it.

This assumption is motivated by the infinite regress argument, according to which, since human capacities are finite, we cannot go through an infinite series of premisses. So a series of premisses in a science cannot be infinite, it must have a beginning, and hence there must be immediately justified knowledge.

Thus, Russell states that, "since human capacity is finite, what is known of a science cannot contain more than a finite number of definitions and propositions," so "every series of definitions and propositions must have a beginning, and therefore there must be undefined terms and unproved propositions. The undefined terms are understood by means of acquaintance. The unproved propositions must be known by self-evidence" (Russell 1992b, 158).

The infinite regress argument, however, is inadequate. Although, by the finiteness of human capacities, we cannot go through an infinite series of premisses, this does not mean that the series of the premisses cannot be infinite, but only that, at each stage, we can only go through a finite initial segment of the series. And yet, as in the analytic method, we can go through longer and longer finite initial segments.

Of course, if the series of the premisses is infinite, there will be no immediately justified premisses, so no knowledge will be definitive. But this does not mean that there will be no knowledge. There would be no knowledge only if the premisses, or hypotheses, occurring in the infinite series were arbitrary. But they need not be arbitrary. As in the analytic method, they must be plausible, namely the arguments for them must be stronger than the arguments against them, on the basis of the existing knowledge. If the hypotheses are plausible, then there will be knowledge, albeit provisional knowledge always in need of further consideration, since new data may always emerge.

That the infinite regress argument is inadequate means that the first assumption of the foundationalist view is unjustified. As in the analytic method, the process of problem solving may consist in formulating hypotheses which are plausible, where each hypothesis is in its turn a problem which must be solved, and is solved by formulating another hypothesis which is plausible, and so on, *ad infinitum*.

3.3 The Second Assumption of the Foundationalist View

Let us now examine the second assumption of the foundationalist view, that immediately justified knowledge is absolutely certain, being based on intuition. The foundationalist view tries to motivate this assumption in several ways. One of the most significant ones is Russell's.

According to Russell, immediately justified knowledge consists, on the one hand, of "the general truths of logic" and, on the other hand, of "the particular facts of sense" (Russell 1993, 78). The latter are "truths of perception," namely truths which are "immediately derived from sensation" (Russell 1997, 113). Both the general truths of logic and the truths of perception are intuitive truths, and hence are absolutely certain. Indeed, "the more we reflect upon these, the more we realize exactly what they are, and exactly what a doubt concerning them really means, the more luminously certain do they become" (Russell 1993, 78). Admittedly, "verbal doubt concerning even these is possible," but "real doubt, in these two cases, would, I think, be pathological" (ibid.).

In view of the illusions of sense, Russell's claim that real doubt about the truths of perception would be pathological may seem odd. But Russell argues that "there are no such things as 'illusions of sense'. Objects of sense, even when they occur in dreams, are the most indubitably real objects known to us" (ibid., 92–93). What is illusory in the illusions of sense "is only the inferences to which they give rise; in themselves, they are every bit as real as the objects of waking life" (ibid., 93).

This view on the truths of perception is not peculiar to Russell, but goes back to the Greeks. Thus Aristotle states that "perceptions are always true, while most of the imaginings turn out to be false" (Aristotle, *De Anima*, Γ 3, 428 a 11). Epicurus states that "our sensations" are "the standards of truth," since "there is nothing which can refute sensations or convict them of error" (Diogenes Laertius, *Vitae Philosophorum*, X, 31). Even objects of sense "occurring to madmen and to people in dreams are true" (ibid., X, 32). Conversely, "falsehood and error always lie in that which is added by our judgment" (ibid., X, 50).

Kant reasserts this view: "The senses do not err; yet not because they always judge correctly, but because they do not judge at all," indeed, "error, and thus also illusion as leading to the latter, are to be found only in judgments" (Kant 1998, A293/B350). Thus, to the human being "the full moon, which he sees ascending near the horizon through a hazy air, seems to be further away, and also larger, than when it is high in the heavens," but this "is an error of the understanding, not of the senses" (Kant 2007, 258). The "illusion is not ascribed to the senses, but to the understanding, whose lot alone it is to render an objective judgment from the appearance" (Kant 2002, 86).

3.4 Frege's and Russell's Foundational Programs for Mathematics

Russell's assumption, that immediately justified knowledge includes the general truths of logic, is the basis of Russell's logicist program: To show that "all pure mathematics deals exclusively with concepts definable in terms of a very small number of fundamental logical concepts," and that "all its propositions are deducible from a very small number of fundamental logical principles" (Russell 2010, xliii). If this could be shown, it would follow that all mathematical knowledge is absolutely certain, because the fundamental logical principles are intuitive truths, and hence are absolutely certain.

Russell's logicist program is an extension of Frege's logicist program, according to which all arithmetical truths are deducible from a very small number of fundamental logical principles, so arithmetic is merely "a further development of logic" (Frege 1984, 145). Logical principles flow from intellectual intuition, which is "the logical source of knowledge" (Frege 1979, 267). On the other hand, geometrical truths are not deducible from logical principles, indeed "our knowledge of them flows from a source very different from the logical source, a source which might be called spatial intuition" (Frege 1980, 37). Namely, Kant's pure sensible intuition. Russell's logicist program is an extension of Frege's logicist program since it assumes that, not only arithmetical truths, but also geometrical truths, are deducible from a very small number of fundamental logical principles.

Frege's logicist program, however, fails because, by Gödel's first incompleteness theorem, there are arithmetical truths which cannot be deduced from logical principles. *A fortiori*, Russell's logicist program fails.

3.5 Russell's Alternative Foundational Program for Mathematics

Russell, however, makes another attempt to show that that all mathematical knowledge is absolutely certain. He argues that the axioms of mathematics are absolutely certain, because their logical consequences are true.

Indeed, Russell states that there is a "close analogy between the methods of pure mathematics and the methods of the sciences of observation" (Russell 1973, 272). In mathematics, "we tend to believe the premisses because we can see that their consequences are true, instead of believing the consequences because we know the premisses to be true. But the inferring of premisses from consequences is the essence of induction" (ibid., 273–274). Thus, "the reason for accepting an axiom" is "always largely inductive" (Whitehead and Russell 1925–1927, I, 59). Hence, "the method in investigating the principles of mathematics is really an inductive method, and is substantially the same as the method of discovering general laws in any other science" (Russell 1973, 274).

This attempt also, however, fails because from a false axiom one can deduce true consequences. In order to assert that an axiom is true, one ought to be able to show that all of its logical consequences are true, but this is generally unfeasible. (For more on this, see Chap. 21).

Russell himself acknowledges that, by considering the logical consequences of an axiom, "infallibility is never attainable, and therefore some element of doubt should always attach to every axiom and to all its consequences" (Whitehead and Russell 1925–1927, I, 59). Thus he admits that his attempt is inadequate with respect to the aim of showing that all mathematical knowledge is absolutely certain.

3.6 Hilbert's Foundational Program for Mathematics

In addition to Russell's and Frege's attempts, other attempts have been made to show that mathematical knowledge is absolutely certain, because it is based on intuition. A very influential attempt has been Hilbert's foundational program.

According to Hilbert, the hard core of mathematics consists of knowledge about "concrete objects that are intuitively present as immediate experience prior to all thought" (Hilbert 1967b, 464). Such knowledge is absolutely certain, because it is based on concrete intuition.

However, mathematics as a whole also includes knowledge about abstract objects, obtained by abstract operations. Such knowledge is problematic, because "abstract operation with general concept-scopes and contents has proved to be inadequate and uncertain" (Hilbert 1996c, 1121).

Therefore, Hilbert states his consistency program: To demonstrate by absolutely reliable means – namely, means based on concrete, sensible intuition – that mathematics as a whole is consistent. For, the "extension by the addition" of abstract objects to concrete objects "is legitimate only if no contradiction is thereby brought about in the old, narrower domain" (Hilbert 1967b, 471). If this could be demonstrated, it would follow that all mathematical knowledge is absolutely certain.

Hilbert's consistency program, however, fails because, by Gödel's second incompleteness theorem, it is impossible to demonstrate by absolutely reliable means that mathematics as a whole is consistent. (For details on Hilbert's consistency program, see, for example, Cellucci 2007, 38–62).

3.7 Russell's Foundational Program for Empirical Knowledge

Russell's assumption that immediately justified knowledge also includes the truths of perception, is the basis of his foundational program for empirical knowledge: To show that all empirical knowledge can be "inferred in some sense" from the truths of perception, "though not necessarily in a strict logical sense" (Russell 1993, 75). If this could be shown, it would follow that, since the truths of perception are intuitive truths and hence are absolutely certain, all empirical knowledge is absolutely certain.

But Russell's foundational program for empirical knowledge fails, because it is based on the assumption that "the 'evidence of the senses' is proverbially the least open to question" (ibid., 74). Then, in order to see that the truths of perceptions are actually truths, we would have no means more reliable than the evidence of the senses itself. Indeed, Russell states: "If the reality is not what appears, have we any means of knowing whether there is any reality at all? And if so, have we any means of finding out what it is like?" (ibid., 16). Such questions "are bewildering," and philosophy cannot answer to them, although it can at least ask them, and "show the strangeness and wonder lying just below the surface even in the commonest things of daily life" (ibid.).

Thus, however, Russell implicitly admits that the evidence of the senses does not provide an adequate basis for asserting that all empirical knowledge is absolutely certain.

3.8 Inadequacy of the Architectural Metaphor

It could be argued that, like the above foundational programs, all other attempts to show that immediately justified knowledge is absolutely certain, being based on intuition, fail.

The attempt to save knowledge from sceptical doubt by appealing to intuition calls to mind baron Munchausen who, in trying to jump on his horse across a swamp, took off badly and fell in up to his neck: "I should, beyond any doubt, have come to an untimely end, had I not, by the force of my unaided arm, lifted up my pig-tail, together with my horse" (Raspe 2005, 56).

We can no more pull ourselves up by intuition, out of the swamps of sceptical doubt, than baron Münchausen could pull himself up by his hair – specifically, his pig-tail – out of the swamp into which he had fallen. Thus the second assumption of the foundationalist view is unjustified.

From the fact that both the first and the second assumption of the foundationalist view are unjustified, it seems fair to conclude that the foundationalist view is untenable.

In fact the architectural metaphor, to which the foundationalist view owes its name, is inadequate. Knowledge is not an edifice built up according to a plan fixed beforehand, but the plan develops as knowledge grows. Moreover, knowledge can be extended not only by adding new floors, namely new knowledge, to the edifice, but also by modifying the existing floors, namely already obtained knowledge. Furthermore, to extend knowledge may require establishing new relations between different edifices, namely, between different systems of knowledge.

3.9 The Weak Foundationalist View

Since the foundationalist view is untenable, some people have proposed to replace it with a milder version, the weak foundationalist view, according to which immediately justified knowledge, while not being absolutely certain, has at least some intrinsic credibility. In the past three centuries, this view has had several proponents, from Reid to Russell himself and Wittgenstein.

For example, Wittgenstein states that "giving grounds" eventually "comes to an end" (Wittgenstein 1969b, § 204). The end consists of propositions such as "'Here is a hand' (namely my own hand)" (ibid., § 52). Such propositions "seem to underlie all questions and all thinking" (ibid., § 415). They are certain, but not absolutely certain, because "one may be wrong even about 'there being a hand here'" (ibid., § 25). This proposition is certain only in the sense that "it simply gets assumed as a truism, never called in question" (ibid., § 87).

Propositions of this kind are accepted as immediately justified knowledge, not because they are absolutely certain, but because they are reasonable, namely because we might "say: 'the reasonable man believes this'" (ibid., § 323). Indeed, "when we say that we know that such and such ..., we mean that any reasonable person in our position would also know it, that it would be a piece of unreason to doubt it" (ibid., § 325). That propositions which seem to underlie all questions and all thinking are reasonable, implies that they may eventually be replaced with others, because "what men consider reasonable or unreasonable alters. At certain periods men find reasonable what at other periods they found unreasonable. And vice versa" (ibid., § 336).

Wittgenstein's view, however, is unjustified because, as a criterion of validity, reasonableness is unreliable. For example, Wittgenstein states: "Suppose some adult had told a child that he had been on the moon. The child tells me the story, and I say it was only a joke" because "no one has ever been on the moon; the moon is a long way off and it is impossible to climb up there or fly there" (ibid., § 106). Of course, "there might be people who believe that that is possible," but "we say: these people do not know a lot that we know," they "are wrong and we know it" (ibid., § 286). That someone has been on the moon, "not merely is nothing of the sort ever seriously reported to us by reasonable people, but our whole system of physics forbids us to believe it" (ibid., § 108).

Wittgenstein's argument is embarrassing. Of course, at the time no one had ever been on the moon, but people who believed that it was possible to go to the moon, and that one day someone would go there, were by no means unreasonable. Our whole system of physics did not forbid them to believe it. In fact, a little later, some men went to the moon without contravening the laws of our whole system of physics, but just on the basis of them. This shows that, as a criterion of validity, reasonableness is unreliable. All the more so because Wittgenstein seems to hold that reasonableness is the only criterion of validity. For, he states that the propositions which seem to underlie all questions and all thinking are propositions "which we affirm without special testing; propositions, that is, which have a peculiar logical role in the system of our empirical propositions" (ibid., § 136).

Therefore, the weak foundationalist is not a viable alternative to the foundationalist view.

3.10 The Alleged Death of Epistemology

That the foundationalist view is untenable has led some people to conclude that epistemology is dead.

Thus Rorty states that, if we "drop the notion of epistemology as a quest" for "truth, we are in a position to ask whether there still remains something for epistemology to be. I want to urge that there does not" (Rorty 1980, 210).

Williams states that "a non-foundational picture will be too trivial to be thought of as an epistemological theory. It is reasonable to take this to mean that epistemology as a serious theoretical enterprise is dead after all" (Williams 1999, 198).

But this is unjustified. For reasons that will be explained in Chap. 8, epistemology cannot be a quest for truth, but this does not mean that that there remains nothing for epistemology to be. As it will be argued in Chap. 9, it can be a quest for plausibility. The purpose of epistemology is not to inquire into the possibility of certain, objective, universal knowledge, because there is no such knowledge, it is rather to inquire into methods to acquire knowledge, fallible knowledge and yet knowledge.

Moreover, a non-foundational picture need not be too trivial to be thought of as an epistemological theory. A non-foundational picture such as the heuristic view, described in Chap. 2, is by no means too trivial to be thought of as an epistemological theory. Indeed, it puts questions about knowledge at the centre of philosophy, reviving the view that philosophy is an enquiry which primarily aims at knowledge and methods to acquire knowledge.

References

Cellucci, Carlo. 2007. *La filosofia della matematica del Novecento*. Rome: Laterza.

Frege, Gottlob. 1979. *Posthumous writings*. Oxford: Blackwell.

———. 1980. *Philosophical and mathematical correspondence*. Oxford: Blackwell.

———. 1984. *Collected papers on mathematics, logic, and philosophy*. Oxford: Blackwell.

Hilbert, David. 1967b. The foundations of mathematics. In *From Frege to Gödel: A source book in mathematical logic, 1879–1931*, ed. Jean van Heijenoort, 464–479. Cambridge: Harvard University Press.

———. 1996c. The new grounding of mathematics: First report. In *From Brouwer to Hilbert*, II, ed. William Ewald, 1117–1134. Oxford: Oxford University Press.

Kant, Immanuel. 1998. *Critique of pure reason*. Cambridge: Cambridge University Press.

———. 2002. *Theoretical philosophy after 1781*. Cambridge: Cambridge University Press.

———. 2007. *Anthropology, history, and education*. Cambridge: Cambridge University Press.

Raspe, Rudolf Erich. 2005. *The adventures of baron Munchausen*. Mineola: Dover.

Rorty, Richard. 1980. *Philosophy and the mirror of nature*. Princeton: Princeton University Press.

Russell, Bertrand. 1971. *Autobiography*. London: Allen & Unwin.

———. 1973. *Essays in analysis*. London: Allen & Unwin.

———. 1992b. *Theory of knowledge: The 1913 manuscript*. London: Routledge.

———. 1993. *Our knowledge of the external world*. London: Routledge.

———. 1997. *The problems of philosophy*. Oxford: Oxford University Press.

———. 2010. *The principles of mathematics*. London: Routledge.

Whitehead, Alfred North, and Bertrand Russell. 1925–1927. *Principia mathematica*. Cambridge: Cambridge University Press.

Williams, Michael. 1999. *Groundless belief*. Princeton: Princeton University Press.

Wittgenstein, Ludwig. 1969b. *On certainty*. Oxford: Blackwell.

Chapter 4
The Limits of Scepticism

Abstract The main motivation of the foundationalist view is to save knowledge from sceptical doubt. This chapter maintains that such motivation is unjustified. Indeed, according to a widespread opinion, absolute scepticism is irrefutable, since no logical argument can be advanced against it. On the contrary, the chapter argues that logical arguments can be advanced against all sceptical doubts raised since antiquity. It would be impossible to show this for all such sceptical doubts in a limited space. Nevertheless, the chapter shows it for some of the main ones, specifically, the sceptical doubts raised by two schools discussed by Aristotle, Sextus Empiricus' indeterminacy doubt, Descartes' dream doubt, Descartes' demon doubt, and Hume's induction doubt. The chapter also argues that, contrary to the widespread opinion that absolute scepticism stands on the side of logic and reason, absolute scepticism stands on the side of mysticism.

4.1 The Question of Sceptical Doubt

In Chap. 3 it has been argued that the foundationalist view is untenable. The aim of this chapter is to argue that, not only the foundationalist view is untenable, but its main motivation, to save knowledge from sceptical doubt, is unjustified.

According to a widespread opinion, absolute scepticism is irrefutable. For example, Russell states that, "if we adopt the attitude of the complete sceptic," then "our scepticism can never be refuted," because, "from blank doubt, no argument can begin" (Russell 1997, 150). Therefore, "against this absolute scepticism, no logical argument can be advanced" (ibid.).

On the contrary, the chapter argues that absolute scepticism is not irrefutable, in fact logical arguments can be advanced against all sceptical doubts raised since antiquity. It would be impossible to show this for all such sceptical doubts in a limited space. Nevertheless, the chapter shows it for some of the main ones, specifically, the sceptical doubts raised by two schools discussed by Aristotle, Sextus Empiricus' intederminacy doubt, Descartes' dream doubt, Descartes' demon doubt, and Hume's induction doubt. The chapter also argues that, contrary to the widespread opinion that absolute scepticism stands on the side of logic and reason, absolute scepticism stands on the side of mysticism.

© Springer International Publishing AG 2017

C. Cellucci, *Rethinking Knowledge*, European Studies in Philosophy of Science 4,
DOI 10.1007/978-3-319-53237-0_4

4.2 The Sceptical Doubts Raised by Two Ancient Schools

Aristotle discusses the sceptical doubts raised by two schools, the first ascribed to Antisthenes, the second ascribed to followers of Xenocrates, or to Menaechmus, or to the youthful Aristotle himself. Both schools assume that, in order to have knowledge based on reason, "one must have knowledge of the primitive premisses" (Aristotle, *Analytica Posteriora*, A 3, 72 b 5). From this assumption, the first school concludes that "there is no knowledge" (ibid., A 3, 72 b 5–6). Instead, the second school concludes that "there is knowledge but there are demonstrations of everything" (ibid., A 3, 72 b 6–7). Including falsehoods.

According to the first school – the one which concludes that there is no knowledge – with respect to the series of the premisses two cases are possible.

1) The series of the premisses does not terminate, so there are no primitive premisses, and "one is led back in an infinite regress" (ibid., A 3, 72 b 8–9). However, "if among the prior items there are no primitive premisses, one cannot know the posterior items because of the prior items," so "there is no knowledge" (ibid., A 3, 72 b 9–10).

2) The series of the premisses "terminates and there are primitive premisses," of which, however, "there is no knowledge because there is no demonstration of them, and this is the only thing by virtue of which there is knowledge" (ibid., A 3, 72 b 11–13). Then, the primitive premisses are mere assumptions. Now, "if it is not possible to have knowledge of the primitive premisses, then it is not possible to have knowledge of what follows from them absolutely and properly, but only on the assumption that they are the case" (ibid., A 3, 72 b 13–15). So, there is no knowledge.

According to the second school – the one which concludes that there is knowledge but there are demonstrations of everything – the series of the premisses and the proposition being proved "proceed in a circle or reciprocally" (ibid., A 3, 72 b 17–18). Then "nothing prevents there being demonstrations of everything" (ibid., A 3, 72 b 16–17). In particular, nothing prevents there being demonstrations of falsehoods. So, there is no knowledge.

Collecting together the cases considered by the two schools, we have the following three possible cases:

(I) The series of the premisses does not terminate, and there is an infinite regress.

(II) The series of the premisses terminates with premisses which are mere assumptions.

(III) The series of the premisses and the proposition being proved proceed in a circle or reciprocally.

According to the two schools, in none of the cases (I)–(III) there is knowledge, so there is no knowledge.

4.3 Aristotle's Argument Against the Two Schools

Aristotle agrees with the two schools that in none of the cases (I)–(III) there is knowledge.

In particular, Aristotle agrees that, when the members of the first school claim that in case (I) there is no knowledge, "they are right, because it is impossible to survey infinitely many items" (Aristotle, *Analytica Posteriora*, A 3, 72 b 10–11). Moreover, if the series of the premisses "does not terminate and there is always something prior to what one looks for, then there will be demonstrations of everything" (ibid., A 22, 84 a 1–2).

Aristotle also agrees that, when the members of the first school claim that in case (II) there is no knowledge, they are right, because mere assumptions need not be true, while, in order to have knowledge, the premisses "must be true, since it is impossible to have knowledge of what is not the case" (ibid., A 2, 71 b 25–26).

Aristotle also agrees that, when the members of the second school claim that in case (III) there is no knowledge, they are right, because "it is impossible for the same thing at the same time to be both prior and posterior to something" (ibid., A 2, 72 b 27–28). Moreover, demonstrations which proceed in a circle or reciprocally "say nothing more than that this is the case if this is the case, and it is easy to prove everything in this way" (ibid., A 3, 72 b 34–35).

But, while agreeing with the two schools that in none of the cases (I)–(III) there is knowledge, Aristotle claims that this does not mean that there is no knowledge. Indeed, none of (I)–(III) is the case, therefore the sceptical doubts raised by the two schools are unjustified.

For, according to Aristotle, "we have knowledge of things through demonstrations," where a demonstration is a deduction which proceeds "from premisses that are true, and primitive, and immediate" (ibid., A 2, 71 b 17–21). And from premisses that are absolutely certain, being based on intuition, since "it is intuition, and not discursive thinking, that apprehends the primitive things," it "is intuition that apprehends the unchanging and first terms in the order of demonstrations" (Aristotle, *Ethica Nicomachea*, Z 11, 1143 a 36–1143 b 3).

That the premisses are true means that, with respect to them, "falsity does not exist, nor error, but only ignorance" (Aristotle, *Metaphysica*, Θ 10, 1052 a 1–2). That the premisses are primitive means that they are principles, because "I call the same thing primitive and principle" (Aristotle, *Analytica Posteriora*, A 2, 72 a 6–7). That the premisses are immediate means that no other proposition is prior to them, because a premiss is "immediate if there is no other proposition prior to it," so "a principle of a demonstration is an immediate proposition" (ibid., A 2, 72 a 7–8). That the premisses are absolutely certain, being based on intuition, means that "the intuition" on which the premisses are based "concerns things about which there is no falsity" (Aristotle, *De Anima*, Γ 6, 430 a 26–27).

But, if we have knowledge of things through demonstrations, and a demonstration is a deduction which proceeds from premisses that are true, primitive, immediate, and absolutely certain, being based on intuition, then none of (I)–(III) is the case.

(I) is not the case, because the premisses of a demonstration are immediate, so no other proposition is prior to them, hence the series of the premisses terminates, therefore there is no infinite regress.

(II) is not the case, because the premisses of a demonstration are true and absolutely certain, being based on intuition, so they are not mere assumptions.

(III) is not the case, because the premisses of a demonstration are principles, so they are immediate premisses, hence there is no other proposition prior to them. Then the series of the premisses and the proposition being proved cannot proceed in a circle or reciprocally, since it is impossible for the same thing at the same time to be both prior and posterior to something.

Since none of (I)–(III) is the case, with respect to the views of two schools Aristotle concludes that "neither of these views is either true or necessary" (Aristotle, *Analytica Posteriora*, A 3, 72 b 7). Therefore, the sceptical doubts raised by the two schools are unjustified.

This is Aristotle's solution of the sceptical doubts raised by the two schools. Clearly Aristotle's solution is based on the foundationalist view, described in Chap. 3. For, it is based on the assumption that we have knowledge of things through demonstrations, where a demonstration is a deduction which proceeds from premisses that are true, primitive, immediate, and absolutely certain, being based on intuition. That premisses are true, primitive, and immediate, implies that they are immediately justified knowledge. Thus there is immediately justified knowledge, and all other knowledge is deduced from it. This is the first assumption of the foundationalist view. On the other hand, that premisses are absolutely certain, being based on intuition, implies that immediately justified knowledge is absolutely certain, being based on intuition. This is the second assumption of the foundationalist view.

4.4 Limits of Aristotle's Solution

That Aristotle's solution of the sceptical doubts raised by the two schools is based on the foundationalist view, implies that his solution is inadequate. For, as argued in Chap. 3, so is the foundationalist view.

Specifically, it is reasonable to claim that in case (II) there is no knowledge. For, in that case, the series of the premisses terminates with premisses which are mere assumptions.

It is also reasonable to claim that in case (III) there is no knowledge. For, in that case, the premisses and the proposition being proved proceed in a circle or reciprocally.

But it is not reasonable to claim that in case (I) there is no knowledge. As we have seen, to support this claim, Aristotle uses two arguments.

The first argument is that the series of the premisses cannot be infinite, because it is impossible to survey infinitely many items. This argument is unjustified because, as already pointed out in Chap. 3, although, by the finiteness of human capacities, we cannot go through an infinite series of premisses, this does not mean that the series of the premisses cannot be infinite, but only that, at each stage, we can only go through a finite initial segment of the series. And yet, as in the analytic method, we can go through longer and longer finite initial segments.

The second argument is that, if the series of the premisses does not terminate and there is always something prior to what one looks for, then there will be demonstrations of everything. This argument is unjustified because, although, if the series of the premisses is infinite, there will be no immediately justified premisses, this does not mean that there will be demonstrations of everything. There would be demonstrations of everything only if the premisses, or hypotheses, occurring in the infinite series were arbitrary. But they need not be arbitrary. As in the analytic method, they must be plausible, namely, the arguments for them must be stronger than the arguments against them, on the basis of the existing knowledge. Now, if the hypotheses must be plausible, then there cannot be demonstrations of everything.

Thus, the claim that in case (I) there is no knowledge, is not reasonable. Therefore, in case (I), the sceptical doubt raised by the first school is unjustified. It is unjustified for a reason different from the one maintained by Aristotle.

4.5 Self-Defeating Character of the Two Schools

Not only, in case (I), the sceptical doubt raised by the first school is unjustified, but all the sceptical doubts raised by the two schools are self-defeating.

The conclusion of the first school, that there is no knowledge, is self-defeating because the conclusion in question is supposed to be knowledge itself. This contradicts the conclusion that there is no knowledge.

The conclusion of the second school, that there is knowledge but there are demonstrations of everything, is self-defeating because, if there are demonstrations of everything, then there is also a demonstration that there is no knowledge, so there is no knowledge. This contradicts the conclusion that there is knowledge.

Nor, to avoid the self-defeating character of the conclusion of the first school that there is no knowledge, one can replace it with the conclusion that there is no knowledge apart from the knowing not to know – the Socratic ‘knowing not to know’. For, if there is no knowledge apart from the knowing not to know, then the conclusion ‘There is no knowledge apart from the knowing not to know’, is not knowledge, therefore it is unjustified.

4.6 Sextus Empiricus' Indeterminacy Doubt

Now we consider Sextus Empiricus' indeterminacy doubt, which Sextus Empiricus states as follows: "Everything is indeterminate" (Sextus Empiricus, *Pyrrhoniae Hypotyposes*, I, 198). For, in any investigation, "to every argument an equal argument is opposed" (ibid, I, 202). So we are led to "suspension of judgment," namely to "a state of the mind on account of which we neither deny nor affirm anything" (ibid., I, 10).

In order to argue that, in any investigation, we are led to suspension of judgment, Sextus Empiricus considers five modes, or forms of argument: the mode from dispute, the mode from infinite regress, the mode from relativity, the mode from assumption, and the mode from reciprocalness. (These modes are often attributed to Agrippa; see Diogenes Laertius, *Vitae Philosophorum*, IX, 88–89). Sextus Empiricus claims that "every matter of investigation can be brought under these modes" (Sextus Empiricus, *Pyrrhoniae Hypotyposes*, I, 169). And he argues that, in each of them, we are led to suspension of judgment. His argument is as follows.

(a) The mode from dispute is that according to which, "with regard to the matter proposed, there has arisen, both among ordinary people and among the philosophers, an unresolved dissension, because of which we are unable to accept something or reject it, and so we are led to suspension of judgment" (ibid., I, 165).

(b) The mode from infinite regress is that in which "what is brought forward as a source of conviction for the matter proposed itself needs another such source, which itself needs another, and so on *ad infinitum*, so that, since we have no point from which to begin to establish anything, suspension of judgment follows" (ibid., I, 166).

(c) The mode from relativity is that in which "the existing thing appears to be such-and-such relative to the judging subject and to the things observed together with it, but as to its real nature we suspend judgment" (ibid., I, 167).

(d) The mode from assumption is that which occurs when people "take as their starting point something which they do not establish but claim to assume simply and without demonstration, in virtue of a concession" (ibid., I, 168). So, since we have no established point from which to start, suspension of judgment follows.

(e) The mode from reciprocalness is that which occurs when "what ought to establish the thing under investigation needs to be made convincing by the thing under investigation; whence, being unable to take either thing in order to establish the other, we suspend judgment about both of them" (ibid., I, 169).

Sextus Empiricus' indeterminacy doubt is not manifestly unjustified. Indeed, his statement that everything is indeterminate is not self-defeating because, when he "says 'Everything is indeterminate', he takes 'is' in the sense of 'appears to me'" (ibid., I, 198). Thus, "when he says this, he is reporting what appears to him about the matters at hand" (ibid., I, 197). Moreover, "by 'everything' he means not whatever exists," but only those matters "which he has considered" (ibid., I, 198).

Nevertheless, that Sextus Empiricus' indeterminacy doubt is not manifestly unjustified, does not mean that it is justified. For, it is reasonable to claim that, in the mode from assumption (d) and in mode from reciprocalness (e), suspension of judgment follows, since such modes correspond to cases (II) and (III), respectively, of the arguments of the two schools discussed by Aristotle. But it is unreasonable to claim that, in the mode from infinite regress (b), suspension of judgment follows, since such mode corresponds to case (I) of the arguments of the two schools discussed by Aristotle and, by what has been argued above, it is unjustified to say that in case (b) suspension of judgment follows. It is also unreasonable to claim that, in the mode from dispute (a), or in the mode from relativity (c), suspension of judgment follows. For, the mode from dispute (a) suggests that any opinion about something can be controverted, and the mode from relativity (c) suggests that any opinion about something is relative to the judging subject and to the things observed together with it. Now, admittedly, any opinion about something can be controverted, and any opinion about something is relative to the judging subject and to the things observed together with it. But some opinions are implausible and can be shown to be implausible, and even if no opinion is absolutely certain, some opinions can be shown to be more plausible than others. Therefore, it is unjustified to claim that in case (a) or in case (c) suspension of judgment follows.

We may then conclude that Sextus Empiricus' indeterminacy doubt, though not manifestly unjustified, is nonetheless unjustified.

4.7 Descartes' Dream Doubt

Now we consider Descartes' dream doubt, which Descartes states as follows: "Admittedly, now I perceive this paper with eyes which are certainly wide awake, this head which I move is not asleep, I consciously and deliberately extend and feel this hand; these things would not happen with such distinctness to someone asleep" (Descartes 1996, VII, 19). Nevertheless, "on other occasions I have been deceived by similar thoughts in my dreams" (ibid.). So, "when I think these things over more carefully, I see so clearly that there are no certain marks by which waking and dreaming can be distinguished that I am stupefied, and this very stupor only reinforces the notion that I am dreaming" (ibid.).

The core of Descartes' doubt is: There are no certain marks by which waking and dreaming can be distinguished; if there are no certain marks by which waking and dreaming can be distinguished, then I am dreaming; therefore I am dreaming.

Descartes' dream doubt is not manifestly unjustified. Already Plato pointed out that it is difficult to say "what evidence could one have to prove, if someone asked now, at the present moment, whether we are asleep and hence our thoughts are a dream, or whether we are awake and really conversing with one another" (Plato, *Theaetetus* 158 b 8–c 1). For, "the two conditions correspond in every circumstance like exact counterparts. There is nothing to prevent us from believing that what we are now conversing about might equally well seem a conversation with one another in sleep. And when in a dream we think we are talking of dreams, there is an extraordinary similarity between the latter and what we talk of when we are talking of waking experiences" (ibid., 158 c 3–7). Then "you see that there is plenty of room for doubt, when we even doubt what is dream and what is waking" (*ibid.*, 158 c 8–d 1).

Nevertheless, that Descartes' dream doubt is not manifestly unjustified, does not mean that it is justified. Indeed, from the fact that it can be argued that there are no certain marks by which waking and dreaming can be distinguished, it does not follow that I am dreaming. For, if there are no certain marks by which waking and dreaming can be distinguished, then there are no certain marks that dreaming is different from waking. In order to infer that I am dreaming, there should be certain marks that dreaming is different from waking, but by Descartes' assumption there are no such marks, so I could as well infer that I am awake.

We may then conclude that Descartes' dream doubt, though not manifestly unjustified, is nonetheless unjustified.

4.8 Descartes' Demon Doubt

Now we consider Descartes' demon doubt, which Descartes states as follows: "I will suppose" that "some malicious demon, supremely powerful and cunning, has employed all his industriousness to deceive me" (Descartes 1996, VII, 22). So "I shall think" that "all external things are merely the delusions of dreams which he has devised to ensnare my judgment. I shall consider myself as not having hands, or eyes, or flesh, or blood, or senses, but as falsely believing that I have all these things" (ibid., VII, 22–23).

Descartes' demon doubt is not manifestly unjustified. Indeed, Descartes states it only instrumentally. He says that "nothing is so conducive to the acquisition of reliable knowledge, as to accustom ourselves beforehand to doubt of all things and especially bodily things" (ibid., VII, 130). Descartes does not want "to imitate the sceptics, who doubt only for doubting's sake, and pretend to be always undecided," on the contrary, his "whole aim is towards certainty, to cast aside the loose earth and sand so as to come upon rock or clay" (ibid., VI, 29). He raises the demon doubt to the only "purpose of completely dispelling all doubts" (ibid., V, 147). There are "reasons which are strong enough to compel us to doubt," but which "are themselves doubtful, and hence are not to be retained later on" (ibid., X, 473–474). Such reasons "are strong so long as we have no others which, by removing the doubt, produce certainty" (ibid., X, 474). The supposition of the malicious demon is one of these reasons.

Nevertheless, that Descartes' demon doubt is not manifestly unjustified, does not mean that it is justified. Descartes himself tries to show that the demon doubt is unjustified by the following argument: "In rejecting everything which we can in any way doubt, and even imagining it to be false, we easily suppose that there is no God, no heaven, no body; and even that we ourselves have no hands, no feet, and indeed no body at all" (*ibid.*, VIII–1, 6–7). And yet, "we cannot for all that suppose that we, who think these things, do not exist. For, it is a contradiction to suppose that what thinks does not, at the very time when it is thinking, exist. And hence this knowledge 'I think, therefore I am' is the first and most certain of all to occur to anyone who philosophizes in an orderly way" (ibid.). So, "observing that this truth 'I think, therefore I am' was so firm and certain that all the most extravagant suppositions of the sceptics were incapable of shaking it, I judged that I could accept it without scruple as the first principle of the philosophy I was seeking" (ibid., VI, 32). Then the "proposition, 'I am, I exist' is necessarily true whenever it is put forward by me, or conceived in my mind" (*ibid.*, VII, 25).

Descartes' argument, however, is inadequate. By accepting 'I think, therefore I am' as the first principle of philosophy, and concluding that the proposition 'I am, I exist' is necessarily true, Descartes thinks to prove existence by thought. But, as Leibniz pointed out, "to say 'I think, therefore I am', is not properly to prove existence by thought, because to think and to be thinking is the same thing; and to say 'I am thinking' is already to say 'I am'" (Leibniz 1965, V, 391). So, to say 'I think' is already to say 'I am'. Therefore, Descartes' argument is begging the question.

Admittedly, in first-order logic we can deduce the conclusion, 'I am', from the premiss, 'I think'. Indeed, 'I think' can be expressed by $A(a)$, and 'I am' by $\exists x(x = a)$. Since, in first-order logic, we have $a = a$, from $A(a)$ we can deduce $a = a \wedge A(a)$, so $\exists x(x = a \wedge A(x))$, hence $\exists x(x = a) \wedge \exists x A(x)$, and therefore $\exists x(x = a)$. Thus, in first-oder logic, we can deduce $\exists x(x = a)$ from $A(a)$.

But, to deduce $\exists x(x = a)$ from $A(a)$ in first-order logic is not properly to prove existence by thought, because first-order logic presupposes that all singular terms denote existing individuals, and hence that in $A(a)$ the singular term a denotes an existing individual, that is, me. Thus, as Leibniz states, the premiss $A(a)$, 'I think', from which the conclusion $\exists x(x = a)$, 'I am', is deduced, presupposes that I exist, so to say 'I am thinking' is already to say 'I am'.

Actually, in first-oder logic we can even prove $\exists x(x = a)$, 'I am', depending on no assumption. For, in first-order logic, we have $a = a$, from which we can deduce $\exists x(x = a)$. Thus, even if I did not exist, in first-order logic it could be proved that I exist. But, once again, this is not properly to prove existence by thought, because first-order logic presupposes that all singular terms denote existing individuals, and hence presupposes that in the proposition $a = a$, from which the conclusion $\exists x(x = a)$ is deduced, the singular term a denotes an existing individual, that is, me.

As an alternative to Descartes' argument, it can be argued that Descartes' demon doubt is unjustified, not trying to prove existence by thought, but using the fact that you cannot deceive someone unless that someone exists.

Indeed, assume *A*: Some malicious demon, supremely powerful and cunning, has employed all his industriousness to deceive me. Since the malicious demon is supremely powerful and cunning and has employed all his industriousness to deceive me, he is deceiving me. But the malicious demon cannot deceive me unless I exist. Therefore, I exist. As Descartes himself acknowledges, if there exists a malicious demon, "supremely powerful and cunning, who is deliberately deceiving me all the time," then "I too undoubtedly exist, if he is deceiving me" (Descartes 1996, VII, 25). Therefore *B*: There is something about which the malicious demon is incapable of deceiving me, namely, that I exist. On the other hand, since the malicious demon is supremely powerful, there is nothing that cannot be brought about by him, just as there is nothing that "cannot be brought about by God" (ibid., V, 224). Therefore $\neg B$: There is nothing about which the malicious demon is incapable of deceiving me. Thus A yields a contradiction, B and $\neg B$. Then, by reduction to the impossible, we conclude $\neg A$: No malicious demon, supremely powerful and cunning, may have employed all his industriousness to deceive me.

This shows that Descartes' demon doubt is unjustified. For, if I suppose that some malicious demon, supremely powerful and cunning, has employed all his industriousness to deceive me, then we get a contradiction.

We may then conclude that Descartes' demon doubt, though not manifestly unjustified, is nonetheless unjustified.

4.9 Hume's Induction Doubt

Finally, we consider Hume's induction doubt, which Hume states as follows: "I have found that such an object has always been attended with such an effect," and from this by induction I infer that "the other objects, which are, in appearance, similar, will be attended with similar effects" (Hume 1975, 34). This conclusion is not absolutely certain, unless induction can be justified by argument. Now, all arguments "may be divided into two kinds, namely, demonstrative" argument, "or that concerning relations of ideas," and probable argument, "or that concerning matter of fact and existence" (ibid., 35). But induction cannot be justified by demonstrative argument, "since it implies no contradiction that the course of nature may change, and that an object, seemingly like those which we have experienced, may be attended with different or contrary effects," and, whatever "implies no contradiction" can "never be proved false by any demonstrative argument" (ibid.). On the other hand, induction cannot be justified by probable argument, hence by induction itself, since all probable arguments "suppose, as their foundation, that the future will resemble the past, and that similar powers will be conjoined with similar sensible qualities," hence to endeavour to justify induction by probable argument "is begging the question" (ibid., 37). Since induction cannot be justified either by demonstrative argument or by probable argument, induction cannot be justified by argument. Therefore, the conclusions inferred by induction cannot be absolutely certain and are only probabilities, probability being "evidence which is still attended with uncertainty" (Hume 1978, 124). But then the conclusions inferred by induction cannot be knowledge. For, knowledge must be absolutely certain, because knowledge is "the assurance arising from the comparison of ideas" (ibid.). There is knowledge only if we can "determine their relations, without any possibility of error" (ibid., 71). Knowledge is based on reasoning which can be carried out with "perfect exactness and certainty" (ibid.). Since the conclusions inferred by induction cannot be knowledge, it follows that induction cannot produce knowledge. Indeed, "knowledge and probability are of" wholly "contrary and disagreeing natures" (ibid., 181).

Hume's induction doubt is not manifestly unjustified because, as Hume maintains, induction cannot be justified by argument, in Hume's sense – in Hume's sense, because induction, and generally non-deductive rules, can be justified by argument in a sense different from Hume's (see Cellucci 2013a, Chap. 18). That induction cannot be justified by argument in Hume's sense was already pointed out, long before the modern era, by the Indian philosophers of the Carvaka school. Indeed, Hume's argument is that "the inductive passage from observed cases to all cases cannot be justified except on the assumption that the nature is uniform and that the future resembles the past, an assumption that amounts to begging the question" (Chakrabarti 2010, 15). Now, this argument is "similar to the Carvaka argument that if the claim of pervasion," namely, of invariant and universal connection, "is justified through inference, one would have to use pervasion itself, inviting either vicious regress or circularity" (ibid.).

Nevertheless, that Hume's induction doubt is not manifestly unjustified, does not mean that it is justified. For, from the fact that induction cannot be justified by argument in Hume's sense, it does not follow that it cannot produce knowledge. Induction can produce knowledge, since it can lead to conclusions that are plausible, and, as it will be argued in Chap. 9, the only knowledge we can have about the world is plausible knowledge. Hume's conclusion that induction cannot produce knowledge depends on the assumption that knowledge must be absolutely certain. But this assumption is unjustified, because knowledge can only be plausible, hence fallible knowledge, and yet knowledge.

We may then conclude that Hume's induction doubt, though not manifestly unjustified, is nonetheless unjustified.

4.10 Scepticism, Mysticism, and the Foundationalist View

What has been said above shows that logical arguments can be advanced against the sceptical doubts raised by the two schools discussed by Aristotle, Sextus Empiricus' indeterminacy doubt, Descartes' dream doubt, Descartes' demon doubt, and Hume's induction doubt. It could be argued that logical arguments can be advanced also against all other sceptical doubts raised since antiquity. Therefore, the claim that, against absolute scepticism, no logical argument can be advanced, is unjustified.

According to a widespread opinion, absolute scepticism stands on the side of logic and reason. But it is not so, actually absolute scepticism stands on the side of mysticism.

As Russell himself acknowledges, at the origin of mysticism there is "the doubt concerning common knowledge, preparing the way for the reception of what seems a higher wisdom" (Russell 1994, 27). Indeed, sceptical doubt leads the mystic to "the belief in insight as against discursive analytic knowledge: the belief in a way of wisdom, sudden, penetrating, coercive, which is contrasted with the slow and fallible study of outward appearance by science relying wholly upon the senses" (ibid.,

26). This belief gives the mystic "the sense of a mystery unveiled, of a hidden wisdom now suddenly become certain beyond the possibility of a doubt," and is, for the mystic, "the gateway to an ampler world" (ibid., 27). For, it gives the mystic "the conception of a reality behind the world of appearance and utterly different from it," a reality which is "thinly veiled by the shows of sense, ready, for the receptive mind, to shine its glory," and which the mystic "knows, with a knowledge beside which all other knowledge is ignorance" (ibid., 27–28). Thus, sceptical doubt is part of "the mystic's initiation" since it leads to the "belief in the possibility of a way of knowledge which may be called revelation or insight or intuition, as contrasted with sense, reason, and analysis, which are regarded as blind guides leading to the morass of illusion" (ibid., 27).

As already pointed out, the main motivation of the foundationalist view is to save knowledge from sceptical doubt. To this purpose, the foundationalist view assumes that knowledge must be ultimately based on intuition, on the grounds that only intuition can rescue knowledge from falling into the morass of illusion. But, on account of the connection between absolute scepticism and mysticism, quite the opposite is the case. Intuition cannot rescue knowledge from falling into the morass of illusion, on the contrary, intuition is a blind guide leading to the morass of illusion, since it cannot safely guide us to anything (see Chap. 10).

The belief that intuition is the ground of the absolute certainty of knowledge is a crucial feature of the foundationalist view, and of any theory of knowledge which assumes that the aim of knowledge is certainty. Therefore, it is any theory of knowledge which assumes that the aim of knowledge is certainty that is a blind guide leading to the morass of illusion.

References

Cellucci, Carlo. 2013a. *Rethinking logic: Logic in relation to mathematics, evolution, and method.* Dordrecht: Springer.

Chakrabarti, Kisor Kumar. 2010. *Classical Indian philosophy of induction: The Nyāya viewpoint.* Lanham: Lexington Books.

Descartes, René. 1996. *Oeuvres*. Paris: Vrin.

Hume, David. 1975. *Enquiries concerning human understanding and concerning the principles of morals*. Oxford: Oxford University Press.

———. 1978. *A treatise of human nature*. Oxford: Oxford University Press.

Leibniz, Gottfried Wilhelm. 1965. *Die Philosophischen Schriften*. Hildesheim: Olms.

Russell, Bertrand. 1994. *Mysticism and logic*. London: Routledge.

———. 1997. *The problems of philosophy*. Oxford: Oxford University Press.

Chapter 5
Philosophy and the Humanistic Disciplines

Abstract According to the heuristic view, philosophy is not essentially different from the sciences. Some people oppose this view, claiming that philosophy is a humanistic discipline which aims to make sense of ourselves and of our activities, contrary to the sciences, which have nothing to say about this. In this chapter it is argued that these claims are unjustified. For, only if we have a global view of the world and our place in it, we may understand who we are and where we are going. To this aim, philosophy must be an investigation about the world and must be able to use the achievements of the present sciences. In fact, humanistic disciplines are disciplines which are not opposed to the sciences, but study human beings alongside with the sciences.

5.1 The View That Philosophy Is a Humanistic Discipline

In Chap. 2 it has been maintained that philosophy is continuous with the sciences, in the sense that it aims at a kind of knowledge which differs from scientific knowledge in no essential respect and is not restricted to any area. Some people oppose this view, claiming that philosophy is a humanistic discipline, by which they mean a discipline essentially different from the sciences.

Thus Williams states that the view that philosophy is continuous with the sciences is a form of scientism, where scientism is "a misunderstanding of the relations between philosophy and the natural sciences which tends to assimilate philosophy to the aims, or at least the manners, of the sciences" (Williams 2006, 182). Scientism assumes that, since the sciences "possess intellectual authority," philosophy must "try to share in it" (ibid., 188). But, according to Williams, this assumption is unjustified, because philosophy is essentially different from the sciences.

Indeed, the sciences describe "the world 'as it is in itself', independent of perspective" (ibid., 184). They provide the kind of representation "that might be reached by any competent investigators of the world, even though they differed from us – that is to say, from human beings – in their sensory apparatus and, certainly, their cultural background" (ibid., 185). Therefore, the sciences give "the 'absolute conception' of the world," namely "a conception of the world that might be arrived at by any investigators, even if they were very different from us" (Williams 1985, 139).

© Springer International Publishing AG 2017

C. Cellucci, *Rethinking Knowledge*, European Studies in Philosophy of Science 4,
DOI 10.1007/978-3-319-53237-0_5

On the contrary, philosophy is "part of a wider humanistic enterprise of making sense of ourselves and of our activities," and, in order to answer many of its questions, "it needs to attend to other parts of that enterprise, in particular to history;" therefore philosophy "should get rid of scientistic illusions," and "should not try to behave like an extension of the natural sciences" (Williams 2006, 197).

Admittedly, "some philosophical subjects have scientific neighbours," but "even in areas where its practices are most relevant, science can be a bad model for philosophy" (Williams 2014, 367). In fact, "there are several features of natural science which, applied to philosophy, may have a baleful effect on it" (ibid.). Hence, the assimilation of philosophy to the aims of the sciences "is a mistake" (Williams 2006, 182). In particular, there is no reason why "the idea that science and only science describes the world as it is in itself, independent of perspective," should mean that "there is no independent philosophical enterprise" (ibid., 184).

5.2 Relation with a Non-analytic Tradition

Although Williams is a significant representative of classical analytic philosophy, his view of philosophy as a humanistic discipline follows a well-established non-analytic philosophical tradition concerning the relation between the sciences and the humanistic disciplines. Dilthey and Husserl are the main representatives of this tradition.

For example, Husserl opposes the view that philosophy is continuous with the sciences, maintaining that this view "decapitates philosophy" (Husserl 1970, 9). It involves that, "in our vital need," philosophy "has nothing to say to us. It excludes in principle precisely the questions" which humanity "finds the most burning: questions of the meaning or meaninglessness of the whole of this human existence" (ibid., 6). But this is a mistake, because philosophy and the sciences have completely different aims. The sciences aim at "objective knowledge of the world, the universe of realities existing in themselves," and have "the intent of knowing being-in-itself through truths in themselves" (ibid., 316–317). Conversely, philosophy is a humanistic discipline in which "theoretical interest is directed at human beings exclusively as persons, at their personal life and accomplishments, and correlatively at the products of such accomplishments" (ibid., 270). In order to deal with these questions, philosophy needs to attend to its own history, because we must "reflect back, in a thorough historical and critical fashion," in order to arrive at "a radical self-understanding: we must inquire back into what was originally and always sought in philosophy" (ibid., 17).

However, although Williams' view of philosophy follows a well-established philosophical tradition concerning the relation between the sciences and the humanistic disciplines, his view is based on somewhat problematic assumptions that we will presently discuss.

5.3 What Is Scientism, Really?

Williams describes scientism as the view that assimilates philosophy to the aims, or at least the manners, of the sciences. This is a common way of describing scientism, but it suffers from the indeterminacy of the expression 'the sciences'. For, since the seventeenth century, new sciences have arisen and continue to arise. As a result, the methods of natural science, or the categories and things recognized in natural science, are not fixed but change over time.

Scientism is more appropriately described as the view that assimilates philosophy to the aims of the present sciences. Scientism, thus meant, arises from overrating the scope of the present sciences. It assumes that the present sciences cover all fields of knowledge, and that they make us know the world as it is in itself, independent of perspective. Several analytic philosophers share this assumption of scientism, even those who, like Williams, are opposed to assimilating philosophy to the aims of the sciences. As we will presently argue, however, this idea is unjustified.

5.4 Sciences and the World as It Is in Itself

Williams assumes that the sciences describe the world as it is in itself. This assumption, however, is unjustified. With Galileo's philosophical revolution, science renounced Aristotle's aim to penetrate the true and intrinsic essence of natural substances, contenting itself with dealing with some of their phenomenal properties mathematical in kind, such as location, motion, shape, size (see Chap. 8). Therefore, its theories are not about the essence of natural substances.

Then, it is unjustified to say that the sciences describe the world as it is in itself, in fact they only describe certain phenomenal properties of the world. It is because of his philosophical revolution that Galileo has an important place, not only in the history of science, but also in the history of philosophy. In particular, without Galileo's philosophical revolution, Kant's distinction between phenomena and things in themselves would not have been so significant.

5.5 Sciences and the Independence of Perspective

Williams assumes that the sciences describe the world independent of perspective. This assumption is widely shared. Indeed, as Putnam points out, "analytic philosophy has become increasingly dominated by the idea that science, and only science, describes the world as it is in itself, independent of perspective" (Putnam 1992, ix–x).

But saying that the sciences describe the world independent of perspective, conflicts with the fact that the representation of the world that the sciences provide is essentially dependent on the human cognitive apparatus. As Calvino states: "Although science interests me just because of its efforts to escape from anthropomorphic knowledge, I am nonetheless convinced that our imagination cannot be anything but anthropomorphic" (Calvino 1988, 90). Indeed, "we can know nothing about what is outside us if we overlook ourselves" (Calvino 2010, 107).

In particular, the representation of the world that the sciences provide is affected by the limitations of our cognitive apparatus – not only our perceptive apparatus but also our conceptual apparatus, since the latter is derived to a large extent from our perceptual apparatus. Such limitations are not accidental because, as Locke pointed out, they are essential to the survival of the species (see Chap. 9).

5.6 Sciences and the Absolute Conception

Williams assumes that the sciences give the absolute conception of the world, where "the notion of an absolute conception can serve to make effective a distinction between 'the world as it is independent of our experience' and 'the world as it seems to us'" (Williams 1985, 139). However, as Putnam points out, to this aim Williams would need "an absolute notion of 'absoluteness'," since only such notion could serve to make effective a distinction between 'the world as it is independent of our experience' and 'the world as it seems to us', but Williams has "only a perspectival notion of absoluteness, not an absolute one" (Putnam 2001, 608).

The question of whether some conception of the world might be absolute, amounts to the question of whether any investigators, even very different from us, would converge upon it. But this will inevitably be a question of interpretation, and such question can be settled only by the most plausible interpretations that can be given of the different communities of investigators. Now, on a Quinean or Davidsonian view of interpretation, interpretations are open to the indeterminacies of translation, and "Williams accepts a Quinean or Davidsonian view of interpretation, and the corollary of semantic indeterminacy" (Blackburn 2010, 253). This means that, within one perspective, there may be convergence upon a given conception of the world, while, within another perspective, there may be none. Then, as Putnam states, Williams has only a perspectival notion of absoluteness, not an absolute one.

5.7 Philosophy as Different from the Sciences

Williams assumes that philosophy is different from the sciences, because it is part of a wider humanistic enterprise of making sense of ourselves and of our activities, from which the sciences are excluded. This assumption is somewhat surprising, because Williams is a significant representative of analytic philosophy, and it is widely held that analytic philosophers "are expected to integrate the results and methods of the sciences with their own philosophizing," to the point that "distinguishing between philosophy per se and the sciences will get more difficult" (Schwartz 2012, 322).

Anyway, Williams' assumption ignores that we cannot make sense of ourselves and of our activities unless we have a global view of the world and our place in it. Only on that basis we can understand who we are and where we are going. This requires that, as argued in Chap. 2, philosophy be an inquiry which primarily aims at acquiring knowledge about the world, including ourselves, and can make use of the achievements of the present sciences, since the latter are what we already know.

Williams' assumption implies that the conclusions reached by the sciences are irrelevant to those reached by the humanistic disciplines, and in particular by philosophy. However, as Zeki maintains, "the path to acquiring knowledge – whether grounded in scientific experimentation or through philosophical or humanistic speculation – must use similar mental processes" (Zeki 2014, 13). And, if "similar brain processes are involved in humanistic and scientific inference, we are led ineluctably to the view that conclusions reached by one are relevant to those reached by the other" (ibid., 14).

Williams' claim that philosophy is different from the sciences is based on the assumption that values and knowledge must be thought of as rigorously separated. But this assumption is unjustified. On the one hand, what we believe ought to be is affected by what we know about the world including ourselves, since knowledge may modify values. On the other hand, values affect theory appraisal, guiding the choice of problems and hypotheses and affecting the evaluation of hypotheses. (For more on this, see below).

5.8 Philosophy and History

Williams assumes that a feature which characterizes philosophy as being a humanistic discipline, different from the sciences, is that, unlike the sciences, philosophy needs to attend to other parts of the humanistic enterprise, in particular to history.

Indeed, Williams states that, even if it is "desirable that scientists should know something about their science's history," this "is not essential to their enquiries" (Williams 2006, 204). On the contrary, philosophers need to know about their discipline's history, because the ignorance of the history of philosophy "stands in the way of our understanding who we are, what our concepts are, what we are up to" (Williams 2014, 412). Philosophy "can get a real hold on its task only with the help of history; or, rather, as Nietzsche put it, philosophising in such a case must itself be historical" (ibid., 409).

The reference to Nietzsche is not surprising, because Williams admired Nietzsche. On the contrary, the claim that philosophy can get a real hold on its task only with the help of history is somewhat surprising, because Williams considered himself an analytic philosopher, and claimed that analytic philosophy "remains the only real philosophy there is" (Williams 2006, 168). Now, analytic philosophy has traditionally proceeded with strongly ahistorical assumptions about the practice of philosophy. Williams himself has done so for most of his life, except in his last book, in which he states that "at a certain point philosophy needs to make way for history, or, as I prefer to say, to involve itself in it" (Williams 2002, 93).

Therefore, Williams' acknowledgment that philosophy can get a real hold on its task only with the help of history, is a change from the traditional practice of analytic philosophy. And it is definitely a change for the better, since philosophy really needs to make use of the experience of the philosophers of the past. As already pointed out in Chap. 2, without such experience, philosophy risks reinventing the wheel, or hunting down trails that are known to be dead ends.

But Williams is wrong in assuming that, while history is fundamental, the sciences are irrelevant to our understanding of who we are, what our concepts are, and what we are up to. Such understanding requires us to acquire knowledge about the world, and the sciences are a constituent part of that aim. Of course, the present sciences may not be sufficient to the task, new knowledge may be required in new areas. For this reason, philosophy must be an inquiry which aims at acquiring knowledge about the world, including ourselves. Nevertheless, the sciences are essential to our understanding of who we are, what our concepts are, and what we are up to.

5.9 Continuity with the Philosophical Tradition

Contrary to the assumption of scientism, philosophy does not reduce to the present sciences. But, on the other hand, it is not opposed to them. As argued in Chap. 2, philosophy and the present sciences are part of a common enterprise, which aims at acquiring knowledge about the world, including ourselves. Philosophy and the present sciences are distinct, but not distant. They are a continuum, not in the sense of scientism, but in the sense that the kind of knowledge at which philosophy aims does not differ essentially from scientific knowledge and is not limited to any area.

The view that philosophy is an inquiry which primarily aims at acquiring knowledge about the world, including ourselves, is continuous with the philosophical tradition. It is simply the view of philosophy that both Plato and Aristotle put forward at the origins of philosophy as a discipline. Thus Plato states that "philosophy is the acquiring of knowledge [*ktesis epistemes*]" (Plato, *Euthydemus*, 288 d 8). Only "the one who is wholeheartedly ready to taste every sort of knowledge, and who is eager to know and is never satisfied, may be justly termed a philosopher" (Plato, *Respublica*, V 475 c 6–8). Aristotle states that "philosophy is rightly called the knowledge of truth" (Aristotle, *Metaphysica*, α 1, 993 b 19–20).

This view of philosophy has been reaffirmed in various ways at the beginning of modern philosophy.

Thus Bacon states: "I have taken all knowledge to be my province" (Bacon 1961–1986, VIII, 109). Indeed, "under philosophy I include all arts and sciences, and in a word whatever has been from the occurrence of individual objects collected and digested by the mind into general notions" (ibid., V, 504).

Descartes states that philosophy aims at "a perfect knowledge of all things that man can know, both for the conduct of his life, and for the preservation of his health, and for the invention of all the arts," indeed acquiring knowledge "is properly called philosophizing" (Descartes 1996, IX–2, 2).

Conversely, the view that philosophy is an inquiry which primarily aims at acquiring knowledge about the world, including ourselves, is not shared by classical analytic philosophy, which has abandoned the aim that Plato and Aristotle originally set for philosophy. But, as we have seen in the Introduction, this has led many scientists to conclude that philosophy is sterile, wan and irrelevant, or even dead. If philosophy wants to be adequate to its past, it must not abandon the aim that Plato and Aristotle originally set for it. Rather, it must be inspired by the hope that it will contribute to that aim.

5.10 Theoretical and Practical Knowledge

It might be thought that the view that philosophy is an inquiry which primarily aims at acquiring knowledge about the world, including ourselves, excludes ethics and politics. But it is not so. This is clear from the fact that Plato and Aristotle, who put forward the view in question, concerned themselves with ethics and politics. Like the sciences, ethics and politics are aimed at knowledge, though practical knowledge rather than theoretical knowledge. For example, when Aristotle states that philosophy is rightly called the knowledge of truth, by 'philosophy' he means both theoretical philosophy and practical philosophy, which includes ethics and politics. In his view, truth is the aim, not only of theoretical knowledge, but also of practical knowledge. For, even "practical philosophers investigate the way in which something is," except that, unlike theoretical philosophers, "they do not aim at knowledge of what is eternal, but of what is relative to a certain circumstance and a certain moment" (Aristotle, *Metaphysica*, α 1, 993 b 22–23). They aim at truth not for its own sake, but only as a means to something else, namely action, because the ultimate aim "of practical knowledge is action" (ibid., α 1, 993 b 21).

Williams himself does not deny that there can be practical knowledge. He only claims that, while the aim of theoretical knowledge is absolute truth, namely truth independent of any perspective, the aim of practical knowledge is relative truth, namely truth dependent on a given perspective. In his view, "ethical knowledge is dependent on 'the local perspectives or idiosyncrasies of enquirers', whereas scientific knowledge may not be" (Williams 2006, xvii).

Admittedly, the logical positivists popularized the view that ethics and politics are essentially different from the sciences because they are concerned with values, whereas the sciences are concerned with facts. Thus Carnap states that "the philosophy of moral values or moral norms" is "not an investigation of facts," its purpose is "to state norms for human action or judgments about moral values," where stating a norm or a value judgment "is merely a difference of formulation" (Carnap 1935, 23). Indeed, a norm "has an imperative form, for instance: 'Do not kill!' The corresponding value judgment would be: 'Killing is evil'" (ibid., 23–24). Since the value statement, 'Killing is evil' has the grammatical form of a declarative sentence, "most philosophers have been deceived by this form into thinking that a value statement is really an assertive proposition, and must be either true or false" (ibid., 24). But, really, "a value statement is nothing else than a command in a misleading grammatical form," so "it is neither true nor false. It does not assert anything and can neither be proved nor disproved" (ibid.). Therefore, facts and values are essentially heterogeneous and, unlike facts, values cannot be investigated.

This view, however, overlooks that, on the one hand, values depend on what we know about the world, including ourselves, and may change as our knowledge changes, so facts affect values. On the other hand, values guide us in choosing scientific problems to work on, and hypotheses to solve them, as well as criteria for evaluating hypotheses, so values affect facts. This is especially clear if, as it will be argued in Chap. 9, the aim of science is to make plausible hypotheses about the world, namely hypotheses such that the arguments for them are stronger than the arguments against them, on the basis of the existing knowledge. Indeed, determining whether the arguments for a hypothesis are stronger than the arguments against it, on the basis of the existing knowledge, requires value criteria. In particular, every scientific revolution leads to a change in values, which produces a change in the choice of scientific problems to work on, and of hypotheses to solve them, as well as of criteria for evaluating hypotheses. Then, it is unjustified to say that facts and values are essentially heterogeneous, and that, unlike facts, values cannot be investigated. As Dewey states, "inquiry, discovery take the same place in morals that they have come to occupy in sciences and nature" (Dewey 2004, 100). If the claim is made that, in ethics and politics, "'values' are involved and that inquiry as 'scientific' has nothing to do with values, the inevitable consequence is that inquiry in the human area is confined to what is superficial and comparatively trivial" (ibid., xvi). In particular, it is unjustified to say that "the justification of evaluations will not be scientific" (Kekes 2015, 260). The justification of evaluations is an intrinsic part of the scientific process.

It might also be thought that the view that philosophy is an inquiry which primarily aims at acquiring knowledge about the world, including ourselves, excludes scholarship, which re-examines philosophy of the past. But it is not so. As pointed out above, without the experience of the philosophers of the past, philosophy risks hunting down trails that are known to be dead ends. Specifically, as we have seen, the view that philosophy is an inquiry which primarily aims at acquiring knowledge about the world, including ourselves, is based on the experience of the philosophy of the past, from Plato and Aristotle to Bacon and Descartes.

5.11 The Humanistic Disciplines Revisited

From what has been said above it follows that, if by a 'humanistic discipline' is meant a discipline essentially different from the sciences, it is unjustified to say that philosophy is a humanistic discipline. Philosophy is not essentially different from the sciences, on the contrary, it is akin to them, being an inquiry which aims at acquiring knowledge about the world, including ourselves. As argued in Chap. 2, in carrying out such an inquiry, philosophy may even give birth to new sciences.

Then, it is unjustified to say that philosophy is a humanistic discipline opposed to the sciences. In particular, it is unjustified to say this on the ground that philosophy as a humanistic discipline can make sense of ourselves and of our activities, while the sciences cannot. Only if we have a global view of the world and our place in it, we may understand who we are and where we are going, and, to this aim, philosophy must be an investigation about the world and must be able to use the achievements of the present sciences.

Humanistic disciplines need not be viewed as opposed to the sciences, they can be viewed as disciplines that study human beings. Not only are there already several scientific disciplines that study human beings, such as psychology, cognitive science, anthropology, sociology, political science, economics, linguistics, but even the more traditional humanistic disciplines, such as art and literature, may contribute to this study. Thus Zeki maintains that "Paul Cézanne's preoccupation, and artistic experimentation, with how color modulates form is but a variant of the neurobiological question of how the separate representations of form and color are integrated in the brain to give us a unitary percept of both" (Zeki 2014, 12). And "one is likely to acquire as much experimentally testable knowledge, for example, from reading Kant on aesthetics or Balzac and Zola on creativity than one would from any presently available scientific text" (ibid., 13). This supports the view that the humanistic disciplines and the sciences are not opposed.

References

Bacon, Francis. 1961–1986. *Works*. Stuttgart Bad Cannstatt: Frommann Holzboog.

Blackburn, Simon. 2010. *Practical tortoise raising and other philosophical essays*. Oxford: Oxford University Press.

Calvino, Italo. 1988. *Six memos for the next millennium*. Cambridge: Harvard University Press.

———. 2010. *Palomar*. New York: Random House.

Carnap, Rudolf. 1935. *Philosophy and logical syntax*. London: Kegan Paul.

Descartes, René. 1996. *Oeuvres*. Paris: Vrin.

Dewey, John. 2004. *Reconstruction in philosophy*. Mineola: Dover.

Husserl, Edmund. 1970. *The crisis of European sciences and transcendental phenomenology*. Evanston: Northwestern University Press.

Kekes, John. 2015. Philosophy: Scientific or humanistic? *Philosophy* 90 : 241–262.

Putnam, Hilary. 1992. *Renewing philosophy*. Cambridge: Harvard University Press.

———. 2001. Reply to Bernard Williams' 'Philosophy as a humanistic discipline'. *Philosophy* 76 : 605–614.

Schwartz, Stephen P. 2012. *A brief history of analytic philosophy: From Russell to Rawls*. New York: Wiley.

Williams, Bernard. 1985. *Ethics and the limits of philosophy*. Cambridge: Harvard University Press.

———. 2002. *Truth & truthfulness: An essay in genealogy*. Princeton: Princeton University Press.

———. 2006. *Philosophy as a humanistic discipline*. Princeton: Princeton University Press.

———. 2014. *Essays and reviews 1959–2002*. Princeton: Princeton University Press.

Zeki, Semir. 2014. Neurobiology and the humanities. *Neuron* 84 : 12–14.

Part II
The Nature of Knowledge

Chapter 6
Knowledge and Naturalism

Abstract In the past century, the prevailing view has been that the main problem of epistemology, or theory of knowledge, is to give a definition of knowledge, and the prevailing definition of knowledge has been that knowledge is justified true belief. This contrasts with the fact that, since antiquity, the definition of knowledge as justified true belief has been known to be subject to counterexamples, and all the variants of this definition that have been put forward are also subject to counterexamples. As an alternative, this chapter argues that the main problem of epistemology is to explain what role knowledge plays in human life. Specifically, unlike justified true belief, knowledge in not merely a state of mind, but rather a response to the environment that is essential for survival. This involves a naturalistic approach to knowledge, according to which knowledge is a natural process, continuous with the biological processes by which life is sustained and evolved, and has a vital role, in the literal sense that life exists only insofar as there is knowledge.

6.1 What Is Knowledge?

After examining the nature of philosophy, this part of the book examines the nature of knowledge. This is essential for a philosophy which aims at knowledge and methods to acquire knowledge.

In the past century, the prevailing view has been that the main problem of epistemology, or theory of knowledge, is to give a definition of knowledge. Thus Russell states that "perhaps the most important and difficult" problem of theory of knowledge is "the definition of knowledge," namely "the question how knowledge should be defined" (Russell 1988, 194).

Moreover, the prevailing definition of knowledge has been that knowledge is justified true belief. This definition is usually motivated by the argument that it "bears its rationale on its face. The belief condition excludes ignorance, the truth condition excludes error, and the justification condition excludes mere opinion" (Williams 2001, 16–17).

The foundationalist view, described in Chap. 3, adopts the definition of knowledge as justified true belief, assuming in particular that, among justified true beliefs, there are some immediately justified and absolutely certain beliefs from which all other knowledge is deduced.

© Springer International Publishing AG 2017

C. Cellucci, *Rethinking Knowledge*, European Studies in Philosophy of Science 4,
DOI 10.1007/978-3-319-53237-0_6

The definition of knowledge as justified true belief is already considered by Plato: "Knowledge is true belief accompanied by a justification" (Plato, *Theaetetus*, 202 c 7–8). According to this definition, "knowledge is true belief" (ibid., 200 e 4). But true belief is not necessarily knowledge. For, true beliefs are unstable and ephemeral, so "they tend not to stay long, they escape from a man's mind, thus they are not worth much unless they are tied down" by justification, that is, by "an account of the reason why" (Plato, *Meno*, 98 a 1–4). In fact, "when a man has a true belief about something but without justification, his mind does think truly of it, but he does not know it; for, if one cannot give or receive a justification of a thing, one has no knowledge of that thing" (Plato, *Theaetetus*, 202 b 8–c 3). Conversely, when true beliefs "are tied down" by justification, they "are stable" and "become knowledge. This is why knowledge is something more valuable than true belief. What distinguishes knowledge from true belief is" in the justification, "in being tied down" (Plato, *Meno*, 98 a 5–8).

That the definition of knowledge as justified true belief is already considered by Plato, has led some people to think that Plato proposes it.

Thus, Chisholm states that "the traditional or classic" definition of knowledge, "and the one proposed in Plato's dialogue, the *Theaetetus*, is that knowledge is justified true belief" (Chisholm 1989, 90).

Gettier states that "Plato seems to be considering some such definition at *Theaetetus* 201, and perhaps accepting one at *Meno* 98" (Gettier 1963, 121, footnote 1).

But Plato considers the definition of knowledge as justified true belief only to criticize and ultimately reject it. Indeed, he makes it quite clear that "neither perception, nor true belief, nor true belief accompanied by a justification, could be knowledge" (Plato, *Theaetetus*, 210 a 9–b 2). For, a justification would have to be knowledge itself. Then, saying that knowledge is true belief accompanied by a justification would amount to saying that knowledge is true belief accompanied by knowledge. But "it would be utterly silly to say" that knowledge is true "belief accompanied by knowledge" (ibid., 210 a 7–8). The circle would be too blatant.

Despite Plato's criticism, in the last century there has been a revival of the definition of knowledge as justified true belief. However, at the beginning of the century, Meinong and Russell had already made it quite clear that this definition is inadequate, by giving counterexamples to it. For instance, one of Meinong's counterexamples is: "Someone, who has tinnitus, thinks to hear the doorbell, and he is right, because accidentally someone is really ringing" (Meinong 1968–1978, V, 619). But that person cannot be said to know that someone is ringing the doorbell. For, it is only accidentally that, when the person in question thinks to hear the doorbell, someone is ringing the doorbell, and "truth must not be accidental" (ibid.). The person's belief that someone is ringing the doorbell is true, and is also justified because, normally, when one hears the doorbell, it is because someone is ringing it. But the person's belief is not knowledge, because what the person hears is only a phantom sound due to his tinnitus.

Several attempts have been made to modify the definition of knowledge as justified true belief so as to avoid counterexamples, but all the variants of this definition that have been formulated are also subject to counterexamples (for details, see, for example, Cellucci 1998, Chap. 4). In particular, the problem raised by Plato remains, because an interpretation of 'justified' strong enough to exclude counterexamples is unlikely to be independent of knowledge itself. Therefore Nozick says that, if asked "what is knowledge," a "reasonable philosopher today might say that, in view of the difficulties thus far encountered, he just does not know," and, "on inductive grounds, that since all previous accounts have fallen so will any one he can formulate" (Nozick 1997, 148).

6.2 A Naturalistic Approach to Knowledge

In view of the difficulties of the prevailing view that the main problem of epistemology is to give a definition of knowledge, and that the definition of knowledge is, 'Knowledge is justified true belief', this chapter puts forward an alternative to it. It argues that the main problem of epistemology is to explain what role knowledge plays in human life, and that knowledge is a response to the environment which is essential for survival.

This is a major change with respect to the prevailing view. For, giving a definition of knowledge does not say anything about the role knowledge plays in human life. Moreover, defining knowledge as justified true belief amounts to reducing knowledge to merely a state of mind of a certain sort. Indeed, as Russell states, according to that definition, knowledge "is a state of mind of a certain sort" (Russell 1988, 194). Conversely, if knowledge is a response to the environment which is essential for survival, then knowledge is not merely a state of mind of a certain sort, hence a subjective mental state. On the contrary, it is an objective phenomenon which plays an essential and pervasive role in the life of all human beings.

To say that knowledge is a response to the environment which is essential for survival, is to adopt a naturalistic approach to knowledge.

Naturalism is expressed by Bacon's motto: "Certe ultra naturam nihil [Certainly, nothing beyond nature]" (Bacon 1961–1986, I, 550). That is, naturalism is the view that all of reality, including our existence and knowledge, must be explained in natural terms because there is nothing beyond nature. Then, a naturalistic approach to knowledge is one according to which knowledge is a natural phenomenon, in particular, a natural process continuous with the biological process by which life is sustained and evolved. For, knowledge has a vital role, in the sense that life exists only insofar as there is knowledge.

A naturalistic approach to knowledge is proposed by Dewey, who states that "knowledge is not something separate and self-sufficing, but is involved in the process by which life is sustained and evolved" (Dewey 2004, 50). Dewey's naturalistic approach, however, has some limitations, because Dewey assumes that "knowledge" means "warranted assertibility" (Dewey 1938, 9). The latter expression designates "the institution of conditions which remove need for doubt" (ibid., 7). This assumption is problematic because, as it will be argued in Chap. 10, knowledge cannot be absolutely certain, so we cannot institute conditions which remove need for doubt. (For more on warranted assertibility, see Chap. 9). Moreover, Dewey assumes that perception is not knowledge, because a "fact 'presented' in perception" is "one directly 'apprehended'," while an object of knowledge is "determined by inference" and hence is "known through a logical way" (Dewey 1911, 396). This assumption is also problematic, because in Chap. 15 it will be argued that perception is based on inference.

6.3 The Biological Role of Knowledge

Since antiquity, the question of explaining what role knowledge plays in human life has had a prominent place on the philosophical agenda.

Particularly significant is Aristotle's view. According to him, human beings "pursue knowledge only for the sake of knowing, and not for any utilitarian end," and they began to pursue it only "when almost all the necessities of life and the things that make for comfort and recreation had been secured" (Aristotle, *Metaphysica*, A 2, 982 b 20–24). Knowledge aims "at no end beyond itself" (Aristotle, *Ethica Nicomachea*, K 7, 1177 b 20). Thus, according to Aristotle, knowledge is an end in itself, and human beings began to pursue it only when almost all the necessities of life had been secured.

But it is not so. Knowledge is not an end in itself, nor human beings began to pursue it only when almost all the necessities of life had been secured. They pursued it from the very beginning, because knowledge has a vital role, vital in the literal sense, because life exists only insofar as there is knowledge. The resources the lack of which hinders the preservation and reproduction of life include not only food and the existence of sexual partners, but also knowledge. Human beings make hypotheses about the environment, on the basis of which they assume behaviours that ensure their survival. Being essential for survival, knowledge has a biological role. In this role, knowledge primarily serves to solve the problem of survival.

Aristotle is right in saying that "all human beings by nature desire knowledge" (Aristotle, *Metaphysica*, A 1, 980 a 21). But he is right in saying so only if by 'nature' one means 'biological nature'. All human beings desire knowledge, not because, as Aristotle states, from knowledge "nothing results apart from contemplation" (Aristotle, *Ethica Nicomachea*, K 7, 1177 b 2). They desire knowledge because knowledge has a vital role, being a certain kind of response to the environment that is essential for survival. Thus, human beings pursue knowledge not simply because they wish but, more important, because they must.

The biological role of knowledge is not limited to human beings, since knowledge is essential for the survival of all organisms. All organisms are cognitive systems, and life itself owes its existence and preservation to cognitive processes. 'All organisms' includes simple ones, such as the prokaryotes, the unicellular organisms that were the first form of life on earth. While not having a nervous system, the prokaryotes have molecules on their cell membrane which act as sensors, and send data about the environment to their DNA. On the basis of the information encoded in their DNA, the prokaryotes interpret the data being received, thus achieving knowledge about the environment. Then they react to this knowledge moving in a liquid environment by means of their flagellum and pili.

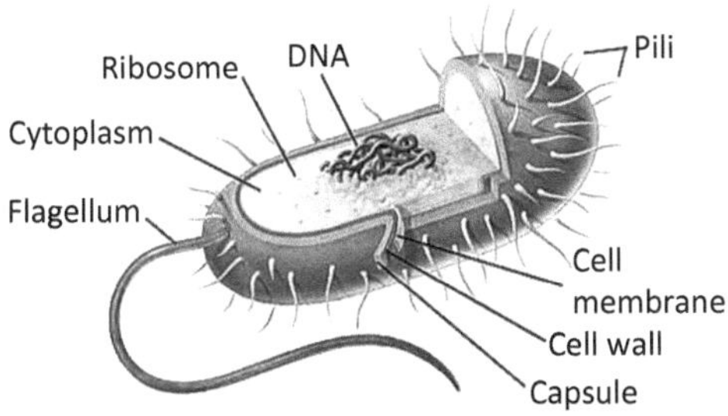

That the prokaryotes are simple organisms does not mean that their amount of DNA is small, since an organism's complexity is not directly proportional to its amount of DNA. Some unicellular organisms have much more DNA than human beings. Rather, an important difference between simple organisms, such as the prokaryotes, and complex ones, such as human beings, is that simple organisms, in a simple and favorable environment, need little knowledge to respond appropriately to the environment and survive. Very rudimentary sensors are sufficient to deal with the stimuli from the environment, and very simple effectors are sufficient to implement appropriate responses. Conversely, complex organisms, in a complex or hostile environment, need a lot of knowledge to respond appropriately to the environment and survive. They need sophisticated sensors and complex effectors, the ability to plan the action predicting its effects, and the ability to learn from experience. By these abilities, an organism may connect what earlier seemed disconnected, and may distinguish what earlier seemed indistinct.

Between the sensors and the effectors, there must be an intermediate system, processing the stimuli received from the environment through sensors, and implementing a response through the effectors. The intermediate system, in simple organisms such as the prokaryotes is the DNA, and in complex organisms is the nervous system. Often the cognitive system is identified with the intermediate system, but this is wrong, because the cognitive system is the whole organism, and also includes processes external to it (see Chap. 17).

Both simple organisms and complex ones pursue knowledge not as an end in itself, but primarily as a means to solve the problem of survival, recognizing those aspects of the environment which make the difference between preserving life and destroying it, and implementing appropriate responses. In particular, human beings pursue knowledge because they are comparatively weak and vulnerable creatures. Being neither though, like sharks, nor numerous and prolific, like ants, human beings must make their evolutionary way in the world mainly by knowledge. It is mainly by knowledge, rather than by sharp teeth or sheer number, that human beings have carved out their niche in evolution's scheme of things.

6.4 Knowledge as a Natural Process

It is often claimed that "knowledge is not a natural phenomenon" (Williams 2004, 194). In view of the biological role of knowledge, this claim is unjustified. As stated above, knowledge is a natural process, continuous with the biological processes by which life is sustained and evolved, and has a vital role, in the literal sense that life exists only insofar as there is knowledge. This depends on the fact that human beings are natural organisms, with cognitive faculties that are the product of evolution. All our knowledge about the world arises from these biologically contingent faculties.

Our cognitive faculties are of several kinds. We have perceptual abilities. We can form concepts in terms of which we think the world. We can reason deductively and non-deductively. We can formulate and evaluate hypotheses about objects and properties that are thought to explain events in the world.

Such cognitive faculties are the basis not only of ordinary everyday knowledge, but also of scientific knowledge. But, while our 'sapient' status started emerging some 10,000 years ago, science has arisen much more recently, because discovering the facts of science, as opposed to the facts of ordinary everyday knowledge, needs much apparatus and professional expertise.

6.5 Knowledge and Consciousness

Against the claim that knowledge is a natural process, Descartes objects that "there can be no knowledge except in a mind" (Descartes 1996, VII, 442). And "there can be nothing in the mind, in so far as it is a thinking thing, of which" the mind "is not conscious" (ibid., VII, 246). Thus knowledge requires consciousness. Now, according to Descartes, only human beings have consciousness, therefore only they can have knowledge, since "animals do not see as we do when we are conscious that we see; but only as we do when our mind is elsewhere" (ibid., I, 413).

Descartes' objection, however, is unjustified, because consciousness is neither necessary nor sufficient for knowledge.

On the one hand, consciousness is not necessary for knowledge, because a significant part of human knowledge is arrived at through processes which are unconscious – they occur too fast and at too low a level in mind to be accessible to direct inspection. Such is perceptual knowledge, which, as it will be argued in Chap. 15, is the result of unconscious inferences.

On the other hand, consciousness is not sufficient for knowledge, because beliefs arrived at consciously, and held to be knowledge, need not be knowledge. For example, Frege arrived consciously at the main axiom of his system, the Basic Law V, and held it to be knowledge, in particular he held it "to be purely logical" (Frege 2013, vii). And yet the Basic Law V was not knowledge, since it led to Russell's paradox (see, for example, Cellucci 2013a, Chap. 10). Since Frege's system was intended to be a realization of Frege's logicist program, described in Chap. 3, this explains Poincaré's sarcastic comment that, after all, logicism "is not sterile, it engenders antinomies" (Poincaré 2013, 483).

6.6 Knowledge and Evolution

It has been stated above that, in its biological role, knowledge primarily serves to solve the problem of survival. Actually, knowledge primarily serves to solve the problem of survival not only of individual organisms, but also of entire species. With respect to species, knowledge helps to solve problems of adaptation.

Suppose that some members of a species develop a new character, for instance a better sensor, which solves a problem of adaptation better than the characters of the other members of the species. Then the members of the species in which the new character is present will tend to have more immediate descendants than those that do not. If the immediate descendants inherit the new character from their parents, the frequency of the latter will increase in the population until, after a certain number of generations, the new character will be present in every member of the species. The new character has a greater adaptive value than the characters of the members of the species in which it is not present, since it increases the ability of the organisms in which that character is present to survive and reproduce.

In the process just described, namely natural selection, knowledge plays an essential role. For, it is not enough that an organism develops, say, a better sensor, the organism must be able to process the data supplied by the sensor in order to implement an effective action for survival.

That knowledge plays an essential role in natural selection, is clear from the fact that organisms that are unable to know that there is an external world have little chance to survive. Organisms that are unable to know that there are other organisms in the external world, have little chance to reproduce. Organisms that are unable to make appropriate hypotheses about the external world are more likely victims of the dangers of the environment. Thus the lack of knowledge is dangerous to the species from an evolutionary point of view.

The essential role played by knowledge in natural selection is acknowledged by Russell, who states that "the whole of our cognitive life is, biologically considered, part of the process of adaptation to facts. This process is one which exists, in a greater or less degree, in all forms of life, but is not commonly called 'cognitive' until it reaches a certain level of development" (Russell 1992a, 160). We "cannot say precisely at what point we pass from mere animal behaviour to something deserving to be dignified by the name of 'knowledge'. But at every stage there is adaptation, and that to which the animal adapts itself is the environment of fact" (ibid.). Therefore, 'knowledge' "must not be defined in a manner which assumes an impassable gulf between ourselves and our ancestors who had not the advantage of language" (ibid., 439).

That, with respect to species, knowledge serves to solve problems of adaptation, means that knowledge is useful not only to avoid the short term threats to the survival of the single organism. The latter can be ensured only for a limited period of time: all the organisms sooner or later die. Other is the case of genes, which contain the instructions of the project of components of organisms. The characters of the project are propagated by promoting the reproduction of genes. With respect to a species, knowledge serves to propagate the characters of the project by promoting the reproduction of genes. Thus knowledge has a biological role, not only with respect to single organisms, but also with respect to entire species. Darwin's theory of evolution needs to be complemented with a theory of knowledge.

6.7 Cultural Role of Knowledge

Of course, in addition to the biological role, knowledge has a cultural role, because in human beings knowledge is a means not only for survival, but also for improving the quality of life, from the routine workings of everyday life to global issues.

The cultural role of knowledge does not reduce to the biological one, because it is based on a non-genetic transmission of information. Indeed, a culture is a body of cognitions that human beings transmit from generation to generation non-genetically, namely not through DNA, but through what they do and communicate. On the other hand, the cultural role of knowledge is not opposed to the biological one, but is continuous with it. As Dewey states, "there is no breach of continuity between operations of inquiry," on which the cultural role of knowledge is based, "and biological operations and physical operations," on which the biological role of knowledge is based, since operations of inquiry "grow out of organic activities," even "without being identical with that from which they emerge" (Dewey 1938, 19).

The continuity between the cultural and the biological role of knowledge is apparent, in particular, from the fact that, even in its cultural role, knowledge can influence biological evolution. The cognitions that form a culture allow organisms to design their environment to some extent, changing the environment so as to make it more suitable to them, and these changes have influence on what organisms survive and reproduce. If several generations of organisms repeatedly modify their environment in the same way, this may lead to changes in the process of natural selection. A simple example of this is given by the inherited culture of pastoralism, which led to the domestication of livestock and the activities associated with the production of milk and its derivatives. This modified the environment of certain human populations for enough generations to select those genes that today make many adults lactose tolerant.

That, even in its cultural role, knowledge can influence biological evolution, holds not only of the human species but also of other species. Some of them are able to modify their environment through their artefacts. Others, while not being able to modify their environment, choose an environment that may affect biological evolution.

In human beings, the biological role of knowledge arose with the human species, perhaps 200,000 years ago. On the other hand, the cultural role of knowledge is much more recent since, as already mentioned, it started emerging only some 10,000 years ago.

6.8 Biological Evolution and Cultural Evolution

In human beings, in addition to the cultural role of knowledge, there is also cultural evolution, namely, in subsequent generations knowledge transmitted non-genetically is modified and increased.

As the cultural role of knowledge is not opposed to the biological one but is continuous with it, so cultural evolution is not opposed to biological evolution but is continuous with it. This depends on the fact that the subjects of cultural evolution are the same as the subjects of biological evolution, namely human beings.

On the other hand, this does not mean that cultural evolution reduces to biological evolution. It does not reduce to biological evolution for at least two reasons.

First, biological evolution is very slow, it takes thousands of unfavourable mutations before a favourable one emerges. Cultural evolution is much faster, being a result of non-genetic interactions between billions of human beings.

Second, certain kinds of organisms are capable of doing things that are not strictly necessary for survival. Such is the case of human beings. In the course of biological evolution, they have been confronted with situations that had not occurred in their evolutionary past. The world changes continually and irregularly, so human beings have to deal all the time with new situations. If their problem solving resources were always strained to the limit, when certain critical situations occurred they might easily fail, and if these failures had frequently occurred in their evolutionary past, we would not be here to tell. To be able to cope with critical situations during times of peak demand, human beings must have excess capacity to spare for other issues at slack times. Thanks to this, in normal circumstances, they may engage in activities that are not immediately necessary for survival. Such activities are a result of cultural evolution.

Then, it is unjustified to say that "from the amoeba to Einstein is only one step" (Popper 1972, 246). Even if, as Russell states, 'knowledge' must not be defined in a manner which assumes an impassable gulf between ourselves and our human ancestors who had not the advantage of language, cultural evolution brings about a significant leap forward with respect to biological evolution. Thus from the amoeba to Einstein is not only one step, but several substantial steps.

Cultural evolution results in a significant difference between human beings and simple organisms, such as the prokaryotes. In their effort to solve the problem of survival, human beings and other organisms are constantly faced with the problem of having only limited resources. But, while simple organisms have little control over their environment, thanks to cultural evolution human beings are able to exert considerable control over it.

Of course, throughout most of their history, human beings were in a condition not very dissimilar from that of simple organisms, such as the prokaryotes. Like them, they had a limited control over their environment and themselves, and hence had to devote a very large part of their efforts to survival. Later on, however, their condition changed, and today, thanks to cultural evolution, they exercise a fairly large control over their environment and themselves. So, they can afford to devote only a relatively small part of their efforts to survival, and engage in activities not directly directed to survival.

For this reason, today human knowledge may seem to serve for purposes other than survival. But it is not so. Even if, thanks to cultural evolution, today human beings can devote only a relatively small proportion of their efforts to survival, in order to survive they must continue to exercise control over their environment and to improve it. Thus, survival remains a primary purpose of knowledge. Even today, knowledge has the biological role of primarily serving to solve the problem of survival, and even cultural evolution may serve to that purpose.

6.9 Objections to the Continuity View

Of course, that cultural evolution does not reduce to biological evolution, does not cancel the fact that it is continuous with biological evolution. Cultural evolution develops on the basis of biological evolution. In this sense we can say that human beings and human knowledge are but a part of nature.

Against the view that cultural evolution is continuous with biological evolution, it has been objected that, through cultural evolution, human beings have developed ways of thinking thanks to which they have made a qualitative leap, freeing themselves from the limitations imposed by their biological makeup. Therefore, human beings are qualitatively superior to all other organisms, which are forced within those limits. There is no biological basis of human thought, the latter is entirely the result of cultural factors. The decisive factor in the qualitative leap is language, which is the key tool of the qualitative superiority of human beings over all other organisms.

Thus Heidegger states that "the fact that physiology and physiological chemistry can scientifically investigate the human being as an organism is no proof that in this 'organic' thing, that is, in the body scientifically explained, the essence of the human being consists" (Heidegger 1998, 247).The "human body is something essentially other than an animal organism. Nor is the error of biologism overcome by adjoining a soul to the human body, a mind to the soul" (ibid.). Just "as little as the essence of the human being consists in being an animal organism can this insufficient definition of the essence of the human being be overcome or offset by outfitting the human being with an immortal soul, the power of reason, or the character of a person" (ibid.). There is an essential difference between the human being and plants and animals, and consists in the fact that plants and animals "lack language" (ibid., 248). In "its essence, language is not the utterance of an organism; nor is it the expression of a living thing. Nor can it ever be thought in an essentially correct way in terms of its symbolic character" (ibid., 248–249). The "human being is not only a living creature who possesses language along with other capacities. Rather, language is the house of being in which the human being ek-sists by dwelling, in that he belongs to the truth of being, guarding it" (ibid., 254).

However, it is unjustified to say that, through cultural evolution, human beings have developed ways of thinking, thanks to which they have made a qualitative leap, freeing themselves from the limitations imposed by their biological makeup. These ways of thinking are made possible by their biological makeup. As a computer is able to execute only the software that its hardware permits it to execute, human beings are able to think only the thoughts that their biological makeup permits them to think. Pinker says that "spiders spin spider webs because they have spider brains, which give them the urge to spin and the competence to succeed" (Pinker 1995, 18). Similarly, we can say that human beings think human thoughts because they have human brains, which give them the urge to think and the competence to succeed.

As to language, "language is no more a cultural invention than is upright posture," it is simply "a distinct piece of the biological makeup of our brains," indeed, human beings "know how to talk in more or less the sense that spiders know how to spin webs" (ibid.). Their alleged superiority is but an anthropocentric conceit. Human beings can do things that other animals cannot do, as other animals can do things that human beings cannot do. Therefore, language should not be seen "as the ineffable essence of human uniqueness but as a biological adaptation to communicate information" (ibid., 19). It is "an evolutionary adaptation, like the eye" (ibid., 24).

As the spider web is an outward projection of the spider brain, language is an outward projection of the human brain. Both the spider web and language reflect the biological makeup of the beings which produces them, and serve biological purposes.

It might be objected that the pursuit of beauty, which is the basis of the supreme creations of art, is evidence for the absolute autonomy of cultural products from the biological matrix. Such objection, however, is not valid. The pursuit of beauty has a biological function since it has an important role in sexual attraction and the reproduction of the species. Indeed, as Etcoff states, beauty is "a biological adaptation," it "provokes pleasure, rivets attention, and impels actions that help ensure the survivals of our genes. Our extreme sensitivity to beauty is hard-wired, that is, governed by circuits in the brain shaped by natural selection," since "in the course of evolution the people" who had such sensitivity "had more reproductive success" (Etcoff 1999, 24).

6.10 Science and Evolution

It has been stated above that the cultural role of knowledge is not opposed to the biological one but is continuous with it. This also applies to scientific knowledge.

Mach states that, although "science apparently grew out of biological and cultural development as its most superfluous offshoot," today "we can hardly doubt that it has developed into the factor that is biologically and culturally the most beneficial. Science has taken over the task of replacing tentative and unconscious adaptation by a faster variety that is fully conscious and methodical" (Mach 1976, 361).

In fact, science provides an essential contribution to solving the survival problem, since it can be viewed as an extension of the activities through which our remotest ancestors solved their survival problem. Such activities, and those underlying science, involve similar cognitive processes.

Our remotest ancestors solved their survival problem, making hypotheses about the location of predators or prey on the basis of clues they found in the environment – such as footprints, crushed or bent grass and vegetation, bent or broken branches or twigs, mud displaced from streams, excrements, and so on. Much in the same way, scientists solve problems about the world, making hypotheses on the basis of clues they find in nature.

6.11 Mathematics and Evolution

Even mathematics can contribute to solve the problem of survival. Mathematical abilities are present not only in human beings, but also in non-human organisms. Such abilities are essential for survival, because only through them organisms can escape from danger, or search for food, or seek out a mate.

The mathematical abilities which are present in non-human organisms can even be quite sophisticated. For example if, standing on a beach by the sea with a dog, you throw a tennis ball in the water diagonally, the dog will not immediately plunge into the sea to swim all the way to the ball, but will run for a while on the shoreline, and only then will dive, swimming up to the ball. This is because the speed with which a dog runs on the shoreline is much higher than that with which the dog swims, therefore the dog will choose to plunge into the sea at a point that minimizes the time needed to reach the ball. This point can be calculated by means of the infinitesimal calculus, and the point actually chosen by the dog is very close to the calculated one (see Pennings 2003). Does this mean that dogs know the infinitesimal calculus? Of course not. Dogs are able to choose the optimal point for diving thanks to natural selection, which gives a greater ability to survive to those organisms that have better judgment. Therefore, the calculation required to determine an optimal point is not made by the dog, but has been made by nature through natural selection. It is thanks to natural selection that dogs are able to solve this infinitesimal calculus problem.

Dogs are able to solve infinitesimal calculus problems even more complex than that. For example, if you toss a frisbee and observe how a dog runs to try to catch it on the fly when it comes down, you will see that the dog does not run in a straight line, but along an arc that ends at the point where the frisbee falls. This problem is more complex than the previous one because, to predict where the frisbee will fall and to determine the direction in which the dog has to run to catch it on the fly, one must take into account both the trajectory and speed of the frisbee and the speed of the dog. The dog does not run in a straight line, but along an arc which ends at the point where the frisbee will fall, because it moves so that the trajectory of the frisbee will appear to it as a straight line (see Shaffer et al. 2004). The dog is able to move so thanks to a complex mathematics that biological evolution has built into its visual and motor system, which leads the dog to move in a way that allows it to keep the frisbee fixed in its visual field. Once again, it is thanks to natural selection that the dog is able to solve this more complex infinitesimal calculus problem.

Rather sophisticated mathematical abilities are present not only in dogs but also in other non-human organisms. And of course in human beings. For example, baseball players, when running to catch the ball, do not run in a straight line but along an arc, and they do so for the same reason as dogs when they run to catch a frisbee.

6.12 Objections to the Dependence of Mathematics on Evolution

Against the claim that rather sophisticated mathematical abilities are present in several non-human organisms, it could be objected that mathematics is the result of processes that human beings carry out consciously, whereas non-human organisms carry out mathematical operations unconsciously, and therefore cannot be said to do mathematics.

But this objection overlooks that no one would deny that human beings, when solving a problem using a computer, do mathematics, and even computerized mathematics programs, when calculating the value of a derivative or an integral, do mathematics. Now, the amount of consciousness of computerized mathematics programs, whatever that might mean, is lower than that of non-human organisms. Therefore, if a computerized mathematics program, when calculating the value of a derivative or an integral, does mathematics, *a fortiori* dogs, when calculating the point where to plunge jump into the water or the point at which the frisbee will fall, do mathematics.

It could be objected that computerized mathematics programs are designed by human beings to do mathematics. Therefore, at the origin of their mathematical abilities there are human beings. But this objection overlooks that, as computerized mathematics programs are designed by human beings to do mathematics, dogs are designed by nature to do mathematics, 'designed' in the sense of natural selection. And human beings who solve a mathematical problem using a computer, or design computerized mathematics programs to do mathematics, have been themselves designed by nature to do mathematics.

It could be also objected that the mathematical abilities of non-human organisms are very specialized, namely suitable for very specific tasks, not for general use.

Admittedly, human beings have very specialized mathematical abilities, deriving from biological evolution. But, thanks to cultural evolution, from a certain point onwards, they have developed mathematical abilities suitable not only for very specific tasks, but also for general use. Moreover, human beings not only have developed mathematical abilities, but have been able to use them to develop mathematics as a discipline, which must be independent of biological evolution, being universal and necessary. In particular, it is thanks to the abilities in question that human beings have been able to design computerized mathematics programs.

But this objection overlooks that, as already pointed out, cultural evolution is a continuation of biological evolution by other means, and is based on it. The subjects of cultural evolution, namely human beings, have abilities that have permitted them to develop mathematics as a discipline, because they were designed by nature to have such abilities, 'designed' in the sense of natural selection. The opposite belief, that cultural evolution is independent of biological evolution, ignores what are the subjects of cultural evolution, and that the latter cannot be separated from the biological nature of these subjects. The subjects of cultural evolution developed mathematics as a discipline thanks to those mathematical abilities for which nature designed them. Therefore, it is unjustified to say that mathematics as a discipline must be independent of biological evolution because it is universal and necessary. Mathematics as a discipline is not universal or necessary, because cultural evolution is not independent of biological evolution, and the latter could have provided us with different mathematical abilities if it had occurred in a different environment. (On this, see also Chap. 18).

6.13 Evolution and Teleology

That mathematics, and cultural evolution generally, are not independent of biological evolution, means that knowledge, while having an immediate aim, namely survival, has no ultimate aim. There is no ultimate aim towards which living things tend, because natural selection has no ultimate aim. In particular, living things are not directed towards an ultimate aim by a divine watchmaker, indeed the watchmaker, natural selection, is blind. Likewise, there is no ultimate aim towards which knowledge tends.

Popper claims that "the growth of our knowledge is the result of a process closely resembling what Darwin called 'natural selection'; that is, the natural selection of hypotheses" (Popper 1972, 261). Like natural selection, "science works by trial (theory making) and by elimination of the errors," and "only the best theories, those which are most fit, survive in the struggle" (Popper 1985, 396). The ultimate aim towards which science tends is truth, because "science aims at truth" (Popper 2000, xxxvi). In fact, theories "get nearer and nearer to the truth" (Popper 1985, 396).

These claims, however, are unjustified. If, as Popper states, the growth of our knowledge is the result of a process closely resembling natural selection, then science cannot be said to tend to an ultimate aim, namely truth. For, natural selection has no ultimate aim. And theories cannot be said to get nearer and nearer to the truth if, as Popper states, "we cannot justify" the belief that our theories "are true; nor can we justify the belief that they are near to the truth" (Popper 2000, 61).

Popper gives a definition of the notion of being near to truth on the basis of which, as Tichý argues, "a false theory can never be" nearer to the truth "than another false theory" (Tichý 1974, 155). This is incongruous because it implies that, among Aristotle's theory of motion and Newton's theory of motion, which are both false, Newton's theory cannot be nearer to the truth than Aristotle's theory.

6.14 Remarks on a Different Naturalistic Approach to Knowledge

It has been stated above that knowledge is a kind of response to the environment which is essential for survival, and that this involves a naturalistic approach to knowledge. However, the naturalistic approach proposed in this book must not be confused with other naturalistic approaches, in particular with Quine's approach.

Quine claims that "the only point of view" the naturalist philosopher "can offer" is "the point of view of our own science" (Quine 1981b, 181). Namely, the only point of view the naturalist philosopher can offer is that of the present sciences. The naturalist philosopher "begins his reasoning within the inherited world theory," namely, within the world theory of the present sciences, and "tries to improve, clarify, and understand the system from within" (ibid., 72). So, according to Quine, the task of the naturalist philosopher is to improve, clarify, and understand the world theory of the present sciences from within.

But Quine's approach seems unrealistic, because improving, clarifying, and understanding the inherited world theory is an integral part of the scientists' work, and scientists are much more competent to the task than philosophers, who do not have the necessary qualification. Moreover, it is limiting to say that the only point of view the naturalist philosopher can offer is the point of view of the present science. As argued in Chap. 2, philosophy deals with problems in areas that the present sciences are unable to handle and, when successful, may even give birth to new sciences. Then, the solutions to philosophical problems cannot be based only on the present sciences. A philosophy dealing with problems that might be solved on the basis only of the present sciences would be reduced to recording the pronouncements of scientists. Moreover, such a philosophy could not be expected to give birth to new sciences if, as Quine claims, epistemology is contained in natural science as a chapter of psychology (see Chap. 1).

6.15 A Theistic Objection to Naturalism

A number of objections have been raised against naturalism, many of them from a theistic perspective.

For example, Plantinga argues that there is a serious conflict between naturalism and science because, from the point of view of naturalism and evolution, the primary purpose of our cognitive faculties "is not that of producing true" beliefs about the world, but rather that of "contributing to survival" (Plantinga 2011, 315–316). So, there is no guarantee that those faculties deliver truth, and hence that they are reliable. Hence, "the probability of our cognitive faculties being reliable, given naturalism and evolution, is low," or, in other terms, "if naturalism and evolution were both true, our cognitive faculties would very likely not be reliable" (ibid., 314). But then, "if I believe both naturalism and evolution, I have a defeater for my intuitive assumption that my cognitive faculties are reliable" (ibid.). And, if I have a defeater for my intuitive assumption that my cognitive faculties are reliable, "then I have a defeater for any belief I take to be produced by my cognitive faculties," including "my belief that naturalism and evolution are true. So my belief that naturalism and evolution are true gives me a defeater for that very belief" (ibid.). Hence, "I cannot rationally accept it. And if one can't accept both naturalism and evolution, that pillar of current science, then there is a serious conflict between naturalism and science" (ibid.). The claim that human beings can achieve scientific knowledge about the world "makes eminently good sense" only "from the perspective of theism," according to which "God creates human beings in his image, a crucial component of which is the ability to know worthwhile and important things about our world" (ibid., 285).

This objection, however, is unjustified. The objection assumes that the aim of science is truth, but this assumption is unjustified, indeed, as it will be argued in Chap. 9, the aim of science is plausibility. Our cognitive faculties contribute to survival by producing plausible hypotheses about the world, and they are reliable because they produce such plausible hypotheses. So, if I believe naturalism and evolution, I do not have a defeater for my assumption that my cognitive faculties are reliable, hence I do not have a defeater for my belief that naturalism and evolution are plausible. Therefore, I can rationally accept that belief, and there is no conflict between naturalism and science, a conflict arises only if one assumes that the aim of science is truth.

Moreover, the objection assumes that the claim that human beings can achieve scientific knowledge about the world makes eminently good sense only from the perspective of theism, according to which God creates human beings in his image. But this assumption is unjustified. Even granting that God creates human beings, surely he does not create them in his image because, while God is supposed to be immortal and omnipotent, alas, human beings are not. What guarantee is there, then, that God creates human beings in his image with respect to the ability to know worthwhile and important things about our world?

References

Bacon, Francis. 1961–1986. *Works*. Stuttgart Bad Cannstatt: Frommann Holzboog.
Cellucci, Carlo. 1998. *Le ragioni della logica*. Rome: Laterza.

———. 2013a. *Rethinking logic: Logic in relation to mathematics, evolution, and method*. Dordrecht: Springer.
Chisholm, Roderick. 1989. *Theory of knowledge*. London: Prentice-Hall.
Descartes, René. 1996. *Oeuvres*. Paris: Vrin.
Dewey, John. 1911. Brief studies in realism. I. *The Journal of Philosophy. Psychology and Scientific Method* 8 : 393–400.
———. 1938. *Logic: The theory of inquiry*. New York: Holt, Rinehart & Winston.
———. 2004. *Reconstruction in philosophy*. Mineola: Dover.
Etcoff, Nancy. 1999. *Survival of the prettiest: The science of beauty*. New York: Anchor.
Frege, Gottlob. 2013. *Basic laws of arithmetic*. Oxford: Oxford University Press.
Gettier, Edmund L. 1963. Is justified true belief knowledge? *Analysis* 23 : 121–123.
Heidegger, Martin. 1998. *Pathmarks*. Cambridge: Cambridge University Press.
Mach, Ernst. 1976. *Knowledge and error: Sketches on the psychology of enquiry*. Dordrecht: Springer.
Meinong, Alexius. 1968–1978. *Gesamtausgabe*. Graz: Akademische Druck- und Verlagsanstalt.
Nozick, Robert. 1997. *Socratic puzzles*. Cambridge: Harvard University Press.
Pennings, Timothy. 2003. Do dogs know calculus? *The College Mathematics Journal* 34 : 178–182.
Pinker, Steven. 1995. *The language instinct*. New York: Harper-Collins.
Plantinga, Alvin. 2011. *Where the conflict really lies: Science, religion, and naturalism*. Oxford: Oxford University Press.
Poincaré, Henri. 2013. *The foundations of science: Science and hypothesis – The value of science – Science and method*. Cambridge: Cambridge University Press.
Popper, Karl Raimund. 1972. *Objective knowledge*. Oxford: Oxford University Press.
———. 2000. *Realism and the aim of science*. London: Routledge.
———. 1985. Evolutionary epistemology. In *Open questions in quantum physics*, ed. Gino Tarozzi and Alwyn van der Merwe, 395–413. Dordrecht: Springer.
Quine, Willard Van Orman. 1981b. *Theories and things*. Cambridge: Harvard University Press.
Russell, Bertrand. 1988. Theory of knowledge. In Bertrand Russell, *Collected papers*, vol. 9, 194–202. London: Routledge.
———. 1992a. *Human knowledge: Its scope and limits*. London: Routledge.
Shaffer, Dennis, Scott Krauchunas, Marianna Eddy, and Michael McBeath. 2004. How dogs navigate to catch frisbees. *Psychological Science* 15 : 437–441.
Tichý, Pavel. 1974. On Popper's definitions of verisimilitude. *The British Journal for the Philosophy of Science* 25 : 155–160.
Williams, Michael. 2001. *Problems of knowledge: A critical introduction to epistemology*. Oxford: Oxford University Press.
———. 2004. Is knowledge a natural phenomenon? In *The externalist challenge*, ed. Richard Schantz, 193–209. Berlin: de Gruyter.

Chapter 7
Knowledge and Reality

Abstract One of the main questions about knowledge is the relation of knowledge to reality. This question is particularly important for the heuristic view of philosophy, according to which philosophy aims at knowledge and methods to acquire knowledge. In the course of the history of philosophy, the question of the relation of knowledge to reality has received several answers. In this chapter, a number of them are briefly discussed and found wanting: direct realism, representative realism, scientific realism, liberalized scientific realism, mathematical structural realism, semantic structural realism, essentialist realism, subjective idealism, and phenomenalism.

7.1 The Relation of Knowledge to Reality

One of the main questions about knowledge is the relation of knowledge to reality. This question is particularly important for the heuristic view of philosophy, according to which philosophy aims at knowledge and methods to acquire knowledge.

In the course of the history of philosophy, the question of the relation of knowledge to reality has received several answers. In this chapter we briefly discuss a number of them and find them wanting: direct realism, representative realism, scientific realism, liberalized scientific realism, mathematical structural realism, semantic structural realism, essentialist realism, subjective idealism, and phenomenalism. An alternative answer to the question of the relation of knowledge to reality is given in Chap. 9.

7.2 Direct Realism

According to direct realism, there are things out there in the world and they are really and directly perceived. We "immediately perceive the external object" and we "perceive it to exist. Nothing can be more absurd than to doubt of it" (Reid 1827, 73). That we immediately perceive the external object means that we perceive it independently of any act of reasoning. Indeed, if our perception of the external object involved an act of reasoning, "the greatest part of men would be destitute of

© Springer International Publishing AG 2017

C. Cellucci, *Rethinking Knowledge*, European Studies in Philosophy of Science 4,
DOI 10.1007/978-3-319-53237-0_7

it; for the greater part of men hardly ever learn to reason; and in infancy and childhood no man can reason" (ibid., 57). This conflicts with the fact that "the information of the senses is as perfect, and gives as full conviction to the most ignorant as to the most learned" (ibid.). On the other hand, that nothing can be more absurd than to doubt that the external object being immediately perceived exists, is clear because the "testimony of our senses carries irresistible conviction along with it" (ibid., 55). So the view of the world that we derive from our senses is to be taken at face value. For, things in the world "are what we perceive them to be" (ibid., 305).

Direct realism, however, is faced with the problem that the same thing may appear differently to different persons, or even to the same person at different times. Moreover, things as they appear to us are not what the sciences tell us they are. For example, objects that appear to be solid and substantial to us, actually consist of mainly empty space, and many of the stars we see in the sky are dead by the time the light from them reaches us. Furthermore, there are sense illusions. Thus the view of the world that we derive from our senses cannot be taken at face value.

7.3 Representative Realism

According to representative realism, there are things out there in the world, but they are not directly perceived. All we perceive are our ideas – representations, or mental images – of things, since "the mind, in all its thoughts and reasonings, hath no immediate object but its own ideas" (Locke 1975, 525). There are certain qualities, called primary qualities, such as size, shape, movement, which are in things, "whether we perceive them or not," and "we have by these an idea of the thing, as it is in itself" (ibid., 140). Thus we know how things are in themselves through our ideas, which represent primary qualities of things.

Representative realism, however, is faced with the problem that, if all we perceive are our ideas of things, then we will have no way of comparing our ideas with things, because we will be unable to go outside our ideas and check them. In particular, we will have no way of verifying that our ideas actually represent primary qualities of things. This will make the world ultimately unknowable.

Moreover, if all we perceive are our ideas of things, this raises the problem: If we have an idea of a thing in our head, who is it that is perceiving that idea? There ought to be a homunculus inside our head perceiving it. But then there ought to be a homunculus inside the head of this homunculus perceiving the idea perceived by the latter. And so on *ad infinitum*. Aristotle did not have to worry about this problem because, as we will see in Chap. 8, he believed that intuition and the object of thought are the same. But representative realism does not have this alternative available. If all we perceive are our ideas of things, then there is need for a subject capable of perceiving such ideas and judging them.

7.4 Scientific Realism

According to scientific realism, the world "is an objective reality that exists independently of human thought" (Sankey 2008, 15). This means that "my own existence will come to an end without the world's coming to an end too" (Popper 1972, 35). The aim of science is "to have true theories about the world, where 'true' is understood in the classical correspondence sense" (Musgrave 1988, 229). That is, "a theory is true if and only if it corresponds to the facts" (Popper 1972, 44). Scientific inquiry "yields knowledge of the truth about the objective reality investigated by scientists" (Sankey 2008, 14). Such truth "is absolute or objective" (Popper 1999, 55). This means that it is not "relative to 'conceptual scheme' or 'paradigm' or 'world-view' or anything else" (Musgrave 1988, 229). Also, "the unobservable theoretical entities of science," such as neutrons or protons, "are real" (Sankey 2008, 43).

Scientific realism, however, is faced with several difficulties. Admittedly, the world is an objective reality that exists independently of human thought. This simply amounts to saying that the world is not man-made.

But to say that the aim of science is to have true theories about the world, that scientific inquiry yields knowledge of the truth about the objective reality investigated by scientists, and that the unobservable theoretical entities of science are real, conflicts with the history of science. The latter offers us many examples of theories that were once considered true but later on have proven to be untenable, and whose unobservable theoretical entities have proven to be not real. Besides, there is no evidence that the present theories, and their unobservable theoretical entities, will fare any better in the future. As Chang points out, "it would be wrong to take" even a statement such as 'Water is H_2O' "as an eternal and unqualified truth," for, "very modern science no longer subscribes to the notion that water is simply H_2O;" indeed, "if we had a simple heap of H_2O molecules, it would not be recognizable as water" (Chang 2012, xvi).

Also, there are theories that are independently confirmable but mutually incompatible, and hence cannot be all true in the sense of the concept of truth as correspondence. Such is the case of general relativity and quantum mechanics, which are very successful but also mutually incompatible. For example, general relativity requires space to be smooth, or at the very least continuous, while quantum mechanics requires space to be chopped up.

Moreover, to say that 'true' is understood in the classical correspondence sense, is faced with the problem that, as it will be argued in Chap. 8, the concept of truth as correspondence does not provide a criterion of truth, namely a way to distinguish, among all propositions, those that are true from those that are false. As a result, we will generally be unable to recognize a truth when we reach one, in particular we will generally be unable to recognize whether a theory is true. So, if the aim of science is to have true theories about the world, where 'true' is understood in the classical correspondence sense, then the aim of science is unachievable.

Furthermore, to say that the truth about the objective reality investigated by scientists is absolute or objective, rather than relative to conceptual scheme or paradigm or world-view or anything else, conflicts with the fact that theories are human constructs and, like all human constructs, depend on our cognitive apparatus and its limitations (see Chap. 9).

These difficulties for scientific realism have been already at least partially pointed out by several scientists.

Thus, Hertz states that in science "we form for ourselves images" of "external objects" which are not true descriptions of things but are simply "our conceptions of things" (Hertz 1956, 1). The images "are not determined without ambiguity" by things, since "various images of the same objects are possible, and these images may differ in various respects" (ibid., 2).

Boltzmann states that "it cannot be our task to find an absolutely correct theory but rather a picture" that "represents phenomena as accurately as possible. One might even conceive of two quite different theories" of the same phenomena, which "in spite of their difference are equally correct" (Boltzmann 1974, 91).

Einstein states that a physicist is "like a man trying to understand the mechanism of a closed watch" (Einstein and Infeld 1966, 31). He "may form some picture of a mechanism which could be responsible for all the things he observes, but he may never be quite sure his picture is the only one which could explain his observations," because "he will never be able to compare his picture with the real mechanism" (ibid.) The picture is "not, however it may seem, uniquely determined by the external world" (ibid.). Thus, while "I admit that science deals with the 'real'," I "am nonetheless not a 'realist'" (Einstein 1998, 652).

7.5 Liberalized Scientific Realism

According to liberalized scientific realism, "the only scientific explanation of the success of science" is that "the theories accepted in a mature science are typically approximately true," and the "terms in mature scientific theories typically refer" (Putnam 1975–1983, I, 73). If a theory accepted in a mature science is not approximately true and its terms do not refer, then "it is a miracle" that the theory "successfully predicts phenomena" (Putnam 1978, 19). Therefore, liberalized scientific "realism is the only philosophy that doesn't make the success of the science a miracle" (Putnam 1975–1983, I, 73).

Since the sixteenth century, several people have supported liberalized scientific realism by essentially the same argument. Thus, Clavius states that the Ptolemaic theory is successful because, "by the assumption of eccentrics and epicycles, not only are all the appearances already known accounted for, but also future phenomena are predicted" (Clavius 1601, 452). The only explanation of the success of the Ptolemaic theory is that the theory is approximately true and its terms – eccentrics and epicycles – refer. For, if the Ptolemaic theory is not approximately true and "the eccentrics and epicycles are fictions," then "it is incredible that we should force the heavens to obey our fictions and to move as we wish or as agrees with our principles" (ibid.). Being a Jesuit, Clavius avoids using the expression 'it is a miracle' and uses instead the expression 'it is incredible', but otherwise Putnam's argument and Clavius' argument are quite similar.

Liberalized scientific realism, however, is unjustified because the argument on which it is based is invalid. While the Ptolemaic theory was successful at Clavius' time, according to contemporary scientific theories it is false, and eccentrics and epicycles do not refer, they are fictions of Ptolemaic astronomers. The history of science offers us many other examples of scientific theories that were once taken to be successful but later on have proven to be untenable, and whose terms do not refer. Thus, contrary to the claims of liberalized scientific realism, the success of a theory does not entail that the theory is approximately true and the terms of the theory refer.

Liberalized scientific realism might try to overcome this problem by assuming that we "need care only about those constituents" of scientific theories "which contribute to successes and which can, therefore, be used to account for these successes" (Psillos 1999, 105). Only those constituents are approximately true. But, as Lyons argues, this assumption "has the potential to fare far worse against the historical argument than the 'naive' holistic versions of realism over which it is thought to be an improvement" (Lyons 2006, 558). For, "each false constituent that is deployed in a key successful prediction constitutes a counterinstance," and, on the other hand, "a particular false constituent stands as a counterinstance each time it is deployed in a successful prediction" (ibid.).

7.6 Mathematical Structural Realism

According to mathematical structural realism, "there exists an external physical reality completely independent of us humans," this "external physical reality is a mathematical structure," and the aim of science "is to find a complete description of it" (Tegmark 2008, 102). Such complete description is provided by mathematics, whose applicability to external physical reality is "a natural consequence of the fact that the latter is a mathematical structure, and we are simply uncovering this bit by bit," so "our successful theories are not mathematics approximating physics, but mathematics approximating mathematics" (ibid., 107).

Mathematical structural realism, however, is faced with several difficulties. The claim that external physical reality is a mathematical structure, conflicts with the fact that only some properties of the world are mathematical in kind. If external physical reality were a mathematical structure, then, by knowing its mathematical properties we would know the essence of natural substances, contrary to Galileo's warning that penetrating the essence of natural substances is an impossible and vain undertaking (see Chap. 8).

Moreover, Tegmark himself admits that the aim of science to find a complete description of the mathematical structure that is our external physical reality, conflicts with the fact that by "the results of Gödel, Church and Turing," there are "questions that can be posed but not answered" and, "for a mathematical structure, this corresponds to relations that are unsatisfactorily defined in the sense that they cannot be implemented by computations that are guaranteed to halt" (ibid., 135).

Tegmark answers this difficulty by assuming that "the mathematical structure that is our external physical reality is defined by computable functions" (ibid., 131). By this assumption, there would be "no physical aspects of our universe that are uncomputable/undecidable, eliminating the above-mentioned concern that Gödel's work makes it somehow incomplete or inconsistent" (ibid., 136).

Tegmark himself, however, admits that this answer is problematic, since "virtually all historically successful theories of physics violate" the assumption that the mathematical structure that is our external physical reality is defined by computable functions, and "it is far from obvious whether a viable computable alternative exists" (ibid., 138). For example, virtually all historically successful theories of physics incorporate "the continuum, usually in the form of real or complex numbers, which cannot even comprise the input to a finite computation since they generically require infinitely many bits to specify" (ibid.).

7.7 Semantic Structural Realism

According to semantic structural realism, "to present a theory is to specify a family of structures, its models" (van Fraassen 1980, 64). This family of structures is specified "directly, without paying any attention to questions of axiomatizability, in any special language" (van Fraassen 1989, 222). So, if a theory is to be identified with anything at all, it "should be identified with its class of models" (ibid.). Here "a model is a mathematical structure" (van Fraassen 2008, 376, footnote 18). More precisely, "a model is a structure plus a function that interprets the sentences in that structure" (van Fraassen 1985, 301). If "a theory is advocated then the claim made is that these models can be used to represent the phenomena, and to represent them accurately," where we say that "a model can (be used to) represent a given phenomenon accurately only if it has a substructure isomorphic to that phenomenon" (van Fraassen 2008, 309).

Semantic structural realism, however, has some serious limitations. A model is a structure and hence a mathematical object, while the phenomenon is not a mathematical object. Van Fraassen himself raises this question: "If the target," that is, the phenomenon, "is not a mathematical object, then we do not have a well-defined range for the function, so how can we speak of an embedding or isomorphism or homomorphism or whatever between that target and some mathematical object?" (ibid., 241). His answer is that we compare the model, not with the phenomenon, but rather with the data model, namely, our representation of the phenomenon. The data model "is itself a mathematical structure. So there is indeed a 'matching' of structures involved; but is a 'matching' of two mathematical structures, namely the theoretical model and the data model" (ibid., 252). This answer, however, is unsatisfactory, because the data model is a mathematical object, while the phenomenon is not a mathematical object, which raises the question of the matching of the data model and the phenomenon. Thus van Fraassen's answer just pushes the problem back one step.

Moreover, even a fiction can have a model, in the sense of a mathematical structure. Therefore, it is not models that can make a distinction between fictions and reality.

Furthermore, semantic structural realism entails that scientific theories, being families of structures, are static things. But, as a matter of fact, scientific theories undergo development. Semantic structural realism has no alternative than treating their development as a sequence of families of models. But then the question arises how the transition from a theory to the next one in the sequence comes about. Semantic structural realism has nothing to say about this, because it does not account for the process of theory formation, which is essential to explain the development of theories and the process of theory change. Therefore, semantic structural realism cannot account for the dynamic character of scientific theories. This is a structural limitation of semantic structural realism.

7.8 Essentialist Realism

According to essentialist realism, there are natural kinds, namely kinds of things in nature, the properties that distinguish natural kinds are their essences, and "the laws of nature describe the essences of the natural kinds" (Ellis 2008, 143). These laws "are metaphysically necessary" (ibid., 144). They are of the form 'All As are Bs', which must be understood as meaning: "Being an A necessitates being a B in this world" (Ellis 2002, 100).

Essentialist realism, however, is faced with the problem that it cannot guarantee that a law of nature is applicable to any individual case. For example, consider the law of inertia: Every body perseveres in its state of rest, or of uniform motion in a right line, unless it is compelled to change that state by forces impressed upon it. From the viewpoint of essentialist realism, this must be understood as meaning: A body's not being compelled to change its state by forces impressed upon it, necessitates that body's persisting in its state of being at rest or of moving uniformly along a straight line. Now, from this law of nature we cannot conclude that any individual body will persist in its state. For, the condition of not being compelled to change its state by forces impressed upon it does not hold for any physical body, because friction, say, is always present. Being frictionless is an essence which does not correspond to any natural kind.

7.9 Subjective Idealism

According to subjective idealism, "houses, mountains, rivers, and in a word all sensible objects," have no existence "distinct from their being perceived," since they are "things we perceived by sense" (Berkeley 1948–1957, II, 42). And they are ideas or sensations, because there is nothing that "we perceive besides our own ideas or sensations" (ibid.). So all our knowledge consists of ideas or sensations. That all sensible objects have no existence distinct from their being perceived, is apparent from the fact that it is "plainly repugnant that any one of these" should "exist unperceived" (ibid.). When sensible objects are perceived by no human mind, they are perceived by "the mind of some eternal spirit" (ibid., II, 43). Specifically, they are perceived by God. So, it can be generally said that all sensible objects have no existence "out of the minds or thinking things which perceive them" (ibid., II, 42). Since all sensible objects have no existence distinct from their being perceived, "their *esse* is *percipi*" (ibid.). Namely, their being is to be perceived.

Subjective idealism is not confronted with the problem of representative realism that, if all we perceive are our ideas of things, then we will have no way of comparing our ideas with things, since we will be unable to go outside our ideas and check them. For, according to subjective idealism, all sensible objects have no existence out of the minds or thinking things which perceive them, so there is no question of going outside ideas. Nevertheless, subjective idealism must resort to God to guarantee that all sensible objects continue to exist when they are perceived by no human mind, so it essentially depends on the unproven assumption of the existence of God.

Moreover, according to subjective idealism, my present knowledge of the tree outside my window, say, consists of certain sensations, my knowledge of that tree in an hour's time will consist of other sensations, and I will have no way of judging that those sensations will be sensations of the same tree. Generally, subjective idealism is unable to account for our knowledge, because much of our knowledge does not consist of sensations, it is the result of inferences, including non-deductive inferences which go much beyond sensations. This contradicts the assumption of subjective idealism that all knowledge consists of sensations.

7.10 Phenomenalism

According to phenomenalism, "the sensations" are "all that we can possibly know of the objects," and also "all that we have any ground for believing to exist" (Mill 1963–1986, IX, 6). Here by 'sensations' are to be meant not only actual sensations, but also the possibilities of sensation, since actual sensations are "fugitive: the possibilities, on the contrary, are permanent" (ibid., IX, 180). An object "may be defined" as "a permanent possibility of sensation" (ibid., IX, 183). Or rather, as a permanent possibility of groups of sensations, since most sensations "occur in fixed groups" (ibid., IX, 6). In addition to groups of sensations, in our experience we recognize certain constant sequences, where their constancy "is a constancy of antecedence and sequence" (ibid., IX, 180). Such constant sequences are sequences of sensations, or rather of groups of sensations, because "in almost all the constant sequences which occur in nature, the antecedence and consequence do not obtain between sensations, but between the groups we have been speaking about" (ibid., IX, 180–181).

While subjective idealism uses God to guarantee that things continue to exist when no human mind perceives them, phenomenalism replaces God with the possibilities of sensation. This, however, is also unsatisfactory. According to phenomenalism, an object may be defined as a permanent possibility of sensation. But a permanent possibility of sensation will have to be defined by reference to the object of which it is supposed to be a permanent possibility of sensation. This involves a circle.

Moreover, phenomenalism claims that in our experience we recognize certain constant sequences, in almost all of which the antecedence and consequence obtain between groups of sensations. But, in order to recognize such sequences, we should be able to recognize certain sensations as members of those groups, and the only basis we have for recognizing that very different sensations are members of such groups are the constant sequences themselves. Thus, on the one hand, we should be able to recognize groupings of sensations prior to being able to recognize constant sequences, and, on the other hand, we should be able to recognize constant sequences prior to being able to recognize groupings of sensations. This involves a circle.

References

Berkeley, George. 1948–1957. *Works*. London: Nelson.

Boltzmann, Ludwig. 1974. *Theoretical physics and philosophical problems: Selected writings*. Dordrecht: Springer.

Chang, Hasok. 2012. *Is water* H_2O? *Evidence, realism and pluralism*. Dordrecht: Springer.

Clavius, Christophorus. 1601. *In sphaeram Ioannis de Sacro Bosco commentarius*. Venice: Sub signo Solis.

Einstein, Albert. 1998. Letter to Eduard Study, 25 September 1918. In Albert Einstein, *Collected papers*, vol. 8, 651–653. Princeton: Princeton University Press.

Einstein, Albert, and Leopold Infeld. 1966. *The evolution of physics*. New York: Simon & Schuster.

Ellis, Brian. 2002. *The philosophy of nature: A guide to the new essentialism*. Chesham: Acumen.

———. 2008. Essentialism and natural kinds. In *The Routledge companion to philosophy of science*, ed. Martin Curd and Stathis Psillos, 139–148. London: Routledge.

Hertz, Heinrich Rudolf. 1956. *The principles of mechanics presented in a new form*. Mineola: Dover.

Locke, John. 1975. *An essay concerning human understanding*. Oxford: Oxford University Press.

Lyons, Timothy D. 2006. Scientific realism and the strategema de Divide et Impera. *The British Journal for the Philosophy of Science* 57 : 537–560.

Mill, John Stuart. 1963–1986. *Collected works*. Toronto: University of Toronto Press.

Musgrave, Alan. 1988. The ultimate argument for scientific realism. In *Relativism and realism in science*, ed. Robert Nola, 229–252. Dordrecht: Springer.

Popper, Karl Raimund. 1972. *Objective knowledge*. Oxford: Oxford University Press.

———. 1999. *All life is problem solving*. London: Routledge.
Psillos, Stathis. 1999. *Scientific realism: How science tracks truth*. London: Routledge.
Putnam, Hilary. 1975–1983. *Philosophical papers*. Cambridge: Cambridge University Press.
———. 1978. *Meaning and the moral sciences*. London: Routledge.
Sankey, Howard. 2008. *Scientific realism and the rationality of science*. Aldershot: Ashgate.
Tegmark, Max. 2008. The mathematical universe. *Foundations of Physics* 38 : 101–150.
van Fraassen, Bas. 1980. *The scientific image*. Oxford: Oxford University Press.
———. 1985. Empiricism in the philosophy of science. In *Images of science: Essays on realism and empiricism*, ed. Paul Churchland and Clifford Hooker, 245–308. Chicago: The University of Chicago Press.
———. 1989. *Laws and symmetry*. Oxford: Oxford University Press.
———. 2008. *Scientific representation: Paradoxes of perspective*. Oxford: Oxford University Press.

Chapter 8
Knowledge and Truth

Abstract Since antiquity, several philosophers and scientists have claimed that the aim of science is truth. This raises the question: What is truth? A popular answer is that a proposition is true if it corresponds to the facts to which it refers. This is the concept of truth as correspondence. However, the concept of truth as correspondence does not provide a criterion of truth, namely a way to distinguish, among all propositions, those that are true from those that are false. This chapter argues that, in fact, no criterion of truth is possible, for any concept of truth. Thus, if the aim of science is truth, then the aim of science is unreachable, since we will generally be unable to recognize a truth when we reach one. In addition to truth as correspondence, the chapter also considers some alternative concepts of truth, namely truth as consistency, truth as systematic coherence, truth as satisfiability, and truth as provability, and finds all of them wanting.

8.1 The Aim of Science and Truth

In order to give a more satisfactory answer to the question of the relation of knowledge to reality, we must consider what the aim of science is.

Since antiquity, several philosophers and scientists have claimed that the aim of science is truth.

Thus, Aristotle states that "the aim of theoretical science is truth" (Aristotle, *Metaphysica*, α 1, 993 b 20–21). Frege states that "all sciences have truth as their goal," so "to discover truths is the task of all sciences" (Frege 1984, 351). Husserl states that the aim of science is "the greatest possible conquest of the realm of truth" (Husserl 2001, I, 18).

In particular, the claim that the aim of science is truth is a basic tenet of scientific realism. Thus, Popper states that "science is the search for truth," and "truth is therefore the aim of science" (Popper 1996, 39). Sankey states that "truth is the aim of science" (Sankey 2008, 13). Musgrave states that "the aim of science" is "to discover the truth" (Musgrave 1996, 19).

However, the claim that the aim of science is truth is problematic, and so is the concept of truth itself. This chapter analyses the problems they raise.

© Springer International Publishing AG 2017

C. Cellucci, *Rethinking Knowledge*, European Studies in Philosophy of Science 4,
DOI 10.1007/978-3-319-53237-0_8

8.2 Truth as Correspondence

The claim that the aim of science is truth raises the question: What is truth? Since antiquity, various concepts of truth have been proposed. In particular, Aristotle discusses two concepts of truth: truth as correspondence, and truth as intuition of the essence. As we will see, he maintains that the former is inadequate, and only the latter is adequate.

According to the concept of truth as correspondence, "to say of what is that it is, and of what is not that it is not, is true," while "to say of what is that it is not, or of what is not that it is, is false" (Aristotle, *Metaphysica*, Γ 7, 1011 b 26–27).

For Aristotle, this concept of truth is inadequate because it does not provide a criterion of truth, namely a way to distinguish, among all propositions, those that are true from those that are false. For, we cannot ascertain whether a proposition corresponds to "any nature existing objectively outside thinking" (ibid., E 4, 1028 a 2). We can only compare the proposition with a representation of some thing existing outside thinking. But a representation is only "an affection of thinking" (ibid., E 4, 1027 b 34–1028 a 1). Thus, we can only compare a proposition with an affection of thinking. Then "the false and the true are not in things" but only "in thinking" (ibid., E 4, 1027 b 25–27). They do not express a real connection between a proposition and the world. So, the concept of truth as correspondence does not provide a criterion of truth, hence it cannot be used in practice, and therefore "must be dismissed" (ibid., E 4, 1027 b 34).

Contrary to this interpretation, Aristotle is often presented as being a supporter of the concept of truth as correspondence, or even its author. Thus Tarski states that "the earliest explanation" of the concept of truth as correspondence "can be found in Aristotle's *Metaphysics*" and this is the concept of truth "put forward by Aristotle" (Tarski 1969, 63). It is the "classical Aristotelian conception of truth" (Tarski 1944, 342).

These claims are unjustified. It is unjustified to say that the earliest explanation of the concept of truth as correspondence can be found in Aristotle's *Metaphysics*. Already Plato states that "the proposition which says of what is that it is, is true, while the one which says of what is that it is not, is false" (Plato, *Cratylus*, 385 b 7–8). Indeed, "the true proposition states the things that are as they are," while "the false one states things different from the things that are" (Plato, *Sophista*, 263 b 4–7).

Moreover, it is unjustified to say that the concept of truth as correspondence is put forward by Aristotle and is the classical Aristotelian conception of truth. As we have seen, Aristotle argues that the concept of truth as correspondence is inadequate.

It may even be disputed whether, by his statement, Aristotle really means to give a definition of the concept of truth as correspondence. Thus Engel says that "it is not obvious that Aristotle is here giving a definition of truth" (Engel 2002, 15). Anyway, his supposed definition of truth expresses "a relation of identity between what we say (or think) and reality, rather than a relation of correspondence. If to say what is true and to say what is (or what is false and what is not) is to say the same thing, truth and being are one and the same thing" (ibid.).

Thus, under Engel's interpretation, Aristotle would not be discussing two different concepts of truth but a single one, namely the concept of truth as intuition of the essence (described below). Indeed, it is with the concept of truth as intuition of the essence that truth and being are one and the same thing.

After Aristotle, it has been repeatedly reaffirmed that the concept of truth as correspondence is inadequate because it does not provide a criterion of truth.

Thus Kant states that the concept of truth as correspondence does not provide a criterion of truth since, on the basis of it, "my cognition, to count as true, is supposed to agree with its object. Now I can compare the object with my cognition, however, it" (Kant 1992, 557). But, "since the object is outside me, the cognition in me, all I can ever pass judgment on is whether my cognition of the object agrees with my cognition of the object" (ibid., 557–558). With the concept of truth as correspondence "it is just as when someone makes a statement before a court and in doing so appeals to a witness with whom no one is acquainted, but who wants to establish his credibility by maintaining that the one who called him as witness is an honest man" (ibid., 558). (For details on Kant's conception of truth, see Capozzi 2013, Chap. 12).

Frege states that the concept of truth as correspondence does not provide a criterion of truth because, on the basis of it, in order to establish if something is true, "we should have to inquire whether it is true that an idea and a reality, say, correspond" (Frege 1984, 353). But "it would only be possible to compare an idea with a thing if the thing were an idea too," and "this is not at all what people intend when they define truth as the correspondence of an idea with something real. For in this case it is essential precisely that the reality shall be distinct from the idea" (ibid.). So "the attempted explanation of truth as correspondence breaks down. And any other attempt to define truth also breaks down" (ibid.).

8.3 Concept of Truth and Criterion of Truth

Against the claim that the concept of truth as correspondence is inadequate because it does not provide a criterion of truth, it might be objected that a concept of truth need not provide a criterion of truth, because truth is independent of our ability to have knowledge of it.

Thus Popper states that "it is wrongly assumed that" the concept of truth as correspondence "is intended to yield a criterion of truth," that is, "a method of deciding whether or not a given statement is true" (Popper 1972, 317). Such concept of truth only plays "the role of a regulative idea. It helps us in our search for truth that we know there is something like truth or correspondence" (ibid., 318).

David states that the argument that the concept of truth as correspondence ought to provide a criterion of truth "is fallacious and has to be abandoned" (David 2004, 374). The "correspondence theory does not entail that we have to know that a belief corresponds to a fact in order to know that it is true" (ibid.). So, "our strategy for how to go about obtaining knowledge does not have to be a strategy of 'comparing' beliefs with facts. Our strategy can be one of making observations and experiments, of deducing logical consequences from what we already know" (ibid.).

This objection, however, is unjustified because if, as Popper says, a concept of truth need not provide a criterion of truth but only plays the role of a regulative idea, then we may be unable to recognize a truth when we reach one, so truth is generally unknowable. Hence, if the aim of science is truth, then the aim of science is unachievable.

Moreover, it is unjustified to say, as David does, that, under the correspondence theory, our strategy for how to go about obtaining knowledge does not have to be a strategy of comparing beliefs with facts. If we go about obtaining knowledge without comparing beliefs with facts, then we will not know that beliefs are true in the sense of the correspondence theory. In particular, it in unjustified to say that our strategy for how to go about obtaining knowledge can be one of making observations and experiments, of deducing logical consequences from what we already know. If we proceed in this way, we obtain knowledge which is plausible, not knowledge which is true in the sense of the correspondence theory.

8.4 Truth as Correspondence and Scientific Realism

Sankey states that the concept of truth as correspondence is the only concept of truth "compatible with realism" (Sankey 2008, 17). If so, then, since the concept of truth as correspondence does not not provide a criterion of truth and hence makes us unable to recognize whether a theory is true, scientific realism makes scientific knowledge generally unachievable. This is a problem for scientific realism, that is declaredly "a position of epistemic optimism, which holds against the sceptic that humans are able to acquire knowledge of the world" (ibid., 3).

Even distinguishing between science and non-science is a problem for scientific realism. For, according to Sankey, not only we may have no rational grounds to believe a proposition that is in fact true, but we may even "rationally believe a proposition that is false" (ibid., 112).

Admittedly, Sankey states that "there can be no reason to accept a claim about reality until some method of inquiry is justified" (ibid., 143). But he also states that, "in order for a claim about reality to justify a method of inquiry there must be reason to accept the claim about reality" (ibid.). Clearly, this leads to a circularity. Sankey admits that "such circularity is surely to be avoided," but the only argument he provides to this effect is that "claims about method and claims about reality fit together in relations of mutual support" (ibid.). Then it is unclear how scientific realism can avoid the circularity.

8.5 Impossibility of a Criterion of Truth

Not only the concept of truth as correspondence does not provide a criterion of truth, but, as Tarski states, a "criterion" of truth "will never be found" (Tarski 1944, 363–364).

Indeed, for any concept of truth, no criterion of truth is possible. This can be seen by slightly modifying the argument by which Frege supports his claim that any attempt to define truth breaks down.

For any concept of truth, in order to have a criterion of truth, "certain characteristics" of truth "would have to be specified. And, in application to any particular case, the question would always arise whether it were true that the characteristics were present" (Frege 1984, 353). To answer this question we would need a criterion of truth. But then "we should be going round in a circle" (ibid.). Therefore, no criterion of truth is possible.

That, for any concept of truth, no criterion of truth is possible, is relevant to the question of whether the aim of science is truth. For, as already stated above, if not only the concept of truth as correspondence does not provide a criterion of truth, but, for any concept of truth, no criterion of truth is possible, we will generally be unable to recognize a truth when we reach one, so the aim of science will be unachievable.

8.6 An Alleged Rehabilitation of Truth as Correspondence

Despite the problems raised by the concept of truth as correspondence, in the past century this concept of truth has had many supporters. In particular, Popper maintains that the "main achievement of Tarski's invention of a method of defining truth" is "the rehabilitation of the notion of truth or correspondence to reality, a notion which had become suspect" (Popper 1972, 59–60).

But this is unjustified. Tarski did not invent a method of defining truth. Indeed, he states that, "if we succeed in introducing the term 'true' into the metalanguage in such a way that every proposition of the form discussed can be proved on the basis of the axioms and rules of inference of the metalanguage, then we shall say that" this way of using the concept of truth is "adequate" (Tarski 1983, 404). Thus, Tarski requires that, if *L*-true is to be an adequate truth predicate for a language *L*, then all instances of the expression

$$(T)\,\text{`}P\text{'}\ \text{is}\ L-\text{true}\ \mathit{iff}\ P,$$

where '*P*' is a name of *L*-sentence *P*, must be theorems of metatheory MT. But Tarski makes it quite clear that "neither the expression (T) itself" nor "any particular instance of the form (T) can be regarded as a definition of truth" (Tarski 1944, 344).

In particular, Tarski did not invent a method of defining truth in the sense of the concept of truth as correspondence. As Putnam points out, the expression (T) does not refer to any extralinguistic reality, only "to purely interlinguistic aspects of the usage of 'true'," as it is clear from the fact that "we might still be able to certify a definition of 'true as a sentence of *L*' to be 'adequate'" even "though the extra-logical constants of *L* are totally uninterpreted" (Putnam 1975–1983, II, 70). Actually, the expression (T) only correlates the language of metatheory MT and language *L*, instead of comparing *L* with extralinguistic reality, which is what 'correspondence' means.

Tarski himself makes it quite clear that (T) is not intended to give a definition of truth in the sense of the concept of truth as correspondence. He reports that, in a group of people, "only 15% agreed that 'true' means 'agreeing with reality', while 90% agreed that a sentence such as 'it is snowing' is true if, and only if, it is snowing" (Tarski 1944, 360). Thus a great majority of people accepted the expression (T) but "seemed to reject the classical conception of truth" (ibid.). That is, the concept of truth as correspondence.

Tarski's requirement, that if *L*-true is to be an adequate truth predicate for a language *L*, then all instances of (T) must be theorems of metatheory MT, presupposes that the axioms of MT be true. But, by Gödel's second incompleteness theorem, it is impossible to demonstrate by absolutely reliable means that the axioms of MT are true. For, if it were possible to demonstrate by absolutely reliable means that the axioms of MT are true, it would be possible to demonstrate by absolutely reliable means that they are consistent, which is excluded by Gödel's second incompleteness theorem.

The condition that the axioms of MT be true is essential because, if the axioms of MT are not true, it may be the case that all instances of (T) are theorems of MT even though some of them are false. Against this, Raatikainen objects that no instance of (T) can be false. He argues that, if some instance of (T) is false, this means that, for some sentence *P*, the equivalence

(1) '*P*' is *L* true *iff* *P*

is false. Then either:

(a) '"*P*" is *L* true' is true *but* *P* is false;

or

(b) '"*P*" is *L* true' is false *but* *P* is true.

According to Raatikainen, "neither case is actually possible" (Raatikainen 2003, 39). Indeed if, as in (a), ' "*P*" is *L*-true' is true, "then certainly *P* is true, not false" (ibid.). And if, as in (b), ' "*P*" is *L*-true' is false, then certainly *P* is false, not true.

Thus no instance of (T) "can be false" and hence all instances of (T) "are in fact necessary" (ibid.).

This argument, however, is invalid because if, as in (a), ' "*P*" is *L*-true' is true, then to infer that certainly *P* is true we would need that the equivalence (1) be true, while by hypothesis it is false. Similarly if, as in (b), ' "*P*" is *L*-true' is false, then to infer that certainly *P* is false we would need that the equivalence (1) be true, while by hypothesis it is false.

8.7 Truth as Intuition of the Essence

As an alternative to the concept of truth as correspondence, Aristotle puts forward the concept of truth as intuition of the essence. According to him, "truth is intuiting and stating" the essence of a thing, because "stating and asserting are not the same" (Aristotle, *Metaphysica*, Θ 10, 1051 b 24–25). Now, stating the essence of a thing means giving a definition of it, since "definition is the discourse which reveals the essence of a thing" (Aristotle, *Topica*, VII, 3, 153 a 15–16). Therefore, truth is intuiting the essence of a thing and giving a definition of it.

By intuiting the essence of a thing we know "what that thing is in itself" (Aristotle, *Metaphysica*, Z 4, 1029 b, 14–15). And "to know a thing is to know its essence," because "any single thing and its pure essence coincide" (ibid., Z 6, 1031 b 19–21). For this reason, according to Aristotle, while the concept of truth as correspondence is inadequate, the concept of truth as intuition of the essence is adequate.

It might be objected that if, as Aristotle assumes, we do have a faculty – intuition – capable of apprehending the essence of things, then, for Aristotle, truth as correspondence must be a genuine concept of truth, because intuition gives us direct access to things existing outside thinking.

This objection, however, is invalid because, if we have a faculty capable of apprehending the essence of things, then truth does not consist in the correspondence of a proposition with some thing existing outside thinking, but rather in the direct apprehension of the essence of the thing. By such apprehension, intuition and the essence of the thing are the same. In fact, Aristotle states that intuition "becomes the object of thought by the act of apprehension and thinking, so that intuition and the object of thought are the same" (ibid., Λ 7, 1072 b 20–21). But, if intuition and the object of thought are the same, then speaking of correspondence is misleading. While truth as correspondence merely involves a congruence between thought and being, truth as intuition of the essence involves an identity of thought and being.

8.8 Truth and Modern Science

The concept of truth as intuition of the essence, however, applies only to Aristotle's science, in which "we have scientific knowledge of a thing when we know its essence" (Aristotle, *Metaphysica*, Z 6, 1031 b 6–7). Indeed, Aristotle's science aimed at penetrating the essence of natural substances. This made impossible a mathematical treatment of nature, because essence is not mathematical in kind. For this reason, Aristotle states that "the method of mathematics is not well suited to physics" (ibid., α 3, 995 a 16–17).

Conversely, the concept of truth as intuition of the essence does not apply to modern science, which arose from Galileo's philosophical revolution: the decision to renounce Aristotle's aim to penetrate the essence of natural substances, dealing only with some of their phenomenal properties mathematical in kind, such as location, motion, shape, or size.

Indeed, Galileo famously states: "Either, by speculating, we seek to penetrate the true and intrinsic essence of natural substances, or we content ourselves with coming to know some of their properties [*affezioni*]" (Galilei 1968, V, 187). Trying to penetrate the essence of natural substances is "a not less impossible and vain undertaking with regard to the closest elemental substances than with the remotest celestial things" (ibid.). Therefore, we will content ourselves with dealing with "some properties of them, such as location, motion, shape, size, opacity, mutability, generation, and dissolution" (ibid., V, 188). While we cannot know the essence of natural substances, "we need not despair of our ability" to come to know such properties "even with respect to the remotest bodies, just as those close at hand" (ibid.).

All the properties in question are mathematical in kind, namely, they are of the same kind as properties dealt with in mathematics. They include mutability, generation, and dissolution, because the latter are changes in size and shape. On the contrary, the essence of natural substances, which is object of Aristotle's science, is not mathematical in kind.

That modern physics originated from Galileo's philosophical revolution is acknowledged also by Newton, who states that "the moderns, rejecting substantial forms and occult qualities, have endeavored to subject the phenomena of nature to the laws of mathematics" (Newton 1962, I, xvii).

The decision to renounce Aristotle's aim to penetrate the essence of natural substances, dealing only with some of their phenomenal properties mathematical in kind, made a mathematical treatment of nature possible, and even necessary, since dealing with phenomenal properties mathematical in kind essentially required mathematics. For this reason, Galileo states that "trying to deal with natural questions without geometry is attempting to do what is impossible to be done" (Galilei 1968, VII, 229). And Kant states that "in any special doctrine of nature there can be only as much proper science as there is mathematics therein" (Kant 2002, 185). For the same reason, since all physical laws are mathematical equations, the very "concept of physical law" was "virtually unknown to antiquity and the middle ages," and "did not arise before the middle of the seventeenth century" (Zilsel 1942, 245).

8.9 Absoluteness Claims

Oddly enough, some supporters of the view that the aim of science is truth seem to take no account of Galileo's philosophical revolution, since they claim that natural science describes the world as it is in itself.

As we have seen in Chap. 5, this is the case of Williams. This is also the case of Dummett who, on the one hand, says that "knowledge consists in the apprehension of the truth of propositions" (Dummett 2010, 17). On the other hand, he maintains that "science is in large part an attempt to answer" the question of "what things are like in themselves, as opposed to how they appear to us" (ibid., 43). Science tries to "arrive at an account of how things are in themselves, not depending at all upon the particular way we experience them or observe them directly or indirectly," as opposed to "a description mediated by how things appear to us" (Dummett 2006, 94). An "early reflection of this effort was Galileo's distinction between primary qualities such as shape, which we perceive as they are in reality," and hence are the things as they really are in themselves, and "secondary qualities such as color, which are propensities to produce in us sensations that may resemble the physical basis of those propensities" (Dummett 2010, 43).

Thus, however, Dummett misinterprets the character of modern science, which does not concern the essence of natural substances, but only some of their phenomenal properties mathematical in kind. Moreover, he misinterprets Galileo's distinction between primary and secondary qualities because, for Galileo, primary qualities depend on how we conceive of things. Indeed, Galileo states that primary qualities are qualities such that, "as soon as I conceive a matter or corporeal substance, I feel compelled by the need to conceive along with it" that it has such qualities, and "I cannot separate it from these conditions by any stretch of imagination" (Galilei 1968, VI, 347). That is, primary qualities are qualities such that, when I conceive a matter or corporeal substance, I cannot conceive it as not having those qualities. Primary qualities are mathematical in kind. On the other hand, secondary qualities are qualities such that, when I conceive a matter or corporeal substance, "I do not feel my mind forced to apprehend it as necessarily accompanied by such conditions" (ibid., VI, 347–348). That is, secondary qualities are qualities such that, when I conceive a matter or corporeal substance, I can conceive it as not having such qualities. Secondary qualities are not mathematical in kind. From this it is apparent that primary qualities are not the things as they really are in themselves, but are dependent on how we conceive a matter or corporeal substance.

It might be objected that Galileo also states that natural "philosophy is written in this very great book that is continually open to us before our eyes (I say the universe)" and "is written in mathematical language" (ibid., VI, 232). If primary qualities are mathematical in kind, this would seem to imply that they are the things as they really are in themselves. Indeed, Husserl claims that for Galileo "nature is, in its 'true being-in-itself', mathematical" (Husserl 1970, 54).

But this objection is invalid because, by saying that the universe is written in mathematical language, Galileo does not mean to say that nature is, in its true being-in-itself, mathematical. Otherwise, by knowing primary qualities, we would know the essence of natural substances, contrary to Galileo's warning that penetrating their essence is an impossible and vain undertaking. By saying that the universe is written in mathematical language, Galileo means to say that scientific knowledge is not about secondary qualities but about primary qualities, which however are not the things as they really are in themselves, but are only phenomenal properties of natural substances.

8.10 Alternative Concepts of Truth

As stated above, because of Galileo's philosophical revolution, the concept of truth as intuition of the essence does not apply to modern science. Admittedly some people, such as Ellis 2002, have proposed to return to some kinds of essentialism, but, as Khalidi points out, "essentialism encounters some fundamental problems which constitute obstacles to integration with science" (Khalidi 2009, 86). Indeed, essentialism is incompatible with the basic assumptions of modern science.

On the other hand, the concept of truth as correspondence does not apply to modern science, for the reason stated by Aristotle, Kant and Frege. Therefore, none of the two concepts of truth considered by Aristotle applies to modern science.

One may ask if some other concept of truth might apply to modern science. The answer is negative, because all alternative concepts of truth that have been proposed are inadequate. It would be impossible to discuss all of them in the limited space of this chapter. However, we will consider four concepts of truth which seem especially relevant to distinct parts of modern science: Hilbert's concept of truth as consistency, Joachim's concept of truth as systematic coherence, Tarski's concept of truth as satisfiability, and Tait's concept of truth as provability.

8.11 Truth as Consistency

According to Hilbert's concept of truth as consistency, a sentence is true if and only if it is consistent with a specified set of other sentences. Indeed, Hilbert states that, "if the arbitrarily given axioms do not contradict one another with all their consequences, then they are true" (Hilbert 1980a, 39). In particular, a sentence which is not in contradiction with arbitrarily given axioms is true, so consistency is "the criterion of truth" (ibid.).

But the concept of truth as consistency is inadequate, because two mutually inconsistent sentences can be both consistent with the same set of other sentences. For example, the continuum hypothesis and its negation are both consistent with Zermelo-Fraenkel's axioms of set theory, and hence, by the concept of truth as consistency, they are both true. But since they are mutually inconsistent, this is impossible.

Moreover, by a corollary of Gödel's first incompleteness theorem, for any consistent, sufficiently strong, formal system *S*, there is a consistent extension *T* of *S* in which some false sentence is demonstrable (see, for example, Cellucci 2013a, Chap. 12). From the corollary it follows that the axioms of *T*, though consistent, cannot be said to be true – if they could be said to be true, only true sentences would be demonstrable in *T*. Thus, contrary to Hilbert's claim, consistency is not a sufficient condition for truth. Actually, that consistency is not a sufficient condition for truth, had been made quite clear well before Gödel. For example, Pascal warns that "contradiction is not an indication of falsehood and the absence of contradiction is not a sign of truth" (Pascal 1995, 61). A similar warning is made by Kant (see Chap. 20).

8.12 Truth as Systematic Coherence

According to Joachim's concept of truth as systematic coherence, a proposition is true if and only if it forms a whole with a specified set of other propositions, such that "all its constituent elements reciprocally involve one another, or reciprocally determine one another's being as contributory features in a single concrete meaning" (Joachim 1906, 66). This is the concept "of truth as 'systematic coherence'" (ibid., 65). Systematic coherence "must not be confused with the 'consistency' of formal logic," because "a piece of thinking might be free from self-contradiction, might be 'consistent' and 'valid' as the formal logician understands those terms, and yet it might fail to exhibit that systematic coherence which is the character of a significant whole" (ibid., 76).

But the concept of truth as systematic coherence is inadequate, for example, because the propositions of a fable form a systematically coherent whole, though being a fiction. So, truth as systematic coherence need not have anything to do with reality at all.

Joachim answers this objection by saying that his concept of truth is "intended to describe the nature of truth as an ideal, as the character of an ideally complete experience" (ibid., 78). Human knowledge "is clearly not a significant whole in this ideally complete sense," so his concept of truth is "an ideal, and an ideal which can never as such, or in its completeness, be actual as human experience" (ibid., 79).

But this answer is inadequate because, if the concept of truth is an ideal that can never be actual as human experience, then such concept is humanly transcendent, and hence cannot be rationally propounded. Nor can one use the argument that the concept of truth as systematic coherence does not imply that a proposition which forms a systematically coherent whole with a specified set of other propositions is true – it only implies that a proposition which forms a systematically coherent whole with a specified set of other true propositions is true. By this argument, the latter propositions are true if and only if they form a systematically coherent whole with a specified set of other true propositions, and so on. This leads to an infinite regress, which is not suited to the concept of truth.

8.13 Truth as Satisfiability

According to Tarski's concept of truth as satisfiability, a sentence is true if and only if there is a domain such that "every infinite sequence of objects" in the domain "satisfies the sentence" (Tarski 1983, 200).

But the concept of truth as satisfiability is inadequate, because even a fiction can be true in terms of it. Indeed, even for a fiction there can be a domain such that every infinite sequence of objects in the domain satisfies the sentence, hence, that a sentence is true in terms of this concept of truth does not entail that it is true in the real world. So, truth as satisfiability need not have anything to do with reality at all.

The scientific realists who assume the concept of truth as satisfiability, are also bound to assume that domains are isomorphic or at least homomorphic to the real world. But since isomorphism or homomorphism is a relation between mathematical objects, this involves assuming that the real world is a mathematical object. Such assumption is problematic because only some properties of the world are mathematical in kind.

Moreover, according to the concept of truth as satisfiability, in order to know that a sentence is true we must demonstrate that there exists a domain such that every infinite sequence of objects in the domain satisfies the sentence. This must be demonstrated in some mathematical theory, and ultimately in set theory, and presupposes that the axioms of set theory be true. But, by Gödel's second incompleteness theorem, it is impossible to demonstrate by absolutely reliable means that the axioms of set theory are true.

8.14 Truth as Provability

According to Tait's concept of truth as provability, a proposition is true if and only if it is provable from given axioms, so there exists "an objective criterion for truth, namely provability from the axioms" (Tait 2001, 32).

But the concept of truth as provability is inadequate because it raises the question: In what sense are the axioms true? This question has no satisfactory answer. As Tait acknowledges, in order to consider the axioms to be true, "we ought to require at least that the axioms be consistent" (ibid., 22). But, by Gödel's second incompleteness theorem, generally we cannot know whether the axioms are consistent. And it is no way out to say that "consistency is just something that, ultimately, we must take on faith" (ibid., 23). If so, then the fact that provability from the axioms is an objective criterion of truth would depend on faith.

Moreover, by Gödel's first incompleteness theorem, already in the case of arithmetic, there is a sentence A such that neither A nor its negation $\neg A$ is provable from the axioms. Which of A and $\neg A$ is true by the criterion of provability from the axioms? Tait states that the sentence A is "indeterminate; but that doesn't mean that there can be no natural grounds, if only we seek them, on which to determine it" (ibid., 31). Then, however, A would not be determined by the criterion of provability from the axioms.

All the more so because Tait states that "to say that there may be natural grounds on which to decide" A "one way," that is, to accept A, "does not mean that there may not also be natural grounds upon which to decide it the other way" (ibid.). That is, it does not mean that there may not also be natural grounds to accept $\neg A$. This amounts to saying that there is no objective criterion of truth.

8.15 Truth and Mythology

It has been argued above that four concepts of truth, which seem especially relevant to distinct parts of modern science, are inadequate. It could be argued that all other alternative concepts of truth that have been proposed are inadequate. From this, it seems fair to conclude that the view that the aim of science is truth is inadequate.

In fact, truth is a remnant of Aristotelian science, which was aimed at knowing the essence of natural substances, and for which, therefore, truth, and specifically truth as intuition of the essence, was central. Other is the case of modern science, which originated from Galileo's philosophical revolution: the decision to renounce Aristotle's aim to penetrate the essence of natural substances, dealing only with some of their phenomenal properties mathematical in kind, such as location, motion, shape, or size.

But, prior to being a remnant of Aristotelian science, truth is a remnant of Greek mythology. Thus, Aesop states that Prometheus "decided one day to sculpt a statue of Truth," then he put the statue "in the kiln," and when it "had been thoroughly baked, he infused" it "with life: sacred Truth walked with measured steps" (Aesop, *Fables* 530. *Prometheus and Truth*). Pindar calls "truth the daughter of Zeus" (Pindar, *Olympian odes* 10, 3–4). Bacchylides says that "truth is from the same city as the Gods; she alone lives with the Gods" (Bacchylides, Fragment 57).

Greek philosophy grew out of Greek mythology and progressively got rid of many elements of the mythical tradition. Nevertheless, it retained some of them. Truth is one of such leftover elements. Thus Parmenides presents the Goddess as saying to a young man that she will reveal him "the unshaken heart of persuasive truth," which is opposed to "the opinions of mortals, where truth cannot reside" (Parmenides 28 B 1 Diels–Kranz). In particular, she will reveal him the way of truth, which is "the path of persuasion, since it follows truth" (Parmenides 28 B 2 Diels–Kranz).

The reason why philosophy, at least some philosophy, makes reference to truth even today, is that the search for truth responds to a deep need of human beings. They seek truth because they look for secure and stable reference points to shelter themselves, against the storm and confusion of life. But this is an illusion, because insecurity and instability are intrinsic constituents of human life and cannot be eliminated.

Like all concepts of a mythical origin, truth is unsuitable for all serious scientific uses since, as argued above, a criterion of truth will never be found. Therefore, as even Hempel admits, "the conception of science as a search for truth is strictly untenable" (Hempel 2000, 77). This raises the need to replace the view that the aim of science is truth with an alternative view.

References

Capozzi, Mirella. 2013. *Kant e la logica I*. Naples: Bibliopolis.
Cellucci, Carlo. 2013a. *Rethinking logic: Logic in relation to mathematics, evolution, and method*. Dordrecht: Springer.
David, Marian. 2004. Theories of truth. In *Handbook of epistemology*, ed. Ilkka Niiniluoto, Matti Sintonen, and Jan Woleński, 331–414. Dordrecht: Springer.
Dummett, Michael. 2006. *Thought and reality*. Oxford: Oxford University Press.
———. 2010. *The nature and future of philosophy*. New York: Columbia University Press.
Ellis, Brian. 2002. *The philosophy of nature: A guide to the new essentialism*. Chesham: Acumen.
Engel, Pascal. 2002. *Truth*. Chesham: Acumen.
Frege, Gottlob. 1984. *Collected papers on mathematics, logic, and philosophy*. Oxford: Blackwell.
Galilei, Galileo. 1968. *Opere*. Florence: Barbera.
Hempel, Carl Gustav. 2000. *Selected philosophical essays*. Cambridge: Cambridge University Press.
Hilbert, David. 1980a. Letter to Frege, 29 December 1899. In Gottlob Frege, *Philosophical and mathematical correspondence*, 38–41. Oxford: Blackwell.
Husserl, Edmund. 1970. *The crisis of European sciences and transcendental phenomenology*. Evanston: Northwestern University Press.
———. 2001. *Logical investigations*. London: Routledge.
Joachim, Harold. 1906. *The nature of truth*. Oxford: Oxford University Press.
Khalidi, Muhammad. 2009. How scientific is scientific essentialism? *Journal for General Philosophy of Science* 40 : 85–101.
Kant, Immanuel. 1992. *Lectures on logic*. Cambridge: Cambridge University Press.
———. 2002. *Theoretical philosophy after 1781*. Cambridge: Cambridge University Press.
Musgrave, Alan. 1996. Realism, truth and objectivity. In *Realism and anti-realism in the philosophy of science*, ed. Robert S. Cohen, Risto Hilpinen, and Qiu Renzong, 19–44. Dordrecht: Springer.
Newton, Isaac. 1962. *Mathematical principles of natural philosophy – The system of the world*. Berkeley: University of California Press.
Pascal, Blaise. 1995. *Pensées and other writings*. Oxford: Oxford University Press.
Popper, Karl Raimund. 1972. *Objective knowledge*. Oxford: Oxford University Press.
———. 1996. *In search of a better world*. London: Routledge.
Putnam, Hilary. 1975–1983. *Philosophical papers*. Cambridge: Cambridge University Press.
Raatikainen, Panu. 2003. More on Putnam and Tarski. *Synthese* 135 : 37–47.
Sankey, Howard. 2008. *Scientific realism and the rationality of science*. Aldershot: Ashgate.
Tait, William Walker. 2001. Beyond the axioms: The question of objectivity in mathematics. *Philosophia Mathematica* 9 : 21–36.
Tarski, Alfred. 1944. The semantic conception of truth and the foundations of semantics. *Philosophy and Phenomenological Research* 4 : 341–376.
———. 1969. Truth and proof. *Scientific American* 220 : 63–77.
———. 1983. *Logic, semantics, metamathematics*. Indianapolis: Hackett.
Zilsel, Edgar. 1942. The genesis of the concept of physical law. *The Philosophical Review* 51 : 245–279.

Chapter 9
Knowledge, Plausibility, and Common Sense

Abstract As an alternative to the view that the aim of science is truth, this chapter maintains that the aim of science is plausibility, and specifically to make plausible hypotheses about the world, namely hypotheses such that the arguments for them are stronger than those against them, on the basis of the existing knowledge. The chapter argues that this meets all the difficulties of the view that the aim of science is truth. It also argues that plausibility is different from truth, probability and warranted assertibility, but is to a certain extent related to Aristotle's *endoxa*. Moreover, the chapter discusses the relation between scientific knowledge and common sense knowledge, arguing that, rather than being opposed to common sense knowledge, scientific knowledge gives an explanation of our common sense knowledge, providing an interpretation of it.

9.1 Plausibility in Place of Truth

An alternative to the view that the aim of science is truth is that the aim of science is plausibility. Specifically, the aim of science is to make plausible hypotheses about the world, namely hypotheses such that the arguments for them are stronger than the arguments against them, on the basis of the existing knowledge.

The concept of truth is inessential to modern science, the concept of plausibility is all that is needed for it. Scientific theories do not deal with the essence of natural substances, but only with some of their phenomenal properties, and deal with them by making plausible hypotheses. Therefore the concept of truth, which is proper to Aristotelian science, must be replaced with that of plausibility. A scientific theory is not a set of truths, but rather a set of problems about the world and plausible hypotheses that permit us to solve them. The only knowledge we can have about the world is plausible knowledge.

That the aim of science is not truth but plausibility avoids the problem of scientific realism and liberalized scientific realism, discussed in Chap. 7, that the history of science offers us many examples of scientific theories that were once taken to be successful but that later on have proven to be untenable. From the empirical success of a scientific theory one may infer that the theory is plausible, not that it is true, and while truth is an absolute concept, plausibility is a relative one. A theory that is plausible at a certain stage may become implausible at a later stage, and conversely

© Springer International Publishing AG 2017

C. Cellucci, *Rethinking Knowledge*, European Studies in Philosophy of Science 4,
DOI 10.1007/978-3-319-53237-0_9

a theory that is implausible at a certain stage may become plausible at a later stage, when the arguments for the theory are stronger than those against it.

Sankey states that, "if a theory is subjected to a battery of demanding tests, consistently yielding accurate predictions in a range of different circumstances, such performance under test is to be accorded evidential weight with regard to the truth of the theory" (Sankey 2008, 143). But this conflicts with the fact that, as we have seen in Chap. 8, Sankey also states that we can rationally believe a proposition that is false. The performance in question cannot be accorded evidential weight with regard to the truth of the theory, but only with regard to its plausibility.

9.2 Plausibility as Different from Truth

Plausibility is different from other concepts to which it is sometimes assimilated, specifically, it is different from truth, probability, or warranted assertibility.

Plausibility is different from truth. Peirce states that "what we mean by the truth" is "the opinion which is fated to be ultimately agreed to by all who investigate" (Peirce 1931–1958, 5.407). This amounts to saying that truth is plausibility. But it is not so, between truth and plausibility there are important differences.

Indeed, truth is an absolute concept, because if a proposition is true, it will remain true forever. Conversely, plausibility is a relative concept, because new data may always emerge which change the balance between the arguments for a hypothesis and the arguments against it. Thus, a hypothesis which is taken to be plausible at one stage may become implausible at a later stage.

Moreover, as argued in Chap. 8, a criterion of truth will never be found. Conversely, a criterion of plausibility exists, and is given by the plausibility test procedure that will be described in Chap. 12.

Furthermore, although a concept of truth can be made mathematically precise through Tarski's expression (T), considered in Chap. 8, this does not provide a means which allows us to distinguish true sentences from false sentences. Conversely, although the concept of plausibility cannot be made mathematically precise, it provides a means which allows us to distinguish plausible propositions from implausible ones.

In addition, as already argued in Chap. 8 with reference to truth as systematic coherence or truth as satisfiability, truth need not have anything to do with reality at all. Conversely, plausibility has to do with reality, because it refers to our experience of reality.

9.3 Plausibility as Different from Probability

Plausibility is different from probability. Pólya states that we can "use the calculus of probability to render more precise our views on plausible reasoning" (Pólya 1954, II, 116). For, "the calculus of plausibilities obeys the same rules as the calculus of probabilities" (Pólya 1941, 457). But it is not so, between plausibility and probability there are important differences.

Indeed, plausibility involves a comparison between the arguments for a hypothesis and the arguments against it, so plausibility is not a mathematical concept. Conversely, probability is a mathematical concept.

This is made quite clear already by Kant, who states that "plausibility is concerned with whether, in the cognition, there are more grounds for the thing than against it" (Kant 1992, 331). So, probability is not a mathematical concept. Conversely, "probability is a fraction, whose denominator is the number of all the possible cases, and whose numerator contains the number of winning cases" (ibid., 328). So, probability is a mathematical concept, and indeed "there is a mathematics of probability" (ibid., 331). (For more on Kant's distinction between plausibility and probability, see Capozzi 2013, Chap. 7, Sect. 5, and Chap. 15).

Moreover, there are hypotheses which are plausible, but conversely, in terms of the classical concept of probability, have zero probability. On the other hand, there are hypotheses which are not plausible, but conversely, again in terms of the classical concept of probability, have a non-zero probability. The same holds on other concepts of probability (see Cellucci 2013a, Chap. 20).

9.4 Plausibility as Different from Warranted Assertibility

Plausibility is different from warranted assertibility. Rescher states that, "in addition to demonstrative reasoning that establishes knowledge there is also the plausible (sub-demonstrative) reasoning that establishes mere credibility or what later philosophers called 'warranted assertability'" (Rescher 2007, 128). Thus, plausible reasoning is the one that establishes warranted assertibility. But it is not so, plausible reasoning is the one that establishes plausibility, and between warranted assertibility and plausibility there are important differences.

Indeed, Dewey states that the "use of a term" such as "warranted assertibility" that "designates a potentiality rather than an actuality involves recognition that all special conclusions of special inquiries are parts of an enterprise that is continually renewed, or is a going concern" (Dewey 1938, 9). This enterprise converges to the ideal limit of truth since, as Peirce says, truth is "the ideal limit towards which endless investigation would tend to bring scientific belief" (ibid., 345, footnote 6). On the contrary, if the aim of science is plausibility, then science is not an enterprise that converges to the ideal limit of truth. For, as stated above, a theory that is plausible at a certain stage may become implausible at a later stage, and conversely a theory that is implausible at a certain stage may become plausible at a later stage when the arguments for the theory are stronger than those against it. Thus a theory can only be plausible, it can never be said to be true.

Moreover, to say that science is an enterprise that converges to the ideal limit of truth, conflicts with the history of science. As already pointed out in Chap. 7, the latter offers us many examples of scientific theories that were once taken to be successful but that, according to contemporary scientific theories, are false. And there is no evidence that scientific theories which today are taken to be successful will not be, according to future scientific theories, false. Even a statement such as 'Water is H_2O' cannot be taken to be an eternal and unqualified truth. Now, how could a sequence of false scientific theories be said to converge to the ideal limit of truth? Conversely, to say a theory can only be plausible does not conflict with the history of science because, as pointed out above, while truth is an absolute concept, plausibility is a relative one.

9.5 Plausibility and *Endoxa*

While plausibility is different from truth, probability, or warranted assertibility, it is to a certain extent related to Aristotle's *endoxa*.

Aristotle states that "those things are *endoxa* which seem so to everyone, or to the majority, or to the wise, and either to all of them, or to the majority, or to the most known and reputed among them" (Aristotle, *Topica*, A 1, 100 b 21–23). This statement, however, is somewhat ambiguous because, in terms of it, the superficial or biased beliefs of the majority might be considered to be *endoxa*. Therefore, the statement should be complemented by Aristotle's statement that, "for every thesis, we must examine the arguments for it and the arguments against it" (ibid., Θ 14, 163 a 37–b 1). For, "going through the objections on either side, we shall more readily discern the true as well as the false in any subject" (Aristotle, *Topica*, A 2, 101 a 35–6). Indeed, "if the objections are answered and the *endoxa* remain, we shall have proved the case sufficiently" (Aristotle, *Ethica Nicomachea*, Z 1, 1145 b 6–7).

With this interpretation, *endoxa* are things which seem acceptable to everyone, or to the great majority, or to the wise, etc., on the basis of an examination of the arguments for and against them, from which the former turn out to be stronger than the latter. On these grounds, it seems fair to conclude that plausibility is to a certain extent related to Aristotle's *endoxa*. Striker even uses 'plausible' "to translate the Greek *endoxon*" (Striker 2009, 77).

However, that plausibility is to a certain extent related to Aristotle's *endoxa* does not mean that it is the same as *endoxa*. Indeed, according to Aristotle, *endoxa* are continuous with truth. For, he states that "an ability to aim at *endoxa* is a characteristic of one who also has a similar ability in regard to the truth" (Aristotle, *Rhetorica*, 1355 a 17–18). On the contrary, plausibility, as defined here, is an alternative to truth. Therefore, plausibility is not the same as *endoxa*.

9.6 Knowledge and Our Ways of Apprehending the World

It has been stated above that the aim of science is plausibility, and specifically to make plausible hypotheses about the world. In fact, plausible hypotheses are our ways of apprehending the world. On the other hand, the only knowledge we can have about the world is through our ways of apprehending it.

That the only knowledge we can have about the world is through our ways of apprehending it, implies that the laws of nature do not exist in the world but only in the minds of human beings. As Kant states, the laws of nature do not exist "in the appearances, but rather exist only relative to the subject in which the appearances inhere" (Kant 1998, B164).

It also implies that the mathematics used in formulating the laws of nature does not exist in the world. Thus, ellipses do not exist in the world, they are only concepts in terms of which human beings account for the observational data about the positions of planets. A light ray through a slit does not know anything about Fourier transforms, the latter are only a tool by which human beings account for the resulting light diffraction.

Saying that the mathematics used in formulating the laws of nature exists in the world may lead to odd conclusions. For example, Russell claims that "there is no transition from place to place, no consecutive moment or consecutive position, no such thing as velocity except in the sense of a real number which is the limit of a certain set of quotients," so "we must entirely reject the notion of a state of motion" (Russell 2010, 480). Thus, however, Russell mistakes a mathematical model – Weierstrass's discretization of continuity – for external physical reality. This leads him to the odd conclusion that there is no transition from place to place, no consecutive moment or consecutive position, no velocity, but only sequences of pairs of real numbers and limits of such sequences.

But mathematical models are not to be mistaken for the external physical reality, they are simply means by which we make the world understandable to ourselves. To claim that the laws of nature, or the mathematics used in formulating them, exist in the world is to confuse the means by which we make the world understandable to ourselves with the world itself. The laws of nature and the mathematics used in formulating them have a place only within the frame of scientific theories. Mathematics is not the essence of the world, but only a wholly human way of seeing and thinking about it, it simply offers us heightened ways to see and think about the world.

9.7 Limitations of the Human Cognitive Apparatus

It has been stated above that the only knowledge we can have about the world is through our ways of apprehending it. Now, our ways of apprehending the world essentially depend on our cognitive apparatus, which is a result of biological evolution. This implies that we cannot have full knowledge of all aspects of the world. As Heraclitus says, “nature loves to hide” (Heraclitus 22 B 123 Diels–Kranz).

Indeed, having evolved in response to certain challenges of the world, our cognitive apparatus can only deal with those aspects of the world from which such challenges arose, therefore, it is subject to limitations. Indeed, our perceptual apparatus is subject to limitations. Thus, unlike goldfishes we are incapable of detecting infrared radiations, unlike bees we are incapable of detecting ultraviolet radiations, unlike bats we are incapable of ultrasound echolocation. But our conceptual apparatus too is subject to limitations. Wittgenstein denies that, “if such-and-such facts of nature were different, people would have different concepts” (Wittgenstein 1958, II, xii.230). But it is not so, because our conceptual apparatus is derived to a large extent from our perceptual apparatus, thus from facts of nature.

Since our perceptual apparatus and our conceptual apparatus are subject to limitations, the purpose of knowledge cannot be to give a complete picture of the world. Rather, the purpose of knowledge is to make hypotheses that work in solving the problems with which we are confronted, starting with that of survival.

9.8 The Relevance of Such Limitations

That we cannot have full knowledge of all aspects of the world, because our cognitive apparatus is subject to limitations, is essential for life. For, while a cognitive apparatus like ours, which is subject to certain limitations, can deal with those aspects of the world that are essential to our survival, a cognitive apparatus not subject to those limitations would be unable to deal with them.

This is made quite clear already by Locke, who states that, despite their limitations, our senses are adequate “to provide for the conveniences of living” which “are our business in this world” (Locke 1975, 302). Conversely, “were our senses alter’d, and made much quicker and acuter, the appearance and outward scheme of things would have quite another face to us,” and “would be inconsistent with our being, or at least well-being, in the part of the universe which we inhabit” (ibid.). For example, if vision “were in any man 1000, or 100000 times more acute than it is by the best microscope,” that man could not “take in but a very small part of any object at once, and that too only at a very near distance” (ibid., 303). He “could not see things he was to avoid, at a convenient distance; nor distinguish things he had to do with by those sensible qualities others do” (ibid.). Then vision would be inadequate to provide for the conveniences of living. Therefore, the limitations of our cognitive apparatus are essential for life.

9.9 Knowledge and Things in Themselves

That we cannot have full knowledge of all aspects of the world is related to Kant's view that "what the things may be in themselves I do not know, and also do not need to know, since a thing can never come before me except in appearance" (Kant 1998, A277/B333). For, "we can cognize objects only as they appear to us (to our senses), not as they may be in themselves" (Kant 2002, 70).

As Kant makes it quite clear, 'objects as they appear to us' and 'objects as they may be in themselves' do not designate two different realms of being, but only two different ways of considering the same things. The distinction between them arises from the fact that "the same objects can be considered from two different sides, on the one side as objects of the senses and the understanding for experience, and on the other side as objects that are merely thought at most for isolated reason striving beyond the bounds of experience" (Kant 1998, Bxviii–Bxix, footnote). Thus "all objects that are given to us can be interpreted in two ways, on the one hand, as appearances, on the other hand, as things in themselves" (Kant 1999, 199, footnote). But "we cannot apply any of our concepts of the understanding" to things in themselves, so the concept of 'thing in itself' "remains empty for us, and serves for nothing but to designate the boundaries of our sensible cognition" (Kant 1998, A289/B345). That we can cognize objects only as they appear to us (to our senses), not as they may be in themselves, is just another way of saying that we can only know that "which can be brought into connection with our actual perceptions" (Kant 2002, 143). Hence, it is just another way of saying that we cannot have full knowledge of all aspects of the world, since our cognitive apparatus can only deal with the phenomenal aspects of the world, and in fact not all of them.

Kant's view that we can cognize objects only as they appear to us, not as they may be in themselves, has been criticized by several people. For example, Plotkin states that knowledge is essential to survival, so, "that we survive at all is proof that knowledge is possible" (Plotkin 1997, 240). Now, knowledge is "correspondence to the things-in-themselves" (ibid.). Hence, "if Kant were correct, if living things could never know the things in themselves, then life would never survive" (ibid., 241). Then, "since we and the myriad other forms of life do survive, Kant must be wrong" (ibid.). Plotkin's argument, however, is based on the assumption that, if living things could never know the things in themselves, then life would never survive. This assumption is unwarranted. Surely, the prokaryotes did not survive because they knew the things in themselves!

As Hanna points out, there is a sharp contrast between Kant's view and scientific realism, because the latter "in most of its forms is committed to microphysical noumenal realism" (Hanna 2006, 49). Specifically, it is committed to the view that "the essential properties of individual material substances, natural kinds, events, processes, and forces" are all "reducible to the microphysical" and "mind-independent properties of metaphysically ultimate objects or things-in-themselves" (ibid.).

Contrary to scientific realism, knowledge is not about things in themselves, but only about phenomenal aspects of the world, and indeed only some of them. Some supporters of scientific realism "portray anti-realists as denying" that "the existence and nature of the stars are in various ways independent of our minds" (Devitt 2013, 112). But this is a misunderstanding. What anti-realists deny is not that the existence and nature of things in the external world are independent of our mind, but only that our minds have knowledge of properties of metaphysically ultimate objects or things in themselves.

9.10 Common Sense Knowledge and Scientific Knowledge

Our knowledge of the world includes common sense knowledge. The latter is usually described as what everyone knows. Obviously, the question immediately arises as to what is the relation between scientific knowledge and common sense knowledge.

According to a widespread opinion, scientific knowledge is opposed to common sense knowledge. If something fits with scientific knowledge, then it is not common sense knowledge, and if something fits with common sense knowledge, then it is not scientific knowledge.

Thus Wolpert maintains that "both the ideas that science generates and the way in which science is carried out" are "against common sense" (Wolpert 1992, 1). Conversely, "if something fits in with common sense it almost certainly isn't science" (ibid., 11). In fact, "a significant proportion of British citizens does not think the earth goes round the sun. And I doubt that of those who do believe the earth moves round the sun, even one person in 100,000 could give sound reasons for their conviction" (ibid., ix–x).

An extreme form of the view that scientific knowledge is opposed to common sense knowledge is the opinion that only the world of science is real. Thus, Sellars claims that "the common-sense world of physical objects in space and time is unreal," namely "there are no such things," only the world of science is real, because "science is the measure of all things, of what is that it is, and of what is not that it is not" (Sellars 1997, 83).

Another extreme form of the view that scientific knowledge is opposed to common sense knowledge, but in the opposite direction, is the opinion that only the common-sense world is real. Thus, Husserl claims that "the only real world" is "the one that is actually given through perception, that is ever experienced and experienceable – our everyday life-world" (Husserl 1970, 49). Science works out a "surreptitious substitution of the mathematically substructed world of idealities for the only real world" (ibid., 48–49).

But the view that scientific knowledge is opposed to common sense knowledge is unjustified. Rather than being opposed to common sense knowledge, scientific knowledge gives an account of our common sense knowledge, providing an interpretation of it.

For example, our common sense knowledge tells us that the sun rises each morning, it goes across the sky and sets each evening, and in the night the moon, stars and planets become visible and move in the sky. Ptolemaic astronomy gives an account of our common sense knowledge, interpreting it as follows: all heavenly objects, including the sun, moon, planets and stars, move around the earth in a circular orbit, while the earth is at the center of the universe and is fixed and immovable. Copernican astronomy gives another account of our common sense knowledge, interpreting it as follows: the planets move around the sun, and the moon moves around the earth, in a circular orbit, while the sun is at the center of the universe and is fixed and immovable, just like the stars. Neither Ptolemaic nor Copernican astronomy are opposed to our common sense knowledge, only, they give different accounts of it.

Kuhn claims that Copernican astronomy is a "violation of common sense" because it contravenes "the first and most fundamental suggestions provided by the senses about the structure of the universe" (Kuhn 1957, 43). But it is not so. Simply, Copernican astronomy gives an account of our common-sense knowledge which is more plausible than the account given by Ptolemaic astronomy. Copernican astronomy is not a violation of common sense, it is only an interpretation of our common sense knowledge. In fact, it is an interpretation which itself needs corrections, because the sun is not at the center of the universe and is neither fixed nor immovable, just like the stars. Moreover, planets move around the sun, and the moon moves around the earth, in an elliptic orbit.

9.11 Common Sense Knowledge and Innate Knowledge

In order to better understand why it is unjustified to say that scientific knowledge is opposed to common sense knowledge, we must seek to clarify the nature of common sense knowledge. It has been said above that common sense knowledge is usually described as what everyone knows. Then common sense knowledge cannot be culturally and historically specific knowledge, since the latter changes with culture and history, so it cannot be what everyone knows. Therefore, common sense knowledge cannot consist of beliefs which are widely held within a culture.

Common sense knowledge can only be innate knowledge, namely, knowledge possessed from birth. Actually, human beings have some innate knowledge. Experiments have revealed that "geometrical knowledge arises in humans independently of instruction, experience with maps or measurement devices, or mastery of a sophisticated geometrical language" (Dehaene et al. 2006, 384). Experiments have also "revealed the rudiments of arithmetic" in "young infants before they acquire number words," although they "appear to represent only the first three numbers exactly" (Pica et al. 2004, 499). So "numerical competence can be present in the absence of a well-developed lexicon of number words" (ibid., 503). Experiments have even revealed that "infants have detailed knowledge about how" physical "objects behave and interact from the first weeks of life," and, "as early as 2 months of age, they have initial concepts about continuity, cohesion, and change properties" (Hespos and vanMarle 2012, 25–26). Thus, experiments have revealed that young infants have some innate knowledge of space, number, and physical objects.

Common sense knowledge consists of this and other possible innate knowledge. The latter is presupposed in all interactions of human beings with the world, and science itself starts from it. This clarifies why it is unjustified to say that scientific knowledge is opposed to common sense knowledge. Being culturally and historically specific, scientific knowledge is different from common sense knowledge, nevertheless it starts from common sense knowledge, and gives an account of it. Thus, the view that the aim of science is to make plausible hypotheses about the world can account for the relation between science and common sense knowledge.

9.12 Common Sense Knowledge and Scientific Realism

Unlike the view that the aim of science is to make plausible hypotheses about the world, scientific realism is unable to account for the relation between science and common sense knowledge. Among supporters of scientific realism, one may find two different positions on this relation.

Some supporters of scientific realism maintain that common sense knowledge is false on the basis of science, so scientific realism "invites us to reject commonsense" (Feyerabend 1981–2016, I, xii). The latter is to be completely replaced by science. If "the complete replacement of commonsense" by science "succeeds, then man will have succeeded in ridding himself of prejudices which would have prevented him from ever seeing the world as it is and from figuring out the details of its real structure" (ibid., I, 127–128). For, "science not only produces predictions, it is also about the nature of things" (ibid., I, 3).

This view, however, is untenable. Scientific realism claims that the aim of science is to have true theories about the world, and science really has such theories. Now, the evidential basis for science is given by observation, and observation, deriving either from sense perception or from instrumentation, is based on common sense. So, if common sense must be rejected, observation itself must be rejected. But then the evidential basis for science vanishes, and we have no grounds for claiming that science has true theories about the world. This contradicts the claim of scientific realism that science has true theories about the world. Besides, if we have no grounds for claiming that science has true theories about the world, then we have no grounds for claiming that common sense knowledge is false on the basis of science. So we must accept common sense knowledge instead of science. Then, to say that common sense knowledge is false on the basis of science is self-defeating.

Other supporters of scientific realism maintain that "commonsense beliefs" have "survival value because they are for the most part true. Our species could not have survived if the majority of the commonsense beliefs on which we base our everyday interaction with the world were false" (Sankey 2014, 17–18). A "robust sense of reality provides us with a reasonable degree of practical certainty that things are by and large as they appear to us" (ibid., 15). Science itself "starts from common sense," so "science does not lead to the overthrow of common sense" (ibid., 23).

This view is also untenable. The claim that commonsense beliefs have survival value because they are for the most part true, conflicts with the fact that, as argued in Chap. 8, a criterion of truth will never be found, so human beings will generally be unable to recognize a truth when they reach it. Then, our species could not have survived because the majority of the commonsense beliefs on which we base our everyday interaction with the world are true. For, our species would have been generally unable to recognize that they are true. Besides, the claim that a robust sense of reality provides us with a reasonable degree of practical certainty that things are by and large as they appear to us, conflicts with the fact that common sense knowledge is based on our cognitive apparatus which, as argued above, is subject to limitations. Because of these limitations, things are not by and large as they appear to us.

References

Capozzi, Mirella. 2013. *Kant e la logica I*. Naples: Bibliopolis.

Cellucci, Carlo. 2013a. *Rethinking logic: Logic in relation to mathematics, evolution, and method*. Dordrecht: Springer.

Dehaene, Stanislas, Véronique Izard, Pierre Pica, and Elizabeth Spelke. 2006. Core knowledge of geometry in an Amazonian indigene group. *Science* 311 : 381–384.

Devitt, Michael. 2013. Hilary and me: Tracking down Putnam on the realism issue. In *Reading Putnam*, ed. Maria Baghramian, 101–120. London: Routledge.

Dewey, John. 1938. *Logic: The theory of inquiry*. New York: Holt, Rinehart & Winston.

Feyerabend, Paul K. 1981–2016. *Philosophical papers*. Cambridge: Cambridge University Press.

Hanna, Robert. 2006. *Kant, science, and human nature*. Oxford: Oxford University Press.

Hespos, Susan J., and Kristy van Marle. 2012. Physics for infants: Characterizing the origins of knowledge about objects, substances, and number. *WIREs Cognitive Science* 3: 19–27.

Husserl, Edmund. 1970. *The crisis of European sciences and transcendental phenomenology*. Evanston: Northwestern University Press.

Kant, Immanuel. 1992. *Lectures on logic*. Cambridge: Cambridge University Press.

———. 1998. *Critique of pure reason*. Cambridge: Cambridge University Press.

———. 1999. *Correspondence*. Cambridge: Cambridge University Press.

———. 2002. *Theoretical philosophy after 1781*. Cambridge: Cambridge University Press.

Kuhn, Thomas Samuel. 1957. *The Copernican revolution*. Cambridge: Harvard University Press.

Locke, John. 1975. *An essay concerning human understanding*. Oxford: Oxford University Press.

Peirce, Charles Sanders. 1931–1958. *Collected papers*. Cambridge: Harvard University Press.

Pica, Pierre, Cathy Lemer, Véronique Izard, and Stanislas Dehaene. 2004. Exact and approximate arithmetic in an Amazonian indigene group. *Science* 306 : 499–503.

Plotkin, Henry. 1997. *Darwin machines and the nature of knowledge*. Cambridge: Harvard University Press.

Pólya, George. 1941. Heuristic reasoning and the theory of probability. *The American Mathematical Monthly* 48 : 450–465.

———. 1954. *Mathematics and plausible reasoning*. Princeton: Princeton University Press.

Rescher, Nicholas. 2007. *Dialectics: A classical approach to inquiry*. Frankfurt: Ontos Verlag.

Russell, Bertrand. 2010. *The principles of mathematics*. London: Routledge.

Sankey, Howard. 2008. *Scientific realism and the rationality of science*. Aldershot: Ashgate.

———. 2014. Scientific realism and basic common sense. *Kairos* 10 : 11–24.

Sellars, Wilfrid. 1997. *Empiricism and the philosophy of mind*. Cambridge: Harvard University Press.

Striker, Gisela. 2009. Commentary. In Aristotle, *Prior Analytics*, Book I, ed. Gisela Striker, 67–246. Oxford: Oxford University Press.

Wittgenstein, Ludwig. 1958. *Philosophical investigations*. Oxford: Blackwell.

Wolpert, Lewis. 1992. *The unnatural nature of science*. Cambridge: Harvard University Press.

Chapter 10
Other Questions About Knowledge

Abstract This chapter considers the relations of knowledge to objectivity, certainty, intuition, deduction, and rigour. It argues that knowledge cannot be objective in the sense of being totally independent of any subject, but only in the sense of being as independent as possible of any particular human subject; that knowledge cannot be absolutely certain, since it can only be plausible; that knowledge is not obtained by intuition, not even fallible intuition; that knowledge cannot be obtained merely by deduction, but requires non-deductive reasoning; and that knowledge cannot be obtained by sticking to an abstract ideal of rigour, since what is important is not rigour but fruitfulness, hence the concept of rigour is better replaced with that of fruitfulness.

10.1 Objectivity as Independence of Any Subject

In addition to the questions about knowledge considered in the previous chapters, in this chapter we consider the relations of knowledge to objectivity, certainty, intuition, deduction, and rigour. First we consider the relation of knowledge to objectivity.

Many people who hold that the aim of science is truth, also hold that truth is objective, in the sense of being totally independent of any subject. Therefore, for them, the aim of science is knowledge which is objective, in this sense.

Thus Aristotle states that "it is not because of our truly thinking that you are pale that you are pale, but it is rather because you are pale that we who assert this speak the truth" (Aristotle, *Metaphysica*, Θ 10, 1051 b 6–9).

Frege states that "being true is different from being taken to be true, be it by one, be it by many, be it by all, and is in no way reducible to it" (Frege 2013, xv). For, "truth is something objective," where objective is what is "independent of the judging subject" (ibid., XVII).

Husserl states that "truths are what they are" independently of "whether we have insight into them or not. Since they do not hold in so far as we have insight into them, but we can only have insight into them in so far as they hold, they must be regarded as objective" (Husserl 2001, I, 150).

People holding this view include supporters of scientific realism. Thus Popper states that "truth is absolute and objective" (Popper 1999, 60). Objective is what is

© Springer International Publishing AG 2017

C. Cellucci, *Rethinking Knowledge*, European Studies in Philosophy of Science 4,
DOI 10.1007/978-3-319-53237-0_10

"totally independent of anybody's claim to know" and "is also independent of anybody's belief, or disposition to assent; or to assert, or to act" (Popper 1972, 109). The aim of science is knowledge in this "objective or impersonal sense" (ibid., 286).

However, if the aim of science is knowledge which is totally independent of any subject, then the aim of science will be generally unachievable. Science is a human product, and hence inherently hinges on human cognitive abilities. In order to have knowledge totally independent of any subject, we should be able to give an account of the world completely independent of our biological makeup. But we are unable to give such an account, because the only knowledge we can have about the world is through our ways of apprehending it, which essentially depend on our cognitive apparatus, that is a result of biological and cultural evolution. Therefore, the aim of science cannot be knowledge which is totally independent of any subject, and identifying the aim of science with that kind of knowledge is incompatible with the possibility of science.

10.2 Objectivity as the View from Nowhere

Against the claim that we are unable to give an account of the world completely independent of our biological makeup, it could be objected, as Nagel does, that "limited beings like ourselves can alter their conception of the world so that it is no longer just the view from where they are but in a sense a view from nowhere" (Nagel 1986, 70). This could be achieved in a number of ways.

According to Nagel, I could have a view from nowhere if "auditory and visual experiences could be produced in me not by sound and light but by direct stimulation of the nerves" (ibid., 62).

But it is not so. It is inessential whether information about the world is fed into the brain through our sensory receptors or by direct stimulation of the nerves. In the latter case, the direct stimulation of the nerves would be produced by stimuli simulating those produced by human sensory receptors, and the resulting sensory data would be processed by a human brain. The view of the world thus formed would always be a human one, so it would be a view of the world from a particular perspective. For this reason, the view of the world of a human being is essentially different from that of a bat.

Moreover, according to Nagel, I could have a view from nowhere if I conceived "the world as a place that includes the person I am within it, as just another of its contents – conceiving myself from outside, in other words" (ibid., 63). By so doing, "I can step away from the unconsidered perspective of the particular person I thought I was. Next comes the step of conceiving from outside all the points of view and experiences of that person and others of his species" (ibid.).

But it is not so. For, in order to have a view from nowhere, it would not be enough to conceive from outside all the points of view and experiences of a particular person and others of his species. It would be necessary to conceive from outside all the points of view and experiences of any possible being, even very different from human beings.

Furthermore, according to Nagel, I could have a view from nowhere if I considered that "we are simply examples of mind, and presumably only one of countless possible, if not actual, rational species on this or other planets" (Nagel 1997, 132). There is "a general concept of mind" under which "we ourselves fall as instances – without any implication that we are the central instances" (Nagel 1986, 18). The "basic methods of reasoning we employ are not merely human but belong to a more general category of 'mind'. Human minds now exemplify it, but those same methods and arguments would have to be among the capacities of any species that had evolved to the level of thinking – even if there were no vertebrates, and a civilization of mollusks or arthropods ruled the earth" (Nagel 1997, 140).

But this amounts to assuming that the mind of any possible intelligent being, even very different from human beings, would have the same constitution as the human mind. Now, there is no evidence for this assumption, because we are unable to figure out what the constitution of the mind of any possible intelligent being could be. Also, with this assumption, instead of having a view from nowhere, we would have an entirely anthropocentric view of the world.

As Kant states, "we can accordingly speak of space, extended beings, and so on, only from the human standpoint" (Kant 1998, A26/B42). Outside the human standpoint, "the representation of space signifies nothing at all" (ibid., A26/B42–A27/B43). Therefore, "we cannot judge at all whether the intuitions of other thinking beings are bound to the same conditions that limit our intuition and that are universally valid for us" (ibid., A27/B43). Our form of knowledge is something "which is peculiar to us, and which therefore does not necessarily pertain to every being, though to be sure it pertains to every human being" (ibid., A42/B59).

10.3 Objectivity as Plausibility

The knowledge we can have about the world is not knowledge which is totally independent of any subject, but only knowledge which is as independent as possible of any particular human subject. Specifically, it is knowledge such that most human subjects would agree that the arguments for it are stronger than the arguments against it, on the basis of the existing knowledge. Thus, the only knowledge we can have about the world is plausible knowledge.

This does not diminish the significance of the knowledge we can have about the world, because plausible knowledge is knowledge which is compatible with the existing data. Moreover, plausible knowledge can be rejected or modified at any time, when new data emerge. Therefore, there is nothing arbitrary about it.

To say that the only knowledge we can have about the world is plausible knowledge is to acknowledge that objectivity, meant as independence of any subject, is impossible. Plausible knowledge is objective not in the absolute sense of being totally independent of any subject, but only in the relative sense of being as independent as possible of any particular human subject. This is the only kind of objectivity we can achieve.

Therefore, to say that the only knowledge we can have about the world is plausible knowledge, is to recognize that the aim of science is not knowledge which is objective in an absolute sense, but only knowledge which is plausible, namely, such that the arguments for it are stronger than the arguments against it, on the basis of the existing knowledge.

10.4 Mathematical Knowledge and Plausibility

The claim that the aim of science is not knowledge which is objective in an absolute sense, but only knowledge which is plausible, also applies to mathematical knowledge. The only mathematical knowledge we can have is knowledge which is plausible. It is knowledge which is more and more plausible, and yet only plausible.

This is already pointed out by Hume, who states that "there is no algebraist nor mathematician so expert in his science, as to place entire confidence in any truth immediately upon his discovery of it" (Hume 1978, 180). Admittedly, "every time he runs over his proofs, his confidence encreases; but still more by the approbation of his friends; and is rais'd to its utmost perfection by the universal assent and applauses of the learned world" (ibid.). And yet, "this gradual increase of assurance is nothing but the addition of new probabilities" (ibid.). That is, it is nothing but the addition of "evidence, which is still attended with uncertainty" (ibid., 124). So, it is only the addition of plausibility.

Some people claim that some mathematical knowledge, such as $1 + 1 = 2$, or $2+2=4$, is objective in an absolute sense.

Thus Nagel claims that, "whatever else we may be able to imagine as different, including the possibility that we ourselves should be incapable of thinking that $2+2=4$, none of it tends to confer the slightest glimmer of possibility on that proposition's failing to be true" (Nagel 1997, 55). For "I cannot conceive of $2 + 2$ being equal to 5" (ibid., 63).

But it is not so. As Dehaene points out, "at our scale, the world is mostly made up of separable objects that combine into sets according to the familiar equation $1 + 1 = 2$. This is why evolution has anchored this rule in our genes" (Dehaene 1997, 249). However, "perhaps our arithmetic would have been radically different if, like cherubs, we had evolved in the heavens where one cloud plus another cloud was still one cloud" (ibid.).

Not only we can conceive of $2 + 2$ being equal to 5, but, if the human species had evolved in a different world, conceiving of $2 + 2$ being equal to 5 would have been most natural. In that case, presumably, Nagel would have said that we could not conceive of $2 + 2$ being equal to 4.

10.5 Certainty

Now we consider the relation of knowledge to certainty. Many people who hold that the aim of science is truth, also hold that the aim of science is absolutely certain truth.

Thus Husserl states that the truths which are the aim of science "must carry with them an absolute certainty," that is, an "absolute indubitability" (Husserl 1960, 14). In other words, they must have the peculiarity of "excluding in advance every doubt as 'objectless', empty" (ibid., 16).

However, if the only knowledge we can have about the world is plausible knowledge, then all our knowledge about the world is not absolutely certain. For, a theory that is plausible at a certain stage may become implausible at a later stage, and conversely a theory that is implausible at a certain stage may become plausible at a later stage, when the arguments for the theory are stronger than those against it. As Poincaré says, "in our relative world all certainty is a lie" (Poincaré 1910, VII).

Even mathematical knowledge is not absolutely certain. It is widely believed that mathematics is "the paradigm of certain and final knowledge" (Feferman 1998, 77). In particular, it is believed that, by means of the axiomatic method, "any dispute about the validity of a mathematical proof can always be resolved" (Gowers 2002, 40). But mathematics is not the paradigm of certain and final knowledge. As we have seen in Chap. 3, Frege's, Russell's, and Hilbert's foundational programs for mathematics, which were intended to establish that mathematics is the paradigm of certain and final knowledge, failed. Indeed, as it will be argued in Chap. 16, mathematicians often make mistakes. Thus, even mathematical knowledge is not absolutely certain. In particular, it cannot be maintained that, by means of the axiomatic method, any dispute about the validity of a mathematical proof can always be resolved. By Gödel's second incompleteness theorem, there is no absolutely reliable way of establishing the validity of a mathematical proof. We may conclude, then, that all our knowledge is not absolutely certain.

Against the conclusion that all our knowledge is not absolutely certain, it is sometimes objected that such conclusion is self-defeating, because it implies that the conclusion itself is not absolutely certain. But this objection is invalid. For, saying that the conclusion is not absolutely certain confirms that we can never be absolutely certain of anything, and hence that all our knowledge is not absolutely certain.

This is well illustrated by Peirce's anecdote: "I am a man of whom critics have never found anything good to say" (Peirce 1931–1958, 1.10). Only once "in all my lifetime have I experienced the pleasure of praise" (ibid.). That pleasure "was beatific," but "the praise that conferred it was meant for blame. It was that a critic said of me that I did not seem to be absolutely sure of my own conclusions. Never, if I can help it, shall that critic's eye rest of what I am now writing; for I owe a great pleasure to him; and, such was his evident animus, that should he find that out, I fear the fires of hell would be fed with new fuel in his breast" (ibid.). By his objection, that critic had reaffirmed that "we never can be absolutely sure of anything" (ibid., 1.147).

Admittedly, the search for certainty responds to a deep need of human beings who, in their life dominated by precariousness, feel the necessity to find secure footholds, and look for them in religion, science or philosophy. But precariousness is a constituent part of human life, and there is no ultimate safety from it. All our knowledge, including mathematical knowledge, is not absolutely certain. As Hersh states, "absolute certainty is what many yearn for in childhood, but learn to live without in adult life, including in mathematics" (Hersh 2014, 81). For, "mathematics is human, and nothing human can be absolutely certain" (ibid., 82).

Notwithstanding his support for the foundationalist view, Russell himself ends up admitting this. After decades of fruitless attempts to establish the absolute certainty of mathematics, he says: "I wanted certainty in the kind of way in which people want religious faith. I thought that certainty is more likely to be found in mathematics than elsewhere" (Russell 1971, III, 220). But eventually I found that "the splendid certainty which I had always hoped to find in mathematics was lost in a bewildering maze" (Russell 1995b, 157). Of course, "the demand for certainty is one which is natural to man, but is nevertheless an intellectual vice" (Russell 1950, 26). Uncertainty "is painful, but must be endured if we wish to live without the support of comforting fairy tales," indeed, we must learn "to live without certainty, and yet without being paralyzed by hesitation" (Russell 1945, xiv).

On the other hand, that all our knowledge is not absolutely certain, does not mean that knowledge is impossible. What is impossible is only absolutely certain knowledge. Knowledge would be impossible only if hypotheses were arbitrary, but they are not arbitrary, since they must be plausible. Therefore, we can have knowledge, albeit fallible knowledge. Moreover, we can have fallible but rational knowledge. Popper claims that knowledge has "its basis in an irrational decision," therefore we must admit "a certain priority of irrationalism" (Popper 1945, II, 218). But is is not so. Knowledge has its basis in a wholly rational procedure, the analytic method, so we need not admit any priority of irrationalism.

10.6 Intuition

Now we consider the relation of knowledge to intuition. What is intuition? The latter has been traditionally viewed as that mode of knowledge through which knowledge immediately relates to objects.

Thus Kant states that, "in whatever way and through whatever means a cognition may relate to objects, that through which it relates immediately to them" is "intuition" (Kant 1998, A19/B33).

Wittgenstein states that "what one means by 'intuition' is that one knows something immediately which others only know after long experience or after calculation" (Wittgenstein 1976, 30).

But one can hardly say that intuition is a mode of knowledge through which knowledge immediately relates to objects. This does not hold even for objects as simple as small numbers. As Dieudonné states: "I doubt that anyone seriously has an intuition of an integer greater than ten (by which I mean, an immediate intuition)" (Dieudonné 1975, 40).

Intuition has also been traditionally viewed as a mode of knowledge through which knowledge grasps objects infallibly.

Thus Plato states that, only "about those who can see each of these things in itself" through intuition, we can say that "they have knowledge" (Plato, *Respublica*, V 479 e 6–7). Therefore, knowledge can be obtained only through intuition, and knowledge is "what is infallible" (ibid., V 477 e 7).

But one can hardly say that intuition is a mode of knowledge through which knowledge grasps objects infallibly. There are plenty of examples where intuition leads to weird conclusions – what Poincaré calls "monsters" (Poincaré 2013, 435).

For example, intuition tells us that the area of any plane surface is different from zero since it is enclosed by a line. But a counterexample is provided by the Sierpiński Carpet, which is constructed as follows. Take a solid filled square, divide it into nine equal-sized squares, and remove the middle square. Then take the remaining eight squares, divide each of them into nine equal-sized squares, and remove the middle square from each group of nine. Then take the remaining sixty four squares, divide each of them into nine equal-sized squares, and remove the middle square from each group of nine.

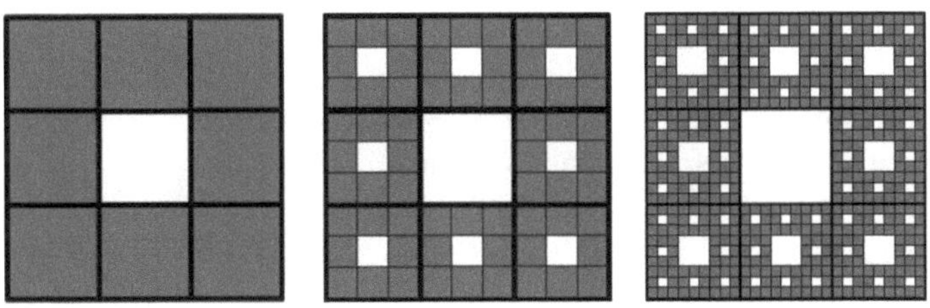

Repeat this construction *ad infinitum*. The area of the resulting plane surface will be zero, but will be enclosed by an infinite line.

As another example, intuition tells us that we cannot decompose a ball into a finite number of non-overlapping pieces and then reassemble them in a different way, to yield two balls identical to the original ball, so that the volume is doubled in the end. But, using the axiom of choice, we can prove that the volume is actually doubled in the end, and also that the reassembly process involves only moving the pieces continuously into place, without changing their shape and without running into one another (see Wilson 2005).

As a further example, Frege formulates his logicist program, according to which all arithmetical truths are deducible from a very small number of fundamental logical principles, based on the assumption that logical principles flow from intellectual intuition, which is infallible (see Chap. 3). In particular, he assumes that the main axiom of his system, the Basic Law V, flows from intellectual intuition, because "one thinks in accordance with it if, e.g., one speaks of extensions of concepts" (Frege 2013, VII). Frege has such a confidence in intellectual intuition as to state that it is "from the outset unlikely that" his system "could be built on an insecure, defective basis" (ibid., XXVI). And he added: "I could only acknowledge it as a refutation" if "someone proved to me that my basic principles lead to manifestly false conclusions. But no one will succeed in doing so" (ibid.). Frege, however, was bad prophet. By showing that the Basic Law V leads to paradox, Russell did succeed in doing exactly so. As Gödel says, "by analyzing the paradoxes to which Cantor's set theory had led," Russell "freed them from all mathematical technicalities, thus bringing to light the amazing fact that our logical intuitions (i.e. intuitions concerning such notions as: truth, concept, being, class, etc.) are self-contradictory" (Gödel 1986–2002, II, 124).

10.7 Fallible Intuition

It could be objected that, although intuition is not a mode of knowledge through which knowledge grasps objects infallibly, this does not mean that intuition has no role in knowledge. Intuition is indispensable to knowledge, in particular to mathematical knowledge. Every process of reasoning in mathematics must be ultimately based on an intuition of some logical principle of deductive inference, such as modus ponens. Otherwise there would be an infinite regress of deductive inferential justifications.

Thus Gödel states that intuition is indispensable to mathematical knowledge because, "in whatever way" mathematics "is built up, one will always need certain "axioms (i.e., deductively unprovable propositions)" for which "there exists no other rational (and not merely practical) foundation except" that they "can directly be perceived to be true" by "an intuition of the objects falling under them" (Gödel 1986–2002, III, 346–347). In particular, such is the case of "modus ponens and complete induction" (ibid., III, 347). In this regard, "the similarity between mathematical intuition and a physical sense is striking. It is arbitrary to consider 'This is red' an immediate datum, but not so to consider the proposition expressing modus ponens or complete induction" (ibid., III, 359).

This objection, however, raises the question: If intuition is fallible, how can we know whether an intuition that a proposition is true is right or wrong? We cannot know it by intuition, because intuition is fallible, we can know it only by reasoning. But if reasoning must be ultimately based on an intuition of some logical principle of deductive inference, how can we know whether such logical principle is true or false? We cannot know it by intuition, because intuition is fallible, we can know it only by reasoning. And so on. Thus we will have an infinite regress of deductive inferential justifications. Therefore, the objection implies that we cannot know whether an intuition is right or wrong, which makes the role of intuition in knowledge highly problematic.

Kripke claims: "I think" that "something's having intuitive content" is "very heavy evidence in favor of anything, myself. I really don't know, in a way, what more conclusive evidence one can have about anything, ultimately speaking" (Kripke 1980, 42). Contrary to Kripke's claim, intuition cannot be evidence for anything. Asserting that a proposition is true on the basis of intuition amounts to saying: I feel it is so; therefore it must be so. But two different persons may say 'I feel it is so' about opposite propositions, which leaves no criterion to choose between them.

In view of the limitations of intuition, one can hardly say that intuition is a ground of knowledge. As Frege admits, "we are all too ready to invoke inner intuition, whenever we cannot produce any other ground of knowledge" (Frege 1960, 19).

10.8 Deduction

Now we consider the relation of knowledge to deduction. Intuition and deduction have been often viewed as complementary modes of knowledge. While intuition has been traditionally viewed as a mode of knowledge through which knowledge immediately relates to objects, deduction has been traditionally viewed as a mode of knowledge which is mediated by inference, consisting of a series of inferences from principles which are known by intuition.

Thus Descartes states that, if we review "all the actions of the intellect by means of which we are able to arrive at a knowledge of things without any fear of being deceived," we "recognize only two: intuition and deduction" (Descartes 1996, X, 368). Intuition permits us to apprehend the simplest of all propositions, which "are known only by intuition" (ibid., X, 370). Deduction permits us to apprehend convoluted propositions starting from the simplest ones, since "we can apprehend" convoluted propositions "only by deducing them from those" (ibid., X, 383).

In addition to being viewed as a mode of knowledge which is mediated by inference, deduction has often been viewed as the essence of reasoning. This has given rise to 'deductivism', the view that all reasoning is either deductive or defective. Deductivism has been very influential, not only in ancient philosophy, noticeably with the Stoics, but also in modern and contemporary philosophy. For example, Popper states that "deductive inference is, like truth, objective, and even absolute" (Popper 2002, 166). As for induction, "there is no such thing" (ibid., 168). Indeed, "every rule of inductive inference ever proposed by anybody would, if anyone were to use it, lead" to "mistakes," so no rule of inductive inference ever proposed "can be taken seriously for even a minute" (ibid., 169).

Hilbert's *Grundlagen der Geometrie* is often presented as the paradigm of deductivism in mathematics, by saying that it "marked an end to an essential role for intuition in geometry" (Shapiro 2000, 151). But this contrasts with the fact that, as it will be argued in Chap. 19, even the very first demonstration in Hilbert's *Grundlagen der Geometrie* is not a deduction from axioms. For, it makes an essential use of properties obtained from a diagram, and the same holds of many other demonstrations therein. Therefore, it seems unjustified to present Hilbert's *Grundlagen der Geometrie* as the paradigm of deductivism.

Deductivism is faced with the difficulty that, by the strong incompleteness theorem for second-order logic, there is no consistent set of rules capable of deducing all second-order logical consequences of any given set of formulas (see, for example, Cellucci 2013a, Chap. 12). Thus deduction is not strong enough to obtain all such consequences. Deduction is not even strong enough to demonstrate all logically valid second-order sentences.

10.9 Rigour

Finally, we consider the relation of knowledge to rigour. Many people who maintain that the aim of science is truth, also maintain that rigour is essential to truth, both in mathematics and natural science, and that rigour means axiomatic rigour.

Thus Frege states that, in mathematics, one must aim at "uninterrupted rigour of demonstration and maximal logical precision" (Frege 1984, 237). Rigour means axiomatic rigour, because axiomatic systems show "the ultimate ground upon which rests the justification for holding" a proposition "to be true" (Frege 1960, 3). They "bring to light each axiom, each presupposition, hypothesis, or whatever one may want to call that on which a proof rests" (Frege 2013, VII).

Hilbert states that, both in mathematics and natural science, rigour is "a requirement for a perfect solution to a problem" (Hilbert 2000, 245). The "requirement of logical deduction by means of a finite number of processes is simply the requirement of rigour in reasoning" (ibid., 244). This requirement is satisfied only using the axiomatic method, because "the more subtle parts of mathematics and the natural sciences can be treated with certainty only in this way; otherwise one is only going around in a circle" (Hilbert 1980b, 51).

Popper states that, in natural science, one must "make every new assumption easily recognizable for what it is," and this "is the reason why the form of a rigorous system is aimed at. It is the form of a so-called 'axiomatized system' – the form which Hilbert, for example, was able to give to certain branches of theoretical physics" (Popper 1959, 71).

This view of the relation of knowledge to rigour, however, is faced with the problem that, as it will be argued in Chaps. 13 and 20, to assume that rigour means axiomatic rigour is incompatible with Gödel's incompleteness theorems.

Even independently of this, to assume that rigour means axiomatic rigour does not provide a fruitful approach to rigour. For example, Frege presents his work of rigorization of arithmetic as the culmination of the work of rigorization of analysis started by Cauchy. Indeed he states that, "proceeding along these lines, we are bound eventually to come to the concept of number and to the simplest propositions holding of positive whole numbers, which form the foundation of the whole arithmetic" (Frege 1960, 2).

But the work of rigorization of analysis started by Cauchy was aimed at removing difficulties about questions, such as Fourier series, convergence, or the existence of derivatives and integrals, which hindered the development of mathematics. And this work was not based on axiomatic rigour. In particular, as Laugwitz points out, "Cauchy resorted to analogy when he extended results on continuous functions to his singular integrals. Had he stuck to" axiomatic rigour, "his research work of the 1820s would never have been written," since axiomatic rigour "means rigidity, while reasoning by analogy was fertile" (Laugwitz 2000, 189).

Conversely, Frege's work of rigorization of arithmetic was not aimed at removing any difficulty which hindered the development of mathematics. It was carried out only in the name of an abstract ideal of logical perfection – the demand that "the fundamental propositions of arithmetic should be proved" with "the utmost rigour; for only if every gap in the chain of deductions is eliminated with the greatest care can we say with certainty upon what primitive truths the proof depend" (Frege 1960, 4). If we try to meet this demand, "we very soon come to propositions which cannot be proved so long as we do not succeed in analyzing concepts which occur in them into simpler concepts," and "here it is above all number which has" to be "defined" (ibid., 5).

Since Frege's definition of number was not aimed at removing any difficulty which hindered the development of mathematics but was carried out only in the name of an abstract ideal of logical perfection, mathematicians accorded it a cool reception. This strongly disappointed Frege. He himself speaks of "the despondency that at times overcame" him "as a result of the cool reception, or rather, the lack of reception, by mathematicians of" his "writings" (Frege 2013, XI). In particular, "researchers in the same area, Mr. Dedekind, Mr. Otto Stolz, Mr. von Helmholtz seem not to be acquainted with" his "works. Kronecker does not mention them in his essay on the concept of number either" (ibid., XI, footnote 1).

The same can be said about Couturat's definition of number. Poincaré ridiculed it, in particular he mocked Couturat's definition of the number one as "the number of elements in a class in which any two elements are identical," by saying that, admittedly, to define the number one, Couturat "does not use the word one; in compensation, he uses the word two. But I fear, if asked what is two, M. Couturat would have to use the word one" (Poincaré 2013, 458).

Rigour has a positive role if it responds to actual needs of mathematics or natural science, and is fruitful for research. If it is an end in itself, it is a too tight dress which impedes movement, and is merely pedantry. What is important is not rigour but fruitfulness, so the concept of rigour is better replaced with that of fruitfulness.

References

Cellucci, Carlo. 2013a. *Rethinking logic: Logic in relation to mathematics, evolution, and method.* Dordrecht: Springer.

Dehaene, Stanislas. 1997. *The number sense: How the mind creates mathematics*. Oxford: Oxford University Press.
Descartes, René. 1996. *Oeuvres*. Paris: Vrin.
Dieudonné, Jean. 1975. L'abstraction et l'intuition mathématique. *Dialectica* 29 : 39–54.
Feferman, Solomon. 1998. *In the light of logic*. Oxford: Oxford University Press.
Frege, Gottlob. 1960. *The foundations of arithmetic: A logico-mathematical enquiry into the concept of number*. New York: Harper.
———. 1984. *Collected papers on mathematics, logic, and philosophy*. Oxford: Blackwell.
———. 2013. *Basic laws of arithmetic*. Oxford: Oxford University Press.
Gödel, Kurt. 1986–2002. *Collected works*. Oxford: Oxford University Press.
Gowers, Timothy. 2002. *Mathematics: A very short introduction*. Oxford: Oxford University Press.
Hersh, Reuben. 2014. *Experiencing mathematics: What do we do, when we do mathematics?* Providence: American Mathematical Society.
Hilbert, David. 1980b. Letter to Frege, 22 September 1900. In Gottlob Frege, *Philosophical and mathematical correspondence*, 50–51. Oxford: Blackwell.
———. 2000. Mathematical problems. Appendix to Jeremy Gray, *The Hilbert challenge*, 240–282. Oxford: Oxford University Press.
Hume, David. 1978. *A treatise of human nature*. Oxford: Oxford University Press.
Husserl, Edmund. 1960. *Cartesian meditations: An introduction to phenomenology*. The Hague: Martinus Nijhoff.
———. 2001. *Logical investigations*. London: Routledge.
Kant, Immanuel. 1998. *Critique of pure reason*. Cambridge: Cambridge University Press.
Kripke, Saul. 1980. *Naming and necessity*. Cambridge: Harvard University Press.
Laugwitz, Detlef. 2000. Controversies about numbers and functions. In *The growth of mathematical knowledge*, ed. Emily Grosholz and Herbert Breger, 177–198. Dordrecht: Springer.
Nagel, Thomas. 1986. *The view from nowhere*. Oxford: Oxford University Press.
———. 1997. *The last word*. Oxford: Oxford University Press.
Peirce, Charles Sanders. 1931–1958. *Collected papers*. Cambridge: Harvard University Press.
Poincaré, Henri. 1910. *Savants et écrivains*. Paris: Flammarion.
———. 2013. *The foundations of science: Science and hypothesis – The value of science – Science and method*. Cambridge: Cambridge University Press.
Popper, Karl Raimund. 1945. *The open society and its enemies*. London: Routledge.
———. 1959. *The logic of scientific discovery*. London: Hutchinson.
———. 1972. *Objective knowledge*. Oxford: Oxford University Press.
———. 1999. *All life is problem solving*. London: Routledge.
———. 2002. *Unended quest: An intellectual autobiography*. London: Routledge.
Russell, Bertrand. 1945. *History of Western philosophy*. New York: Simon & Schuster.
———. 1950. *Unpopular essays*. New York: Simon & Schuster.
———. 1971. *Autobiography*. London: Allen & Unwin.
———. 1995b. *My philosophical development*. London: Routledge.
Shapiro, Stewart. 2000. *Thinking about mathematics. The philosophy of mathematics*. Oxford: Oxford University Press.
Wilson, Trevor. 2005. A continuous movement version of the Banach-Tarski paradox: A solution to De Groot's problem. *Journal of Symbolic Logic* 70 : 946–952.
Wittgenstein, Ludwig. 1976. *Lectures on the foundations of mathematics, Cambridge 1939*. Brighton: Harvester Press.

Part III
The Methods to Acquire Knowledge

Chapter 11
A Discourse on Method

Abstract How is knowledge acquired? In the past century it has been generally believed that this question is vacuous, because there is no method for acquiring knowledge, the basic laws of science can be reached only by intuition, then, from them, conclusions are deduced and compared with experience. This belief has had a very negative impact on logic, the method of science, and philosophy, and contrasts with the views on knowledge from antiquity to the nineteenth century. This chapter maintains that the belief in question has no foundation. It also maintains that there is no more foundation in the claim that today there is no longer need for the traditional approach to science, based on the scientific method, because powerful algorithms can explore huge databases and find correlations and regularities therein, providing rules for prediction.

11.1 The Need for Method

After examining the nature of knowledge, this part of the book examines the methods to acquire knowledge.

However, a preliminary question must be addressed: Do we need a method to acquire knowledge, and is there such a method?

Ryle rejects this question, claiming that "preoccupation with questions about methods tends to distract us from prosecuting the methods themselves. We run, as a rule, worse, not better, if we think a lot about our feet" (Ryle 2009, II, 331).

But, as Williamson points out, "philosophizing is not like riding a bicycle, best done without thinking about it – or rather: the best cyclists surely do think about what they are doing" (Williamson 2007, 8).

As a matter of fact, from antiquity to the nineteenth century, several people have maintained that we need a method to acquire knowledge, and that there is such a method.

Thus, Plato states that, proceeding without method, "would be like walking with the blind. But someone who goes about his subject skillfully must not be like the blind" (Plato, *Phaedrus*, 270 d 9–e 2). Hence, we need a method to acquire knowledge. In fact, there is such a method, namely the analytic method, and, "if we are to believe Hippocrates" of Cos, the physician, we cannot learn anything "unless we follow this method" (ibid., 270 c 3–5).

© Springer International Publishing AG 2017

C. Cellucci, *Rethinking Knowledge*, European Studies in Philosophy of Science 4,
DOI 10.1007/978-3-319-53237-0_11

Aristotle states that, in order to be able to acquire knowledge, "one must have been educated in the method by which each thing should be demonstrated" (Aristotle, *Metaphysica*, α 3, 995 a 12–13). This method is the analytic-synthetic method. (On the latter, see Chap. 12).

Bacon states that "the fabric of the universe, by its structure, is like a labyrinth to the mind observing it," and "a way must constantly be made through the forests of experience and particular things, in the uncertain light of the senses" (Bacon 1961–1986, I, 129). In "such difficult circumstances, one must despair of men's unassisted judgment, or even of any casual good fortune," therefore "we need a thread to guide our steps; and the whole path, right from the very first perceptions of senses, has to be made with a sure method" (ibid.).

Descartes states that people who investigate things at random and without method are "like someone who is consumed with such a senseless desire to find a treasure, that he constantly roams the streets to see if by chance he might find one, lost by a passer-by" (Descartes 1996, X, 371). On the contrary, "a method is necessary to investigate the truth of things" (ibid.). In fact, there is such a method, namely the analytic-synthetic method, since the method "is twofold, one by analysis, the other by synthesis" (ibid., VII, 155).

Kant states that "cognition, as science, must be arranged in accordance with a method," because science "requires a systematic cognition, hence one composed in accordance with rules on which we have reflected" (Kant 1992, 630). In fact, there is such a method and is twofold, since scientific method is divided "into synthetic and analytic. The latter is where I go from consequences to grounds, the former where I go from grounds to consequences" (ibid., 511). Thus "analysis proceeds '*ascendendo*', but synthesis proceeds '*descendendo*'" (ibid., 85).

11.2 The Denial of a Logic of Discovery

While, from antiquity to the nineteenth century, several people have maintained that we need a method to acquire knowledge and that there is such a method, in the past century this has been generally denied. It has been almost universally believed that scientific discovery escapes logical analysis, so there is no logic of discovery. The basic laws, or axioms, of science can be reached only by intuition, then, from them, conclusions are deduced and compared with experience. The basic laws, or axioms, and the conclusions together form what is called a scientific theory, so scientific theories are axiomatic theories and the scientific method is the deductive method.

Thus Einstein states that "there is no logical path" to the basic laws of physics, "only intuition, resting on sympathetic understanding of experience, can reach them" (Einstein 2010, 226). The "intuitive grasp of the essentials of a large complex of facts leads the scientist" to "a basic law," then "from the basic law (system of axioms) he derives his conclusion as completely as possible in a purely logically deductive manner" (Einstein 2002, 108). The conclusions deduced from the basic law are then "compared to experience and in this manner provide criteria for the justification of the assumed basic law. Basic law (axioms) and conclusion together form what is called a 'theory'" (ibid.). So, "while the researcher always starts out from facts, whose mutual connections are his aim," he "adapts to the facts by intuitive selection among the conceivable theories that are based upon axioms" (ibid., 109).

Popper states that "there is no method of discovering a scientific theory" (Popper 2000, 6). Logical analysis is possible only for "questions of the following kind. Can a statement be justified? And if so, how?" but "the initial stage, the act of conceiving or inventing a theory seems" neither "to call for logical analysis nor to be susceptible of it" (Popper 2005, 7). Indeed, "there is no such thing as a logical method of having new ideas, or a logical reconstruction of this process," every "discovery contains 'an irrational element', or 'a creative intuition', in Bergson's sense" (ibid., 8). Then, from a basic law obtained by creative intuition, "conclusions are drawn by means of logical deduction" and compared to experience, so the basic law is tested "by way of empirical applications of the conclusion which can be derived from it" (ibid., 9). This is the scientific method, which is then the method of "deductive testing" (ibid.).

This view, however, is unjustified. Saying that the basic laws of science can be reached only by intuition, conflicts with the fact that an unrestrained intuition might generate so many candidate basic laws that it would be impossible to test all of them. There must be some criterion for picking up some basic law among all the candidate basic laws, and for judging it worth testing. This criterion would act as a rule of discovery, so, at least in this sense, there must be a method of discovery. Besides, testing a basic law may itself involve some research work, hence some discovery task, so justification is not independent of discovery.

Moreover, saying that scientific theories are axiomatic theories, conflicts with Gödel's first incompleteness theorem, by which there are laws of a theory that cannot be deduced from the basic laws of the theory.

Furthermore, saying that the scientific method is the deductive method, amounts to assuming that the only logic needed for theory building is deductive logic. This conflicts with the strong incompleteness theorem for second-order logic, by which there is no consistent set of rules capable of deducing all second-order logical consequences of any given set of formulas. Therefore, deductive logic is not sufficient as a means for theory building.

11.3 Discovery and the Romantic Myth of Genius

The view that there is no logic of discovery and the basic laws of science can be reached only by intuition, is a remnant of the Romantic myth of genius, according to which there is no logic of discovery, discovery is based on leaps of intuition and is the result of extraordinary thought processes which are the privilege of the genius. This myth is still alive today.

Thus Kuhn claims that there are no "rules for inducing correct theories from facts," theories are "imaginative posits, invented in one piece for application to nature" (Kuhn 1977, 279). New theories are born through "flashes of intuition" and, "though such intuitions depend upon the experience, both anomalous and congruent, gained with the old" theories, "they are not logically or piecemeal linked to particular items of that experience" (Kuhn 1996, 123). An essential requisite "for the beginning of an episode of discovery" is "the individual skill, wit, or genius" (Kuhn 1977, 173). The "most profound sort of genius in physical science is that displayed by men" who "enunciate a whole new theory that brings potential order to a vast number of natural phenomena" (ibid., 188). It is the kind of "genius that leaps ahead of the facts, leaving the rather different talent of the experimentalist and instrumentalist to catch up" (ibid., 194).

However, the Romantic myth of genius is at odds with facts. Discovery is not the result of extraordinary thought processes, but rather the result of ordinary thought processes that produce an extraordinary outcome. Indeed the analytic method, through which discovery is reached, is entirely based on rational processes.

Instead of genius, discovery requires a prepared mind, as it is clear from the fact that discovery always occurs within a context of prior knowledge. Specifically, in the analytic method hypotheses are obtained from the problem, and possibly other data already available, by some non-deductive rule. Moreover, hypotheses must be plausible, namely, such that the arguments for them are stronger than the arguments against them, on the basis of the existing knowledge. Thus, both finding hypotheses and establishing that they are plausible need acquaintance with the existing knowledge, hence discovery requires a prepared mind. Quine claims that mathematicians make discoveries "by unregimented insight and good fortune" (Quine 1981a, 87). But it is not so because, as Pasteur says, "fortune favours only the prepared mind" (Pasteur 1922–1939, VII, 131). Now, if fortune favours only the prepared mind, it is unjustified to say that mathematicians make their discoveries by unregimented insight and good fortune.

In addition to a prepared mind, discovery requires total absorption in the problem. An example of this is given by Newton who, when "asked how he had been able to discover his mechanical system of the universe," answered: "*Nocte dieque incubando* [By thinking about it day and night]" (Ortega y Gasset 1957, 42). As a result, "Newton's cat grew quite fat munching on all the food" that Newton, when absorbed in a problem, "left untouched" (Bardi 2006, 28). Mathematicians are notorious for their absent-mindedness, but what other people consider absent-mindedness is actually total absorption in the problem. Instead of genius, such total absorption is one of their keys to success.

11.4 Discovery and Serendipity

Another way of denying that there is a logic of discovery is by appealing to serendipity, which is commonly defined as the faculty of making happy and unexpected discoveries by accident. Some people claim that all discoveries are serendipitous, in this sense. Thus Crick states that "chance is the only source of true novelty" (Crick 1982, 58). Simonton states that "chance must be considered the primary basis for scientific creativity" (Simonton 2004, 161). Kantorovich states that "scientific creativity is equated with unintentionality and serendipity" (Kantorovich 1993, 27). For, "science in fact advances by serendipitous steps" (ibid., 157). Now, if all discoveries are serendipitous, in the sense of being made by accident, then there is no logic of discovery.

But the claim that all discoveries are serendipitous in the sense of being made by accident, is unjustified. Already Bacon pointed out this, though of course without reference to the term 'serendipity', which was coined later on in 1754 by Walpole (see Merton and Barber 2004). Indeed, Bacon states that "many useful things have been discovered by chance [*casu quodam*] or happy opportunity [*per occasionem*], by men who were not looking for them, and were engaged on something else" (Bacon 1961–1986, I, 206–207). But undoubtedly "many more things would be discovered by men who were looking and attending to them, and were doing so with method and order, not impulsively or desultorily" (ibid., I, 207).

Thus, however, Bacon somehow gives credit to the belief that many useful things have been discovered by chance or happy opportunity. This belief is problematic, because really no discoveries are made purely by chance or happy opportunity. For, as stated above, discovery requires a prepared mind. On the other hand, Bacon also says that those useful things have been discovered by men who were not looking for them, and were engaged on something else. This suggests an alternative definition of serendipity, according to which serendipity consists in looking for one thing and finding another. As Gillies states, "in serendipity proper, someone fails to discover what he or she was looking for, but discovers something else unexpected instead" (Gillies 2014, 37). For example, Columbus was looking for a sea route to the East Indies obtained sailing west, but he discovered America instead.

Serendipity, defined as looking for one thing and finding another, can be naturally accounted for in terms of the analytic method. For, one of the distinctive features of the analytic method is that the hypotheses formulated to solve a problem may turn out to solve some other problem, instead of the intended one (see Chap. 12).

Gillies also formulates a concept of "additional serendipity," according to which "the researcher does discover what he or she was looking for, but, in addition, discovers something else unexpected" (ibid.). For example, "the basic results of group theory were discovered in a successful investigation of the solubility of polynomial equations, but group theory turned out unexpectedly to provide, in addition, a useful tool for classifying geometries" (ibid., 37–38).

Additional serendipity can also be naturally accounted for in terms of the analytic method. For, one of the distinctive features of the analytic method is that the hypotheses formulated to solve a problem may turn out to solve also some other problem, in addition to the intended one (see Chap. 12).

11.5 Discovery and Deductive Logic

A variant of the view that there is no logic of discovery, scientific theories are axiomatic theories and the scientific method is the deductive method, is put forward by Musgrave. He states that actually "there is a logic of invention (discovery)," but this "is deductive logic," since "the originality or inventiveness" lies in deducing a "conclusion, which in interesting cases is no trivial routine task" (Musgrave 2011, 222). Therefore, "deductive logic is the only logic that we have or need" (Musgrave 1999, 395). In particular, the basic laws are obtained "(as Newton said) by 'deducing them from the phenomena'. Hence, Newton was right that 'deduction from the phenomena' is deduction (not induction, abduction, or any other ampliative process of inference)" (Musgrave 2006, 302).

Musgrave's variant too, however, is unjustified, because there is an algorithm for enumerating all deductions from given premisses. The algorithm can be said "to proceed like Swift's scholar, whom Gulliver visits in Balnibarbi, namely, to develop in systematic order, say according to the required number of inferential steps, all consequences and discard the 'uninteresting' ones" (Weyl 1949, 24). Given enough time and space, the algorithm will enumerate all deductions, from given premises. Therefore, contrary to Musgrave's claim, originality and inventiveness cannot lie in deducing a conclusion.

Moreover, contrary to Musgrave's claim, Newton did not say that scientific laws and theories are obtained by deducing them from the phenomena. What he actually said is that, in experimental philosophy, "particular propositions are deduced from the phenomena, and afterwards rendered general by induction," and this is the way "the laws of motion and of gravitation were discovered" (Newton 1962, II, 547). So, according to Newton, what is deduced from the phenomena are not scientific laws but singular propositions, namely the descriptions of observations, which are then rendered general by induction. Therefore, in Newton's view, it is not deduction, but induction, which leads to basic laws. (For more on this, see Chap. 13).

11.6 The Denial of Method

In the past century, not only it has been almost universally held that there is no method to acquire knowledge, but some people have even claimed that there is no scientific method at all.

Thus Popper states that "scientific method does not exist" (Popper 2000, 5). Not only there is no method of discovering a scientific theory, but also "there is no method of ascertaining the truth of a scientific hypothesis, that is, no method of verification" (ibid., 6). Therefore, despite his claim that there is a scientific method and consists in the method of deductive testing, somewhat inconsistently Popper concludes that "no scientific method exists" (ibid.). And he adds: "I ought to know, having been, for a time at least, the one and only professor of this nonexistent subject within the British Commonwealth" (ibid., 5).

Popper's claim that no scientific method exists depends on the fact that he makes two tacit assumptions: 1) Only deduction is rational; 2) A method of justification must be algorithmic. Since, as it will be argued in Chap. 12, deduction is non-ampliative and hence cannot yield new knowledge, from 1) Popper infers that discovery must contain an irrational element. On the other hand, since, by the undecidability theorem, there is no algorithm for determining whether a given proposition is true or false, from 2) Popper infers that there is no method of justification.

But assumptions 1) and 2) are unjustified. Assumption 1) is unjustified, because the non-deductive inferences by which hypotheses are obtained in the analytic method, though not yielding absolutely certain hypotheses, are completely rational. Assumption 2) is unjustified, because the method for testing the plausibility of hypotheses, namely the plausibility test procedure which is described in Chap. 12, is a method of justification.

11.7 Method and Rationality

A variant of the view that there is no scientific method at all is put forward by Feyerabend. He claims that "there is no method" (Feyerabend 1981–2016, III, 125). Indeed, there is "not a single rule, however plausible, and however firmly grounded in epistemology, that is not violated at some time or other," and such violations "are necessary for progress" (Feyerabend 1993, 14). The "only principle that does not inhibit progress is: anything goes" (ibid.). This indicates a weakness, not only of methodology, but of reason itself because, "without a frequent dismissal of reason, no progress. Ideas which today form the very basis of science exist only because" they "opposed reason," so "within science reason" must "often be overruled, or eliminated" (ibid., 158). For example, the Copernican theory exists today "only because reason was overruled at some time" (ibid., 116). Indeed, at the time of Galileo, the Copernican theory "was inconsistent with facts" (ibid., 39). But "Galileo uses propaganda. He uses psychological tricks" which are "arguments in appearance only," and "these tricks are very successful: they lead him to victory" (ibid., 65). He turns "an experience which partly contradicts the idea of the motion of the earth" into "an experience that confirms it" (ibid., 71). However, "the experience on which Galileo wants to base the Copernican view is not but the result of his own fertile imagination," it "has been invented" (ibid., 65).

Feyerabend's variant too, however, is unjustified. The claim that, without a frequent dismissal of reason, no progress, is unwarranted. For, a theory that is plausible at a certain stage may become implausible at a later stage, and, conversely, a theory that is implausible at a certain stage may become plausible at a later stage, when the arguments for the theory become stronger than those against it. This was the case of the Ptolemaic theory and the Copernican theory. So science's progress is a rational process.

The Copernican theory does not exists today because reason was overruled by Galileo. Against the earth's motion, the Aristotelians put forward the following argument. Suppose, for argument's sake, that the earth does move. Drop a stone from the top of a tower. If the earth moves, then, "during the time taken by the stone in its fall," the tower, "being carried by the earth's rotary motion, would advance many hundreds of cubits" in the direction of the earth's motion, "and the stone should hit the ground that distance away from the tower's base" (Galilei 1968, VII, 151–152).

But, in fact, the stone does nothing of the kind. Indeed, "the senses" immediately "assure us that the tower is straight and perpendicular, and show us that the stone in falling grazes it without inclining so much as a hairbreadth to one side or the other, and lands at the foot of the tower exactly under the place from which it was dropped" (ibid., VII, 165). Hence, the Aristotelians conclude, the earth does not move.

Against this, Galileo argues that, if the senses assure us that the tower is straight and perpendicular and show us that the stone in falling grazes it, and if the earth moves and consequently carries the tower along with it, then the motion of the stone "would be a compound of two motions," one from above downwards with which the stone "grazes the tower, and another one with which it follows the course of the tower" (ibid.). The result of the compound "would be that the stone would no longer describe a simple straight and perpendicular line, but rather an inclined, and perhaps not straight, one" (ibid.). Therefore, Galileo concludes, "from just seeing the falling stone graze the tower, you cannot affirm with certainty that it describes a straight and perpendicular line, unless you first assumed that the earth is standing still;" for, "if the earth were moving, the stone's motion would be inclined and not perpendicular" (ibid.). But the assumption that the earth is standing still is the conclusion the Aristotelians want to prove, so the argument of the Aristotelians "is assuming as known what it is trying to prove" (ibid.). Namely, it is begging the question.

Then, Galileo does not use propaganda nor uses psychological tricks, instead, he provides an argument against the Aristotelians. His argument is a criticism of the argument of the Aristotelians against the earth's motion, not a counter-argument in favour of the opposite conclusion. As a matter of fact, "eventually, Galileo did formulate several positive arguments for the earth's motion" (Finocchiaro 2010, 129). But the argument above is not such a positive argument. With it, Galileo does not turn an experience which partly contradicts the idea of the motion of the earth into an experience that confirms it. Instead, he points out that the argument of the Aristotelians against the earth's motion is invalid. That the senses assure us that the tower is straight and perpendicular and show us that the stone in falling grazes it, does not mean that the stone actually describes a straight and perpendicular line. If we shut ourselves up in a windowless cabin on some large ship, all motions inside the cabin will appear the same to us, whether the ship is moving or standing still, since "the ship's motion is common to all the things contained in it" (Galilei 1968, VII, 213). Similarly, all motions on the earth, such as that of a stone dropped from the top of a tower, will appear the same to us standing on the earth, whether the earth itself is moving or standing still, since the earth's motion is common to all the things contained it it. That the stone describes a straight and perpendicular line is only an appearance, what the stone actually describes is an inclined, and not straight, line.

Therefore, Galileo does not overrule reason, he acts in accordance with it. It is Feyerabend who claims that reason "does not fit science," and hence "those who admire science and are also slaves of reason" must "make a choice. They can keep science; they can keep reason; they cannot keep both" (Feyerabend 1993, 214). So, rather than Galileo, it is Feyerabend who radically separates and opposes science and reason.

11.8 The Psychology of Discovery

Since the nineteenth century, the view that scientific discovery escapes logical analysis has prompted many people to assert that discovery is a purely subjective, psychological matter, so it cannot be an object of logic but only of psychology.

Thus Frege states that logic cannot be concerned "with the way in which" new propositions "are discovered," but only "with the kind of ground on which their" justification "rests" (Frege 1960, 23). The question of discovery "may have to be answered differently for different persons," so it is a merely subjective, psychological one, only the question of justification "is more definite" (Frege 1967, 5).

This has led several people to replace the logic of discovery with the psychology of discovery. Poincaré played a significant role in this replacement. According to him, mathematical discovery consists "in making new combinations" with concepts "already known" and in selecting "those that are useful and which are only a small minority" (Poincaré 2013, 386). This is the action of the unconscious mind, that selects useful combinations on the basis of the "feeling of mathematical beauty," which is "a true aesthetic feeling that all real mathematicians know" (ibid., 391). Once useful combinations have been selected, it is necessary to submit them to "verification" (ibid., 390). This is the action of the conscious mind.

Poincaré's view, however, is unjustified. If, as Poincaré claims, discovery consists in making new combinations with concepts that are already known and in selecting those that are useful, then ultimately there will be some basic concepts out of which all combinations of concepts will be made. Such basic concepts will provide a universal language for mathematics, namely a language capable of expressing all mathematical concepts, because all mathematical concepts will be definable in terms of them. But this conflicts with Tarski's undefinability theorem, by which there cannot be a theory *T* capable of expressing all mathematical concepts, in particular, the concept of being a true sentence of *T*. Thus, there cannot be a universal language for mathematics. (For more on the limitations of Poincaré's view, see Cellucci 2013a, Chap. 13).

Moreover, the feeling of mathematical beauty, while often useful, can also be unreliable as a means of selection of useful combinations. For example, the feeling of mathematical beauty led Galileo to stick to Copernicus' circular orbits for planets, which contrasted with observations, rejecting Kepler's elliptical orbits, which agreed with them, since to Galileo, who opposed mannerism in favour of Raphael's classicism, Kepler's compressed elliptic orbits appeared unbearable aesthetic deformations. As another example, the feeling of mathematical beauty led Dirac to stick to his own version of quantum electrodynamics, which made predictions that were often infinite and hence unacceptable, rejecting renormalization, which led to accurate predictions.

11.9 An Attempt to Trivialize Discovery

As an alternative to the view that discovery is a purely subjective, psychological matter, and hence cannot be an object of logic but only of psychology, Hilbert attempts to trivialize discovery, at least with regard to mathematics.

To this purpose, on the one hand, Hilbert claims that in mathematics there is no question of discovering axioms, because "the axioms can be taken quite arbitrarily" (Hilbert 2004c, 563). They are only subject to the condition that they must be consistent, namely, to the condition that "the application of the given axioms can never lead to contradictions" (Hilbert 1996a, 1093). On the other hand, Hilbert claims that there is an algorithm for deciding whether or not a given mathematical proposition can be deduced from given axioms. The question of the existence of such an algorithm is the 'decision problem', which "is the best-known and the most discussed; for it goes to the essence of mathematical thought" (Hilbert 1996b, 1113). A "solution of the decision problem is of fundamental importance for the theory of all subjects whose theorems are capable of being logically derived from finitely many axioms" (Hilbert and Ackermann 1928, 73). Indeed, if the axioms can be taken quite arbitrarily, a solution of the decision problem would permit to trivialize the question of discovery. Moreover, it would prove that "there can be no *ignorabimus* in mathematics," which for Hilbert "must remain the ultimate goal" (Hilbert 1905, 249).

Hilbert's attempt to trivialize discovery, however, fails. By Gödel's second incompleteness theorem, it is impossible to show by absolutely reliable means that arbitrarily given axioms are consistent. Moreover, by the undecidability theorem, there is no algorithm for deciding whether or not a given sentence can be deduced from arbitrarily given axioms, so the decision problem has a negative answer.

11.10 Heuristic vs. Algorithmic Methods

It has been stated above that, from antiquity to the nineteenth century, several philosophers have argued that we need a method to acquire knowledge, and that there is such a method.

The word 'method' derives from the Greek *methodos*, which combines *meta*, one of whose meanings is 'toward', with *hodos*, whose primary meaning is 'way'. Hence, *methodos* literally means 'the way toward', the way to be followed in pursuing a certain aim.

Methods can be divided into algorithmic and heuristic. An algorithmic method is a method that guarantees to always produce a correct solution to a problem. Conversely, a heuristic method is a method that does not guarantee to always produce a correct solution to a problem.

A method to acquire knowledge need not be algorithmic, it can be heuristic. While not guaranteeing to always produce a correct solution to a problem, a heuristic method may greatly reduce the search space, namely, the domain within which the solution is sought, thus making a solution feasible when no algorithmic method is available.

That a method to acquire knowledge need not be algorithmic and may be heuristic, was recognized already in antiquity. Mathematics and medicine were the first areas where the need for a method to acquire knowledge arose, and the earliest such method of which we have notice, namely the analytic method of Hippocrates of Chios and Hippocrates of Cos, was a heuristic method (see Chap. 12).

Heuristic methods play an essential role in knowledge, because the world is too complex to be comprehended in its entirety by algorithmic methods, by organisms such as human beings who are limited in space, time and cognitive resources. We can make the world understandable to ourselves only by resorting to heuristic methods.

11.11 Algorithmic Methods, Discovery, and Justification

Contrary to the view that a method to acquire knowledge need not be algorithmic and can be heuristic, several people have claimed that a method to acquire knowledge should be algorithmic, and, from the impossibility of such a method, have concluded that no logic of discovery is possible, discovery can only be an object of psychology.

Thus Lakatos states that, "in the seventeenth or even eighteenth century," it "was hoped that methodology would provide scientists with a mechanical book of rules," namely with an algorithmic method, "for solving problems," but this has turned out to be impossible, so "this hope has now been given up" (Lakatos 1978, I, 103). There can be no logic of discovery, "modern methodologies" consist "merely of a set of (possibly not even tightly knit, let alone mechanical) rules for the appraisal of ready, articulated theories" (ibid.). They are not supposed "to give advice to the scientist" about "how to arrive at good theories" (Lakatos 1971, 174). They only provide "rules governing the (scientific) acceptance and rejection" of ready, articulated "theories" (Lakatos 1978, I, 103). These rules are normative, where, however, "the term 'normative' no longer means rules for arriving at solutions, but merely directions for the appraisal of solutions already there. Thus methodology is separated from heuristics" (ibid., I, 103, footnote 1). Unlike methodology, which consists of such objective rules, heuristics can only have subjective and psychological rules since, "outside the legislative domain of" the "normative rules" of methodology, "there is" only "an empirical psychology" of "discovery" (ibid., I, 103).

This view, however, is unjustified because it is based on the assumption that discovery either is obtained by an algorithmic method or is reached through intuition, *tertium non datur*. This assumption is unfounded because, between the dullness of algorithmic methods and the inscrutability of intuition, there is an intermediate region inhabited by heuristic methods, which are neither algorithmic nor involve intuition. The purpose of a logic of discovery is not to dispense with the need for intelligence by use of an algorithmic method, but rather to expand natural intelligence providing it with heuristic means – means capable of guiding natural intelligence, albeit not infallibly.

Incidentally, some of the people who have used the impossibility of finding an algorithmic method to acquire knowledge as an argument for denying the possibility of a logic of discovery, conversely have affirmed that there is an algorithmic method of testing, or justification.

Thus Carnap, on the one hand, claims that there cannot be a "machine – a computer into which we can put all the relevant observational sentences and get, as an output, a neat system of laws that will explain the observed phenomena;" indeed, for the purpose of discovery, "creative ingenuity is required" (Carnap 1966, 33). On the other hand, he claims that it is "in many cases possible to determine, by mechanical procedures, the logical probability, or degree of confirmation, of" a hypothesis "*h* on the basis of" evidence "*e*" (ibid., 34). So, "the aim of epistemology" can only be "the formulation of a method for the justification of cognitions" (Carnap 2003, 305).

These claims, however, are unjustified. Just as there is no algorithmic method of discovery, there is no algorithmic method of testing. Indeed, by the undecidability theorem, there is not even an algorithmic method for testing whether a formula is logically valid or not. So, as Putnam says, if "there is no logic of discovery," then, "in that sense, there is no logic of testing, either" (Putnam 1975–1983, I, 268). The view that "correct ideas just come from the sky, while the methods for testing them are highly rigid and predetermined, is one of the worst legacies of the Vienna Circle" (ibid.). As a matter of fact, "all the formal algorithms proposed for testing, by Carnap, by Popper, by Chomsky, etc., are, to speak impolitely, ridiculous; if you don't believe this, program a computer to employ one of these algorithms and see how well it does at testing theories!" (ibid.).

11.12 Consequences of the Denial of Method

The view that there is no method to acquire knowledge has had a very negative impact on logic, the method of science, and philosophy.

From Aristotle to Descartes, and even beyond, logic has been a logic of discovery and has been part of the scientific method.

Thus Aristotle states that logic should indicate "how to reach for premisses concerning any problem proposed, in the case of any discipline whatever" (Aristotle, *Analytica Priora*, B 1, 53 a 1–2). Namely, it should indicate "by what method we will find the premisses about each thing" (ibid., A 27, 43 a 21–22).

Descartes states that logic should serve to "discover in general the principles or first causes of everything that exists or can exist in the world" (Descartes 1996, VI, 63–64).

Thus, for Aristotle and Descartes, logic is a logic of discovery, and is part of the analytic-synthetic method which, for them, is the scientific method.

On the contrary, the view that there is no method to acquire knowledge has led to abandon the view that logic is a logic of discovery and is part of the scientific method. Such view has been replaced with the view that there is no logic of discovery, or logical reconstruction of the process of discovery, discovery contains an irrational element and is based on intuition. Moreover, logic has been restricted to deductive logic, and has been separated from the method of science. Thus, Tarski states that the aim of logic is "to perfect and to sharpen the deductive method" (Tarski 1994, xi). There is "little rational justification for combining the discussion of logic and that of the methodology of empirical sciences" (ibid., xiii). As a result, today "logic (classical or otherwise) in philosophy of science is, to put it simply, out of fashion" (Aliseda 2006, 21).

The view that there is no method to acquire knowledge has also led to abandon the view that philosophy aims at knowledge and methods to acquire knowledge. The study of method has been replaced with epistemology, meant as a discipline whose main problem is to give a definition of knowledge, and within epistemology there has been a revival of the definition of knowledge as justified true belief (see Chap. 6). Thus the vigorous, highly relevant discussions on the scientific method in the seventeenth century, have given way to the anemic, largely irrelevant discussions on the definition of knowledge in the twentieth century.

11.13 The Alleged Obsoleteness of Method

Rather than asserting that there is no method to acquire knowledge, a currently widespread view, called 'the big data view', claims that the scientific method has become obsolete. According to this view, there is no longer need for the traditional approach to science based on the scientific method, because powerful algorithms can explore huge databases and find correlations and regularities therein, providing rules for prediction.

The best known formulation of the big data view is by Anderson, who states that, while the "scientific method is built around testable hypotheses," and "this is the way science has worked for hundreds of years," today, "faced with massive data, this approach to science – hypothesize, model, test – is becoming obsolete" (Anderson 2008). We are at "the end of theory," because "the data deluge makes the scientific method obsolete," indeed, "with enough data," no hypothesis is needed, "the numbers speak for themselves" (ibid.). We "can analyze the data without hypotheses about what it might show. We can throw the numbers into the biggest computing clusters the world has ever seen and let statistical algorithms find patterns" (ibid.). Thus, "the new availability of huge amounts of data, along with the statistical tools to crunch these numbers, offers a whole new way of understanding the world. Correlation supersedes causation" (ibid.). Massive data "allow us to say that "correlation is enough" (ibid.). This "kind of thinking is poised to go mainstream" (ibid.).

11.14 The Vindication of Method

The proponents of the big data view often portray Bacon as a precursor. They claim that "Bacon proclaimed that science could discover truths about nature only by empirical testing of all the possible explanations for all the observed phenomena" (Siegfried 2013). For example, science could discover the nature of heat only by recording "all observed facts about all manner of heat-related phenomena" and then performing "experiments to eliminate incorrect explanations" (ibid.). Therefore, "Bacon was a fan of big data. With today's massive computerized collections of massive amounts of data on everything," at last "Bacon's dreams have been realized" (ibid.).

These claims, however, are unjustified. Bacon was not a fan of big data because, according to him, "casual experience which follows only itself," unguided by method, "is merely groping in the dark" (Bacon 1961–1986, I, 203). Without method, human beings "wander and stray in no clear path, but only take their lead from the things they come across," and so "make little progress" (ibid., I, 180). Therefore, human beings must not only "seek to acquire a greater abundance of experiments," but must also develop a method "for carrying on and advancing experience" (ibid., I, 203).

As implicit in Bacon, the view that the data deluge makes the scientific method obsolete is illusory, because data depend on method. The assumption that they are completely independent of method is favoured by the fact that the term 'data' comes from the Latin verb *dare*, which means 'to give'. This suggests that data are raw elements that are given by phenomena. But it is not so, because observation is always selective, every choice of the data is a reflection of an, often unstated, set of assumptions and hypotheses about what we want and expect from the data. Therefore, data are not simple elements that are abstracted from the world in neutral and objective ways. There is always a 'viewpoint' preceding observation and experiment, namely, a hypothesis which guides observation and experiment, and generally data-finding. Data do not speak for themselves, but acquire meaning only when they are interpreted, and interpreting them requires some hypothesis through which to observe them and extract information from them. Science is not about the data, it is about the interpretation of the data, through which alone we can make the world understandable to ourselves.

Systems of data analysis are designed to capture certain kinds of data, and the algorithms used to that purpose are based on some hypothesis and have been refined through testing. Thus, a statistical strategy of identifying patterns within data is based on previous findings and method. Then, it is illusory to think that statistical strategies may automatically discover insights without presupposing any hypothesis or testing. Indeed, the more data, the higher the probability that arbitrary correlations will be found in them. This is admitted even by some supporters of the big data revolution, such as Berman who states that, "for big data projects, holding a prior theory or model is almost always necessary; otherwise, the scientist is overwhelmed by the options" (Berman 2013, 147).

Contrary to the big data view, the scientific approach has by no means become obsolete, it is more important than ever. As Rovelli states, the question of the method of science "is particularly relevant today in science, and particularly in physics," because "in fundamental theoretical physics, it is thirty years that we fail. There hasn't been a major success in theoretical physics in the last few decades" (Rovelli 2014, 215). This "might be in part because of the wrong ideas we have about science, and because methodologically we are doing something wrong, at least in theoretical physics, and perhaps also in other sciences" (ibid.). Thus, the question of the nature of scientific method has not become obsolete, it is more important than ever.

References

Aliseda, Atocha. 2006. *Abductive reasoning: Logical investigations into discovery and explanations*. Berlin: Springer.
Anderson, Chris. 2008. The end of theory: The data deluge makes the scientific method obsolete. *Wired Magazine*, 23 June.
Bacon, Francis. 1961–1986. *Works*. Stuttgart Bad Cannstatt: Frommann Holzboog.
Bardi, Jason Socrates. 2006. *The calculus wars: Newton, Leibniz, and the greatest mathematical clash of all time*. New York: Thunder's Mouth Press.
Berman, Jules. 2013. *Principles of big data: Preparing, sharing, and analyzing complex information*. Amsterdam: Elsevier.
Carnap, Rudolph. 1966. *Philosophical foundations of physics: An introduction to the philosophy of science*. New York: Basic Books.
Carnap, Rudolf. 2003. *The logical structure of the world – Pseudoproblems in philosophy*. Chicago: Open Court.
Cellucci, Carlo. 2013a. *Rethinking logic: Logic in relation to mathematics, evolution, and method*. Dordrecht: Springer.
Crick, Francis. 1982. *Life itself: Its origin and nature*. New York: Simon & Schuster.
Descartes, René. 1996. *Oeuvres*. Paris: Vrin.
Einstein, Albert. 2002. Induction and deduction in physics. In Albert Einstein, *Collected papers*, Vol. 7, 108–109. Princeton: Princeton University Press.
———. 2010. *Ideas and opinions*. New York: Crown.
Feyerabend, Paul K. 1981–2016. *Philosophical papers*. Cambridge: Cambridge University Press.
———. 1993. *Against method*. London: Verso.
Finocchiaro, Maurice A. 2010. *Defending Copernicus and Galileo: Critical reasoning in the two affairs*. Dordrecht: Springer.
Frege, Gottlob. 1960. *The foundations of arithmetic: A logico-mathematical enquiry into the concept of number*. New York: Harper.
———. 1967. *Begriffsschrift*, a formula language, modeled upon that of arithmetic, for pure thought. In *From Frege to Gödel: A source book in mathematical logic, 1879–1931*, ed. Jean van Heijenoort, 5–82. Cambridge: Harvard University Press.
Galilei, Galileo. 1968. *Opere*. Florence: Barbera.
Gillies, Donald. 2014. Serendipity and mathematical logic. In *From a heuristic point of view*, ed. Cesare Cozzo, and Emiliano Ippoliti, 23–39. Newcastle upon Tyne: Cambridge Scholars Publishing.
Hilbert, David. 1905. *Logische Prinzipien des mathematisches Denkens*. Unpublished manuscript. Göttingen: Bibliothek, Mathematischen Institut, Universität Göttingen.
———. 1996a. On the concept of number. In *From Kant to Hilbert: A source book in the foundations of mathematics*, ed. William Ewald, Vol. II, 1092–1095. Oxford: Oxford University Press.

———. 1996b. Axiomatic thought. In *From Kant to Hilbert: A source book in the foundations of mathematics*, ed. William Ewald, Vol. II, 1107–1115. Oxford: Oxford University Press.

———. 2004c. Grundlagen der Geometrie. In *David Hilbert's lectures on the foundations of geometry 1891–1902*, ed. Michael Hallett and Ulrich Majer, 540–602. Berlin: Springer.

Hilbert, David, and Wilhelm Ackermann. 1928. *Grundzüge der theoretischen Logik*. Berlin: Springer.

Kant, Immanuel. 1992. *Lectures on logic*. Cambridge: Cambridge University Press.

Kantorovich, Aharon. 1993. *Scientific discovery: Logic and tinkering*. Albany: State University of New York Press.

Kuhn, Thomas Samuel. 1977. *The essential tension: Selected studies in scientific tradition and change*. Chicago: The University of Chicago Press.

———. 1996. *The structure of scientific revolutions*. Chicago: The University of Chicago Press.

Lakatos, Imre. 1971. Reply to critics. In *PSA 1970: In memory of Rudolf Carnap*, ed. Roger C. Buck and Robert S. Cohen, 174–182. Dordrecht: Springer.

———. 1978. *Philosophical papers*. Cambridge: Cambridge University Press.

Merton, Robert K., and Elinor Barber. 2004. *The travels and adventures of serendipity*. Princeton: Princeton University Press.

Musgrave, Alan. 2006. Responses. In *Rationality and reality: Conversations with Alan Musgrave*, ed. Colin Cheyne and John Worrall, 293–333. Dordrecht: Springer.

———. 1999. How to do without inductive logic. *Science & Education* 8: 395–412.

———. 2011. Popper and hypothetico-deductivism. In *Inductive logic*, ed. Dov Gabbay, Stephan Hartmann, and John Woods, 205–234. Amsterdam: North-Holland.

Newton, Isaac. 1962. *Mathematical principles of natural philosophy – The system of the world*. Berkeley: University of California Press.

Ortega y Gasset, José. 1957. *On love: Aspects of a single theme*. New York: Grenwich Editions/ Meridian Books.

Pasteur, Luis. 1922–1939. *Oeuvres*. Paris: Masson.

Poincaré, Henri. 2013. *The foundations of science: Science and hypothesis – The value of science – Science and method*. Cambridge: Cambridge University Press.

Popper, Karl Raimund. 2000. *Realism and the aim of science*. London: Routledge.

———. 2005. *The logic of scientific discovery*. London: Routledge.

Putnam, Hilary. 1975–1983. *Philosophical papers*. Cambridge: Cambridge University Press.

Quine, Willard Van Orman. 1981a. *Mathematical logic*. Cambridge: Harvard University Press.

Rovelli, Carlo. 2014. Science is not about certainty. In *The universe*, ed. John Brockman, 214–228. New York: Harper.

Ryle, Gilbert. 2009. *Collected papers*. London: Routledge.

Siegfried, Tom. 2013. Rise of big data underscores need for theory. *ScienceNews*, 3 December.

Simonton, Dean Keith. 2004. *Creativity in science: Chance, logic, genius, and Zetigeist*. Cambridge: Cambridge University Press.

Tarski, Alfred. 1994. *Introduction to logic and to the methodology of deductive sciences*. Oxford: Oxford University Press.

Weyl, Hermann. 1949. *Philosophy of mathematics and natural science*. Princeton: Princeton University Press.

Williamson, Timothy. 2007. *The philosophy of philosophy*. Oxford: Blackwell.

Chapter 12
The Methods of Knowledge

Abstract This chapter examines the methods to acquire knowledge. The methods considered are the analytic method, the analytic-synthetic method, the material axiomatic method, and the formal axiomatic method. The chapter also examines the original formulation of these methods, and the relations between them. Since deductive and non-deductive rules play a crucial role in the methods in question, the chapter discusses which of these rules are ampliative, in the sense that the conclusion is not contained in the premisses.

12.1 The Analytic Method

In Chap. 11 it has been argued that we need a method to acquire knowledge, and that the claim that there is no such a method is unjustified. This chapter examines the methods to acquire knowledge.

A method to acquire knowledge that has been widely used since antiquity is the analytic method, also known as the method of analysis.

As already stated in Chap. 2, the analytic method is the method according to which, to solve a problem, one looks for some hypothesis that is a sufficient condition for solving the problem, namely, such that a solution to the problem can be deduced from the hypothesis. The hypothesis is obtained from the problem, and possibly other data already available, by some non-deductive rule, and must be plausible, that is, such that the arguments for the hypothesis are stronger than the arguments against it, on the basis of the existing knowledge. But the hypothesis is in its turn a problem that must be solved, and is solved in the same way. That is, one looks for another hypothesis that is a sufficient condition for solving the problem posed by the previous hypothesis, it is obtained from the latter, and possibly other data already available, by some non-deductive rule, and must be plausible. And so on, *ad infinitum*. Thus solving a problem is a potentially infinite process. If, in the course of this process, a hypothesis turns out to be implausible, one analyses the reasons why such hypothesis is implausible and so arrives at some alternative hypothesis.

In the analytic method there are no principles, everything is a hypothesis. The problem and the other data already available are the only basis for solving a problem. Solving a problem is a potentially infinite process. This is a basic feature of the

© Springer International Publishing AG 2017
C. Cellucci, *Rethinking Knowledge*, European Studies in Philosophy of Science 4,
DOI 10.1007/978-3-319-53237-0_12

analytic method, according to which there is no completed knowledge, every answer to a problem will raise new problems. An inquiry into a problem is the opening of the way to an unlimited deepening of the problem, being a continuous proceeding from the problem not to a conclusive answer, but to more and more fundamental problems.

In Chap. 11 we have seen that, since the nineteenth century, the view that scientific discovery escapes logical analysis has prompted many people to assert that discovery is a purely subjective, psychological matter, so it cannot be an object of logic but only of psychology. This claim is unjustified. For, there is nothing subjective or psychological about the analytic method. The non-deductive rules through which the method is implemented are as objective and non-psychological as the deductive rules, and can be justified like the deductive rules (see Cellucci 2013a, Chap. 18).

In the analytic method, logic plays a double role, since the analytic method involves both non-deductive and deductive rules. Non-deductive rules serve to obtain hypotheses, deductive rules serve to deduce solutions from hypotheses. Thus the analytic method comprises a double movement, an upward movement from problems to hypotheses, and a downward movement from hypotheses to problems. Then the analytic method can be schematically represented as follows.

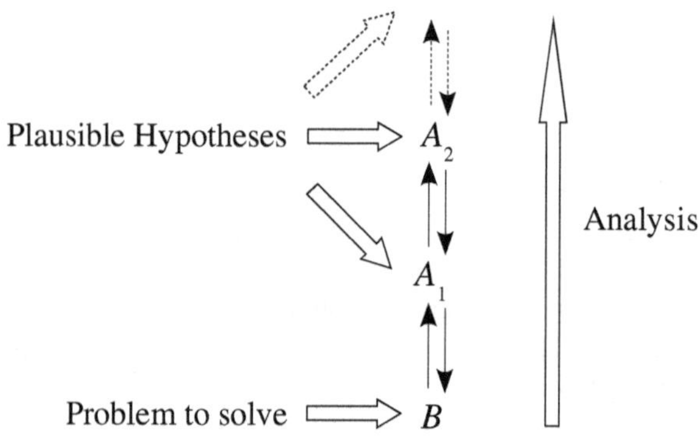

12.2 Original Formulation of the Analytic Method

The analytic method was already used by the mathematician Hippocrates of Chios and the physician Hippocrates of Cos, but was first explicitly formulated by Plato.

It is sometimes claimed that "Plato discovered the method of analysis" (Doxiadis and Sialaros 2013, 373). That is, the analytic method. The evidence offered for this claim is Proclus' statement that the analytic method is "a method which Plato, as they say, communicated to Leodamas" (Proclus 1992, 211.21–22). And Diogenes Laertius' statement that Plato "was the first to explain to Leodamas of Thasos the method of solving problems by analysis" (Diogenes Laertius, *Vitae Philosophorum*,

III, 24). But, in fact, neither Proclus nor Diogenes Laertius asserts that Plato discovered the method of analysis. Proclus merely asserts that Plato communicated, and Diogenes Laertius that Plato was the first to explain, the method of analysis to Leodamas. So the claim that Plato discovered the method of analysis is unjustified.

Actually, Plato did not discover the analytic method, the latter was already in use before Plato. Simply, Plato gave a formulation of the analytic method, based on the previous uses of the method by Hippocrates of Chios and Hippocrates of Cos.

Plato gave such a formulation by saying that, to solve a problem, you "should take refuge in certain hypotheses, and consider the truth of matters in them" (Plato, *Phaedo*, 99 e 5–6). Specifically, "on each occasion I assume the hypothesis which I judge to be the strongest, and I lay down as true whatever seems to me to agree with it", while "I put down as not true whatever does not seem to me to agree with it" (ibid., 100 a 3–7). Once you had assumed a hypothesis, you wouldn't go on until "you had investigated its consequences, to see whether they agreed or disagreed with one another" (ibid., 101 d 4–5). Even if the consequences agreed with each other, this would not mean that the hypothesis is conclusively justified, but only that it does not lead to contradiction. Therefore, you would "give an account of the hypothesis itself," and "you would give such an account in the same way, positing another hypothesis, whichever should seem best of the higher ones, and so on until you came to something adequate" (ibid., 101 d 5–e 1). But the hypothesis thus reached will in turn be a problem that eventually will have to be solved, it will be solved by positing a new hypothesis. And so on, *ad infinitum*. Thus solving a problem is "an infinite task" (Plato, *Parmenides*, 136 c 7).

Husserl claims that the ancients did not "grasp the possibility of an infinite task" (Husserl 1970, 21). From Plato's formulation of the analytic method it is apparent that this claim is unjustified.

Plato formulates the analytic method as the method of philosophy. For example, to solve the problem whether virtue is teachable, Plato formulates the hypothesis: "Virtue is science" (Plato, *Meno*, 87 c 5). The hypothesis is a sufficient condition for solving the problem. For, clearly science and only science is teachable, since "it is plain to everyone that men cannot be taught anything but science" (ibid., 87 c 1–3). From this, by the hypothesis, it follows that "virtue is teachable" (ibid., 88 c 8–9). This solves the problem. But the hypothesis is in its turn a problem that must be solved. And so on, *ad infinitum*.

As another example, to solve the problem whether the soul is immortal, Plato formulates the hypothesis: "Each of the ideas exists" (Plato, *Phaedo*, 102 b 1). Thus, "there exists the beautiful itself, the good itself, the great itself, and so on" (ibid., 100 b 5–7). The hypothesis implies that all things participate in ideas, that is, "if anything else is beautiful besides the beautiful itself, it is beautiful for no other reason than because it participates in that beautiful; and this applies to everything" (ibid., 100 c 4–6). Now, clearly no idea can be its opposite, since "the opposites do not admit each other" (ibid., 104 b 7–8). From this, by the implication of the hypothesis mentioned above, it follows that no thing can participate in an idea and its opposite. But the soul participates in the idea of life, since the soul is "that which must be in a body to make it alive" (ibid., 105 c 9–10). Moreover, the opposite of life "is death" (ibid., 105, d 9). Therefore, the soul cannot participate in the idea of death, hence "the soul is immortal" (ibid., 105 e 6). This solves the problem. But the hypothesis is in its turn a problem that must be solved. And so on, *ad infinitum*.

Plato does not indicate how hypotheses are obtained. The formulation of the analytic method given above remedies this limitation, specifying that they are obtained by non-deductive rules.

12.3 Origin of the Analytic Method

As already stated, Plato did not discover the analytic method, before Plato the latter was already used by the mathematician Hippocrates of Chios and the physician Hippocrates of Cos.

Hippocrates of Chios used the analytic method to solve mathematical problems. For example, to solve the problem of doubling the cube – namely the problem of finding the side of the cube double of a given cube – Hippocrates of Chios formulates the hypothesis: Given any two straight lines, a and b, we can always find two other straight lines, x and y, which are the mean proportionals in continued proportion between a and b, namely, such that $a: x = x: y = y: b$. The hypothesis is a sufficient condition for solving the problem. For, if a is the side of the given cube, then by the hypothesis there are two straight lines, x and y, such that $a: x = x: y = y: 2a$. Then, by compounding the ratios, we have $(a: x)^3 = (a: x)\,(x: y)\,(y: 2a) = (a: 2a)$, hence $a^3: x^3 = 1{:}2$, therefore $x^3 = 2a^3$. So x will be the side of the cube double of the given cube of side a. Thus "the cube will be doubled" (Eutocius, *Commentarii in Libros de Sphaera et Cylindro. Ut Menechmus*, 104.15). This solves the problem. But the hypothesis is in its turn a problem that must be solved. And so on, *ad infinitum*.

Hippocrates of Chios used the analytic method also to solve the problem of the quadrature of certain lunules. For example, to solve the problem of the quadrature of the lunule obtained by circumscribing about a right-angled isosceles triangle a semicircle, and about the base a segment of a circle similar to those cut off by the sides, Hippocrates of Chios formulates the hypothesis: "Similar segments of circle are as the squares on their chords" (Simplicius, *In Aristotelis Physicorum Libros Quattuor Priores Commentaria*, I 2, 61.6–7).

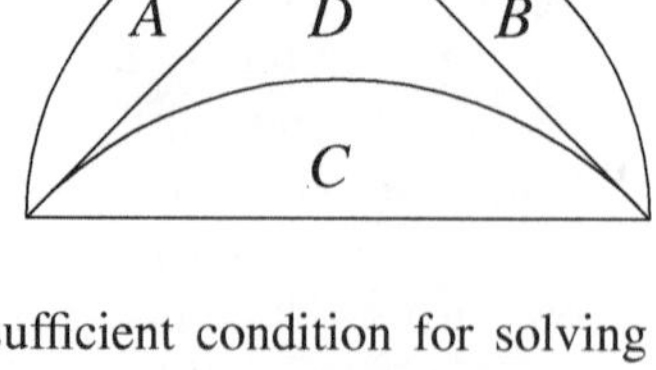

The hypothesis is a sufficient condition for solving the problem. For, by the Pythagorean theorem, the square on the base is equal to the sum of squares on the sides. From this, by the hypothesis, it follows that the segment about the base is equal to the sum of the segments about the sides, that is, $C = A + B$. Now, the lunule consists of the sum of the segments about the sides plus the part of the triangle above the segment about the base, that is, lunule $= A + B + D$. On the other hand, the triangle consists of the segment about the base plus the part of the triangle above the segment about the base, that is, triangle $= C + D$. Since $C = A + B$, from this it follows: lunule $= A + B + D = C + D =$ triangle. Therefore, we may conclude that "the lunule is equal to the triangle" (ibid., I 2, 62.6–7). This solves the problem. But the hypothesis is in its turn a problem that must be solved. And so on, *ad infinitum*.

On the other hand, Hippocrates of Cos used the analytic method to solve medical problems. For example, to solve the problem why the human being is affected and altered by different foods, Hippocrates of Cos formulates the hypothesis: The substances which are in the food "are also in the human being" (Hippocrates 2005, 93). The hypothesis is a sufficient condition for solving the problem. For, when the substances which are in the human being are well "mixed and blended with one another," they do not "cause the human being pain," and conversely, when one of the substances which are in the human being "separates off and comes to be on its own," it "causes the human being pain" (ibid.). So the human being is affected and altered by the substances which are in him. From this, by the hypothesis, it follows that the human being is affected and altered by the substances which are in the food. Since different foods contain different substances, one may then conclude that the human being is affected and altered by different foods. This solves the problem. But the hypothesis is in its turn a problem that must be solved. And so on, *ad infinitum*.

12.4 Analytic Method and Infinite Regress

Since, with the analytic method, solving a problem is a potentially infinite process, it may be objected that, with such method, no solution to a problem is absolutely guaranteed since no hypothesis is absolutely justified, therefore the analytic method cannot yield knowledge.

This is Aristotle's objection to the analytic method, which is based on his argument, already considered in Chap. 4, that, if the series of the premisses does not terminate and there is an infinite regress, then there is no knowledge. (See also his argument against the view that the method of mathematics is the analytic method, considered in Chap. 20). Gödel uses essentially the same argument, when he says that one cannot acquire knowledge "by trying to give" explicit "proofs for axioms, since for that one obviously needs" other "concepts and axioms holding for them. Otherwise one would have nothing from which one" could "prove" (Gödel 1986–2002, III, 383).

This objection, however, is unjustified because, that, with the analytic method, no solution is absolutely guaranteed since no hypothesis is absolutely justified, does not mean that such method cannot yield knowledge. It can yield knowledge, albeit fallible knowledge. The analytic method could not yield knowledge only if the hypotheses were arbitrary. But they are not arbitrary, since they must be plausible. Therefore, the analytic method can yield knowledge. On the other hand, the knowledge it yields is fallible, in the sense that it cannot be ruled out that it may lead to error. New data may always emerge with which the hypotheses on which knowledge is based may turn out to be incompatible.

Lewis claims that “to speak of fallible knowledge, of knowledge despite uneliminated possibility of error, just sounds contradictory” (Lewis 1996, 549). But even mathematical knowledge is fallible, since it cannot be ruled out that it might lead to error. *A fortiori*, this applies to non-mathematical knowledge. Thus, if Lewis were right, knowledge would be impossible.

12.5 The Open-Ended Character of Rules of Discovery

The non-deductive rules by means of which hypotheses are obtained in the analytic method are rules of discovery. Of course, obtaining hypotheses is not a sufficient condition for discovery, hypotheses must also be plausible. Nonetheless, obtaining hypotheses is a necessary condition for discovery, and in that sense non-deductive rules may be said to be rules of discovery.

The rules of discovery include induction, analogy, generalization, specialization, metaphor, metonymy, definition, diagrams. (For a statement of these rules, see Cellucci 2013a, Chaps. 20 and 21). However, the rules of discovery are not a closed set, given once and for all, but rather an open set that can always be extended as research develops. Descartes even states that every solution that is found can be used “as a rule for finding other ones” (Descartes 1996, VI, 20–21).

Each extension of the rules of discovery by new non-deductive rules is a development of the analytic method, which grows as such new rules are added. Thus, as Bacon states, “the art of discovery may grow with discoveries” (Bacon 1961–1986, I, 223).

12.6 Non-ampliativity of Deductive Rules

That, in the analytic method, hypotheses are obtained by non-deductive rules, rather than by deductive rules, depends on the fact that non-deductive rules are ampliative, namely, the conclusion is not contained in the premisses. On the contrary, deductive rules are non-ampliative, namely, the conclusion is contained in the premisses. That deductive rules are non-ampliative has been maintained by several people.

Thus Aristotle states that “a deduction is an argument in which, certain things being posited, something different from them results by necessity, because these things are so. By ‘because these things are so’ I mean to say that it results through these, and by the expression ‘it results through these’ I mean to say that no term is required from outside for the necessity to come about" (Aristotle, *Analytica Priora*, A 1, 24 b 18–22).

Kant states that deductive logic "teaches us nothing at all about the content of cognition," so "using it as a tool (organon) for an expansion and extension of its information, or at least the pretension of so doing, comes down to nothing but idle chatter" (Kant 1998, A61/B86).

Mill states that, in deductive inference, the conclusion is "merely a repetition of the same, or part of the same, assertion, which was contained in the first" (Mill 1963–1986, VII, 158). Therefore, "there is in the conclusion no new truth, nothing but what was already asserted in the premisses, and obvious to whoever apprehends them" (ibid., VII, 160).

De Morgan states that deductive inference is subject "to the great rule of all search after truth, that nothing is to be asserted as a conclusion, more than is actually contained in the premises" (De Morgan 1835, 99).

Peirce states that, in deductive inference, "certain facts are first laid down in the premisses," and "part or all of them" will be thrown "into a new statement" that "will be the conclusion of" the deductive "inference" (Peirce 1931–1958, 2.680). Therefore, deductive inference "is evidently entirely inadequate to the representation" of inference "which goes out beyond the facts given in the premisses" (ibid., 2.681).

Poincaré states that deduction "remains incapable of adding anything to the data that are given it," the data are contained in the premisses, "and that is all we should find in the conclusions" (Poincaré 2013, 31).

Wittgenstein states that, "if one proposition follows from another, then the latter says more than the former, and the former less than the latter" (Wittgenstein 2002, 5.14). Thus "there can never be surprises in logic" (ibid., 6.1251). Indeed, "in logic process and result are equivalent. (Hence the absence of surprise)" (ibid., 6.1261).

12.7 Objections to the Non-Ampliativity of Deductive Rules

Against the claim that deductive rules are non-ampliative, several objections have been raised. We consider two of them.

Frege objects that, although the conclusions of deductions "are in a way contained covertly in the whole set" of premisses "taken together," this "does not absolve us from the labour of actually extracting them and setting them out in their own right" (Frege 1960, 23). Indeed, "what we shall be able to infer from" the premisses "cannot be inspected in advance; here, we are not simply taking out of the box again what we have just put into it. The conclusions we draw" from the premisses may "extend our knowledge, and ought therefore, on Kant's view, to be regarded as synthetic" (ibid., 100–101). The conclusions are contained in the premisses, "but as plants are contained in their seeds, not as beams are contained in a house" (ibid., 101).

This objection, however, is unjustified because, as already mentioned in Chap. 11, there is an algorithm for enumerating all deductions from given premisses. Given enough time and space, a suitably programmed Turing machine will be able to grind out all the deductions from the premisses one by one. If a conclusion can be deduced from them, then the Turing machine, operating in this blind fashion, will sooner or later find a deduction of it. So, contrary to Frege's claim, extracting the conclusions from the premisses is a purely mechanical task. It can be performed by a Turing machine, and hence requires no labour.

Moreover, it is misleading to say that conclusions are contained in the premisses as plants are contained in their seeds. For, plants can develop from seeds only by absorbing water from the soil and harvesting energy from the sun, hence, by using something which is not contained in the seeds. On the contrary, conclusions can be deduced from premisses without using anything not contained in the premisses.

In relation to Frege's argument, a distinction must be made between objective novelty and psychological novelty. The conclusions of deductions may be psychologically surprising, and hence may have psychological novelty, because we are incapable of making even comparatively short deductions without the help of processes external to us. (For an example of this, see the solution of Carroll's problem in Chap. 17). But this does not mean that the conclusions of deductions extend our knowledge, and hence have objective novelty. Frege's claim that they extend our knowledge mistakes psychological novelty for objective novelty, so it is a form of psychologism.

Dummett objects that, if deductive rules were non-ampliative, then, "as soon as we had acknowledged the truth of the axioms of a mathematical theory, we should thereby know all the theorems. Obviously, this is nonsense" (Dummett 1991, 195). Therefore, deduction must have the power "to yield knowledge that we did not previously possess" (ibid.).

This objection, however, is unjustified because, as stated above, we are incapable of making even comparatively short deductions without the help of processes external to us. For this reason, we do not know all the theorems that can be deduced from the axioms of a mathematical theory as soon as we acknowledge that the axioms are true. This has nothing to do with ampliativity.

For other objections against the claim that deductive rules are non-ampliative, see Cellucci 2013a, Chap. 17.

12.8 The Plausibility Test Procedure

In the analytic method, hypotheses must be plausible. In order to show that a hypothesis is plausible we may use the following plausibility test procedure:

1) Deduce conclusions from the hypothesis.

2) Compare the conclusions with each other, in order to see that the hypothesis does not lead to contradictions.

3) Compare the conclusions with other hypotheses already known to be plausible, and with results of observations or experiments, in order to see that the hypothesis is compatible with them.

If the hypothesis passes the plausibility test procedure, then for the moment it is approved. Indeed, 'plausible' comes from the Latin *plausibilis*, which derives from *plaudere* that means 'to applaud', 'to approve'. The hypothesis, however, is approved only for the moment, because new data can always emerge with which the hypothesis may turn out to be incompatible.

On the other hand, if the hypothesis does not pass the plausibility test procedure, it is not rejected outright, but is put on a waiting list, subject to further investigation, since nothing is definitively plausible or definitively implausible.

12.9 Inference Rules, Plausibility, and Experience

The plausibility of a hypothesis obtained by some non-deductive rule is not ensured by that rule but only by experience. For, non-deductive rules are not plausibility preserving, namely, such that if the premisses are plausible so is the conclusion. This depends on the fact that the arguments for the conclusion go beyond the arguments for the premisses. Since non-deductive rules are not plausibility preserving, the plausibility of a hypothesis obtained by some non-deductive rule cannot be ensured by that rule, but only by a comparison with the existing knowledge, and hence by experience.

Even the plausibility of a hypothesis obtained by some deductive rule is not ensured by that rule but only by experience. For, while deductive rules are truth preserving, they are not plausibility preserving. For example, let A and B be two rival scientific hypotheses. In order to be accepted, A and B must both be plausible. They can both be plausible because the arguments for and against A need not coincide with the arguments for and against B. On the other hand, being rival, A and B must be mutually incompatible. This means that certain arguments for A must be arguments against B, and vice versa. Then, although A and B are both plausible, their conjunction, $A \wedge B$, will not be plausible. Therefore, the conjunction introduction rule, 'From A and B infer $A \wedge B$', is not plausibility preserving. Since deductive rules are not plausibility preserving, the plausibility of a hypothesis obtained by some deductive rule is not ensured by that rule, but only by a comparison with the existing knowledge, and hence by experience.

12.10 Inference Rules and Usefulness

That deductive rules are non-ampliative does not mean that they are not useful. By making explicit all or part of what is contained in the premisses, they facilitate the comparison of the premisses with the existing knowledge.

However, for a deductive inference to be useful, we must know that its premisses and conclusion are plausible. Knowing that its premisses are plausible is not enough because, as argued above, deductive rules are not plausibility preserving.

Since, for a deductive inference to be useful, one must know that both its premisses and conclusion are plausible, the usefulness of deductive rules essentially depends on a comparison with the existing knowledge, thus on something external to inference.

Similarly, that non-deductive are ampliative does not ensure that any specific non-deductive rule is useful. For a non-deductive rule to be useful, one must know that both its premisses and conclusion are plausible. Knowing that its premisses are plausible is not enough, because non-deductive rules are not plausibility preserving.

Since, for a non-deductive inference to be useful, one must know that both its premisses and conclusion are plausible, the usefulness of non-deductive rules essentially depends on a comparison with the existing knowledge, thus on something external to inference.

12.11 Basic Features of the Analytic Method

The analytic method has certain basic features, which can be summarized as follows.

1) *The hypotheses to solve a problem need not belong to the field of the problem, but may belong to other fields.* So, the search for a solution to a problem is not carried out in a closed, predetermined space.

2) *The hypotheses to solve a problem are not global but local.* They are not general principles, good for all problems, but are aimed at a specific problem. As Suppes states, "like our own lives and endeavors," hypotheses "are local and are designed to meet a given set of problems" (Suppes 1993, 54).

3) *Being local, the hypotheses to solve a problem can be efficient.* Indeed, being aimed at a specific problem, they can take care of the peculiarities of the problem. This may be essential for the feasibility of a solution.

4) *Different problems will generally require different hypotheses.* This follows from the fact that hypotheses are aimed at a specific problem, so they depend on the peculiarities of the problem.

5) *The same problem may be solved using different hypotheses.* A hypothesis is a window through which you look upon a problem. But the problem can be looked at through other windows from different angles, that reveal heretofore unsuspected aspects, leading to other solutions of the problem. When a problem seems to be solvable only by a single hypothesis, one should worry because it might mean that the solution is wrong or that the problem is ill-posed. Indeed, if one cannot view a problem from different angles, the problem is probably not very interesting. Different hypotheses may establish different relations between the problem and problems of other fields, thus showing the problem in a new perspective.

6) *The hypotheses formulated to solve a problem may turn out to solve some other problem, instead of the intended one.* When the hypotheses formulated to solve a problem fail to solve it, in any case they solve something else, namely, some other problem. The problem they solve may be an insignificant one, but sometimes may have an interest of its own. Whatever the hypotheses solve, as long as it is interesting, qualifies as a significant problem.

7) *The hypotheses formulated to solve a problem may turn out to solve also some other problem, in addition to the intended one.* Indeed, the hypotheses formulated to solve a problem may involve concepts not occurring in that problem, so they may have implications also for other problems, in addition to the intended one.

8) *Solving a problem is both a process of discovery and a process of justification.* It is a process of discovery, since it involves finding hypotheses by non-deductive rules. It is a process of justification, since it involves establishing that the hypotheses thus found are plausible.

9) *Solving a problem essentially requires acquaintance with the existing knowledge.* Hypotheses are obtained from the problem, and possibly other data already available, by some non-deductive rule. Moreover, hypotheses must be plausible, namely, such that the arguments for them are stronger than the arguments against them, on the basis of the existing knowledge. Thus, both finding hypotheses and establishing that they are plausible essentially require acquaintance with the existing knowledge.

10) *Solving a problem is a dynamic process.* First a problem is formulated. Then a hypothesis is obtained by some non-deductive rule and is shown to be plausible. Afterwards, a new hypothesis is obtained to solve the problem posed by the previous hypothesis. And so on.

11) *Solving a problem yields something new.* Being obtained from the problem, and possibly other data already available, by some non-deductive rule, a solution is not contained in the problem or in the other data already available. It possesses novelty with respect to them, because non-deductive rules are ampliative. This explains why solving a problem can yield something new.

12) *Intuition has no role in solving a problem.* It has no role in the formulation of hypotheses, because the latter are obtained from the problem, and possibly other data already available, by some non-deductive rule, so not by intuition but by inference. It has no role in the justification of hypotheses, because their plausibility is established by means of the plausibility test procedure, so not by intuition but by inference.

13) *Hypotheses are neither true nor certain, but this does not diminish the value of the analytic method.* That hypotheses are neither true nor certain follows from the fact that the analytic method is a heuristic method, not an algorithmic one, so it cannot guarantee to yield true or certain knowledge, but only plausible knowledge. This, however, is no limitation, because there is no source of knowledge capable of guaranteeing truth or certainty, plausible knowledge is the best we can achieve, and, on the other hand, without plausible knowledge there is no knowledge at all. As Pólya states, "if you take a heuristic conclusion as certain, you may be fooled and

disappointed; but if you neglect heuristic conclusions altogether you will make no progress at all" (Pólya 1971, 181).

14) *Solving a problem is a potentially infinite process, so no solution is final.* For, no hypothesis is absolutely justified. Any hypothesis which provides a solution to a problem is liable to be replaced with another hypothesis when new data emerge, so every solution is always provisional. As Kant states, "any answer given according to principles of experience always begets a new question which also requires an answer" (Kant 2002, 141–142).

12.12 Analytic Method and Abduction

It might be thought that there is a relation between the analytic method and abduction, because Peirce claims that "abduction is the process of forming explanatory hypotheses. It is the only logical operation which introduces any new idea" (Peirce 1931–1958, 5.172). In fact, "all ideas of science come to it by way of abduction" (ibid., 5.145). Therefore, "abduction must cover all operations by which theories and conceptions are engendered" (ibid., 5.590).

But Peirce formulates abduction as follows: "The surprising fact, C, is observed; but if A were true, C would be a matter of course; hence, there is reason to suspect that A is true" (ibid., 5.189). Then abduction cannot introduce any new idea, because the conclusion A already occurs in the premiss, 'If A were true, C would be a matter of course', so A is already known. Peirce himself states that "A cannot be abductively inferred, or if you prefer the expression, cannot be abductively conjectured until its entire content is already present in the premiss, 'If A were true, C would be a matter of course'" (ibid.). Therefore, "quite new conceptions cannot be obtained from abduction" (ibid., 5.190). (On the problems with Peirce's concept of abduction, see Frankfurt 1958; Hoffmann 1999; Cellucci 2013a, Chap. 18).

Unlike abduction, the analytic method introduces genuinely new hypotheses. Therefore, the analytic method must not be confused with abduction.

An example of this confusion is the assertion that "the method of reasoning used by Sherlock Holmes is abduction" (Patokorpi 2007, 171). This conflicts with Sherlock Holmes' own assertion that "in solving a problem of this sort, the grand thing is to be able to reason backward," namely "analytically" (Doyle 2003, 115). Most people can only "reason forward," namely "synthetically," but "there are few people" who, "if you told them a result, would be able to evolve from their own inner consciousness what the steps were which led up to that result. This power is what I mean when I talk of reasoning backwards, or analytically" (ibid., 115–116). This "science of deduction and analysis is one which can only be acquired by long and patient study, nor is life long enough to allow any mortal to attain the highest possible perfection in it" (ibid., 16).

The science of deduction and analysis to which Sherlock Holmes refers is the analytic method, in which analysis is the upward movement from problems to hypotheses, and deduction is the downward movement from hypotheses to problems.

The assertion that life is not long enough to allow any mortal to attain the highest possible perfection in such a science, is one with which Plato would completely agree because, as we will see below, Plato maintains we can attain such perfection only after death.

12.13 Fortune of the Analytic Method

Although this is not generally realized, the analytic method was not used only in ancient Greece, but has been used up to the present-day.

An example is provided by the solution of Fermat's problem: Show that there are no positive integers x, y, z such that $x^n + y^n = z^n$ for $n > 2$. According to the current prevailing view, the problem was solved by Wiles and Taylor, but this is not literally correct. Actually, the problem was solved by Ribet using the following hypothesis, called the Conjecture of Taniyama and Shimura: Every elliptic curve over the rational numbers is modular. Indeed, Ribet showed: "Conjecture of Taniyama and Shimura ⇒ Fermat's Last Theorem" (Ribet 1990, 127). Therefore, Ribet showed that the Conjecture of Taniyama and Shimura is a sufficient condition for solving Fermat's problem. In particular, Ribet showed that, if there existed positive integers x, y, z such that $x^n + y^n = z^n$ for $n > 2$, they would yield an elliptic curve over the rational numbers which would not be modular, thus contradicting the Conjecture of Taniyama and Shimura.

But the Conjecture of Taniyama and Shimura was in its turn a problem that had to be solved. It was solved by Wiles and Taylor using hypotheses from various fields of mathematics, ranging from differential geometry to complex analysis. Thus what Wiles and Taylor solved was not Fermat's problem but the problem posed by the Conjecture of Taniyama and Shimura. The process through which Fermat's problem was solved can then be schematically represented as follows.

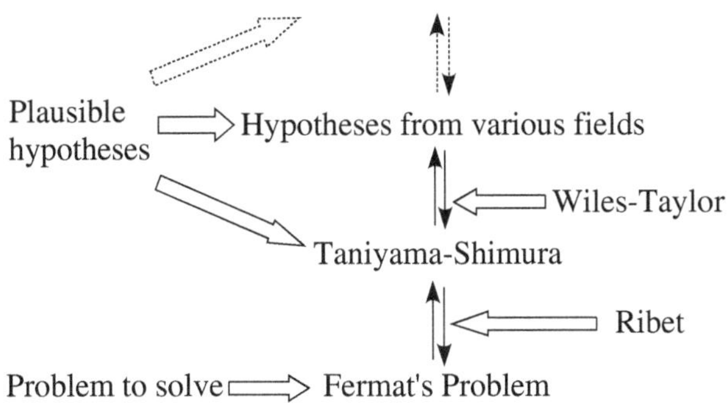

Most mathematicians would object that Fermat's problem was not solved by Ribet since he used a hypothesis, the Conjecture of Taniyama and Shimura, that at the time had not been demonstrated yet. But Wiles and Taylor solved the Conjecture using hypotheses from various mathematics fields, which in turn depended on other hypotheses, and ultimately on the axioms of set theory. Now, the axioms of set theory are a hypothesis that has not been demonstrated yet. So, if Ribet did not solve Fermat's problem since he used a hypothesis that at the time had not been demonstrated yet, then Wiles and Taylor did not solve Fermat's problem since they used a hypothesis, the axioms of set theory, that to this very day has not been demonstrated yet.

12.14 The Analytic-Synthetic Method

Another method to acquire knowledge which has been widely used since antiquity as an alternative to the analytic method, is the analytic-synthetic method, also know as the method of analysis and synthesis.

The analytic-synthetic method is the method according to which, to solve a problem, one looks for some hypothesis that is a sufficient condition for solving the problem, namely, such that a solution to the problem can be deduced from the hypothesis. The hypothesis is obtained from the problem, and possibly other data already available, by some non-deductive rule and must be plausible, namely, such that the arguments for the hypothesis are stronger than the arguments against it, on the basis of the existing knowledge. If the hypothesis thus found is not a principle, one looks for another hypothesis that is a sufficient condition for solving the problem posed by the previous hypothesis, it is obtained from the latter, and possibly other data already available, by some non-deductive rule, and must be plausible. And so on, until one arrives at some hypothesis which is a principle. Principles must be true. When one arrives at some hypothesis which is a principle, the process terminates. This is analysis. At this point, one tries to see whether, inverting the order of the steps followed in analysis, one obtains a deduction of the solution of the problem from the principle arrived at in analysis. This is synthesis.

The analytic-synthetic method can be schematically represented as follows.

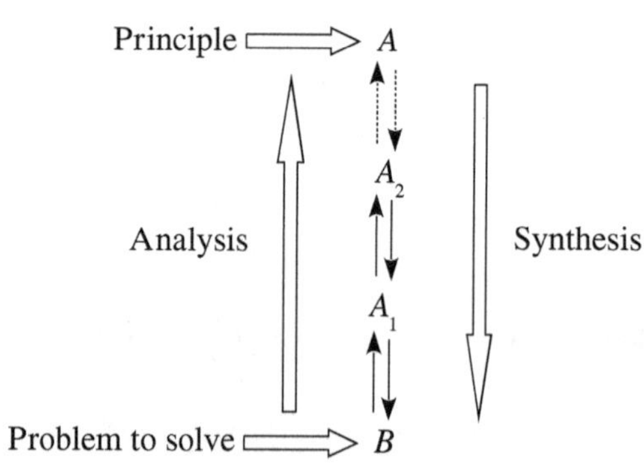

12.15 Original Formulation of the Analytic-Synthetic Method

The analytic-synthetic method was originally formulated by Aristotle. Hintikka and Remes claim that "the only extensive general description of the method of analysis and synthesis to be found in the surviving ancient literature is given by Pappus of Alexandria" (Hintikka and Remes 1976, 254). But it is not so. Aristotle gives an extensive general description of the method of analysis and synthesis.

Indeed, Aristotle states that "the process of knowledge proceeds from what is more knowable and clearer to us to what is clearer and more knowable by nature" (Aristotle, *Physica*, A 1, 184 a 16–18). So "we must necessarily proceed in this way, namely, starting from the things which are less known by nature but clearer to us, towards those which are clearer and more knowable by nature" (ibid., A 1, 184 a 18–21). Now, what is more knowable and clearer to us is the problem to solve, while what is clearer and more knowable by nature are the principles. They are "most knowable" by nature because, "by reason of these, and from these, all other things are known, but these are not known by means of the things subordinate to them" (Aristotle, *Metaphysica*, A 2, 982 b 2–4). Therefore, by saying that we must proceed from what is more knowable and clearer to us to what is clearer and more knowable by nature, Aristotle means to say that we must proceed from the problem to solve to principles.

Specifically, given the problem to solve, we must find "the necessary hypotheses through which the syllogisms come about" (Aristotle, *Topica*, Θ 1, 155 b 29). We will find them "either by syllogism" – namely, by Aristotle's procedure for finding the premisses of a syllogism given the conclusion, see Cellucci 2013a, Chap. 7 – "or by induction" (ibid., Θ 1, 155 b 36). When the hypotheses are found, we "should not put forward these" hypotheses "right away, but rather should stand off as far above them as possible" (ibid., Θ 1, 155 b 29–30). That is, we should find other hypotheses from which the previous hypotheses can be deduced. And so on, until we arrive at some hypothesis which is a principle. Principles must be true, otherwise we would not "have scientific knowledge of what follows from them, absolutely and properly" (Aristotle, *Analytica Posteriora*, A 3, 72 b 14).

When we arrive at some hypothesis that is a principle, the process terminates. This is analysis. But analysis is not always successful, so, "if we meet with an impossibility, we give up," but, "if the thing appears possible we try to do it" (Aristotle, *Ethica Nicomachea*, Γ 3, 1112 b 24–27). That is, we try to see whether, by inverting the order of the steps followed in analysis, we may obtain a deduction of the solution of the problem from the principle arrived at in analysis. This is synthesis.

This procedure serves not only to acquire scientific or mathematical knowledge, but also to solve problems in practical fields which require deliberation to reach the intended end, such as medicine. In such fields "we deliberate not about ends, but about means to ends" (ibid., Γ 3, 1112 b 11–12). For example, doctors do not deliberate whether they are to heal, but take the end they want to achieve, namely healing, and "consider how and by what means it is to be attained" (ibid., Γ 3, 1112 b 15–16). Then they consider "by what other means" these "means will be achieved," and so on, "until they arrive at the first cause, which is last in the order of discovery" (ibid., Γ 3, 1112 b 18–20). This is analysis. At that point, they try to see whether, by inverting the order of the steps followed in analysis, starting from the first cause arrived at in analysis, they may achieve the intended end, namely healing. This is synthesis, or construction. Thus, "he who deliberates seems to carry out an inquiry and an analysis in the way described as though he were carrying out an analysis in a geometrical construction" (Aristotle, *Ethica Nicomachea*, Γ 3, 1112 b 20–21). Since the synthesis, or construction, starts from the first cause, which is last in the order of discovery, "what is last in the order of analysis is first in the order of construction" (*ibid.*, Γ 3, 1112 b 23–24).

From Aristotle's formulation of the analytic-synthetic method it is clear that there is no justification for Hintikka and Remes' claim that the only extensive general description of the method of analysis and synthesis to be found in the surviving ancient literature is given by Pappus of Alexandria. Moreover, unlike Aristotle's formulation, Pappus' description of the method of analysis and synthesis is somewhat problematic because it makes apparently incompatible assertions about the direction of analysis (for details, see Cellucci 2013a, Chap. 5).

12.16 Difference Between the Analytic and the Analytic-Synthetic Method

The analytic method and the analytic-synthetic method are often confused with each other. For example, Menn states that both "Plato and Aristotle" say that "there are two stages of argument, first to the *archai*," namely, the principles, "(contrary to the 'natural' order of things) and then from the *archai* (following the 'natural' order)" (Menn 2002, 193). Thus, according to Menn, both Plato and Aristotle hold that analysis is an upward movement to the principles.

But it is not so. For Plato, analysis is an upward movement to plausible hypotheses, not to principles, therefore the analytic method and the analytic-synthetic method are essentially different. In the analytic method, analysis is a means of obtaining unknown hypotheses. On the contrary, in the analytic-synthetic method, analysis is a means of obtaining demonstrations of propositions from already known principles.

Admittedly, Plato states that ultimately analysis arrives at "the unhypothetical principle of everything" (Plato, *Respublica*, VI 511 b 5–6). But he specifies that, through analysis, we can arrive at the unhypothetical principle of everything only after death, because "as long as we have the body and our soul is contaminated by such an evil, we will never adequately gain the possession of what we desire, and that, we say, is truth" (Plato, *Phaedo*, 66 b 5–7). For this reason, Plato makes the somewhat paradoxical statement that the best way to achieve knowledge is to "practise nothing other than dying and being dead" (ibid., 64 a 5–6). (For more on this, see Cellucci 2013a, Chap. 3).

As long as we are in life, there remains an essential difference between analysis in the analytic method and analysis in the analytic-synthetic method. As Lakatos puts it, in the analytic method analysis is "a venture into the unknown," while, in the analytic-synthetic method, analysis is only "an exercise in mobilizing and ingeniously connecting the relevant parts of the known" (Lakatos 1978, II, 100). Hypotheses, which in the analytic method are "daring and often falsified conjectures," in the analytic-synthetic method "harden into auxiliary theorems" (ibid.).

12.17 Analytic-Synthetic Method and Intuition

In the analytic-synthetic method, intuition plays an essential role. As already mentioned, in such method principles must be true. Then the question arises how principles "become known" to be true, and "what is the state which gets to know them" (Aristotle, *Analytica Posteriora*, B 19, 99 b 17–18).

Now, principles cannot become known to be true by demonstration, nor can the state which gets to know them be discursive thinking, otherwise principles would be demonstrable. But principles "are indemonstrable," otherwise there would be an infinite regress, "so it will not be scientific knowledge but intuition that is concerned with the principles" (Aristotle, *Magna Moralia*, A 34, 1197 a 22–23). Thus, according to Aristotle, principles become known to be true by intuition, and the state which gets to know them is intuitive thinking.

However, while principles become known to be true by intuition, they are not discovered by intuition, they are discovered by the analytic-synthetic method. Indeed, in its first application, the analytic-synthetic is used to discover principles. Once principles have been discovered, the method will be used to discover deductions of given conclusions from the principles. On the other hand, principles become known to be true only by intuition, because although "the hypotheses are the end" of analysis, it is not "argument that teaches us the principles" (Aristotle, *Ethica Nicomachea*, Z 8, 1151 a 16–18).

Intuition and analysis are essentially different. While intuition belongs to intuitive thinking, analysis belongs to discursive thinking. Moreover, while intuition is supposed to be infallible, analysis is fallible, because hypotheses found by means of it may not be true.

12.18 The Material Axiomatic Method

A byproduct of the analytic-synthetic method is the material axiomatic method. It is a byproduct of the analytic-synthetic method, because it can be seen as what results from the latter when the analytic part of the method is omitted and only the synthetic part is retained.

Indeed, the material axiomatic method is the method according to which, in order to justify and teach an already acquired proposition of a given field, one deduces the proposition from the principles, or axioms, of that field. The axioms must be true, in the sense that there must be a system of things, specified in advance, for which the axioms are true.

The purpose of the material axiomatic method is not to obtain new knowledge, but only to justify and teach an already acquired proposition, because deduction produces its teaching through what is already known.

The material axiomatic method can be schematically represented as follows.

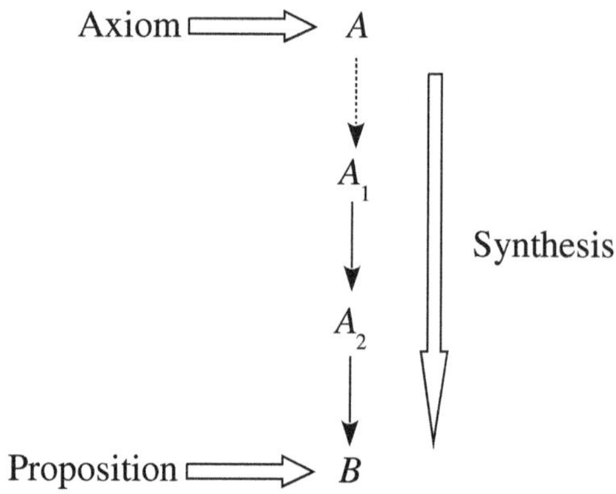

12.19 Original Formulation of the Material Axiomatic Method

The material axiomatic method was already used by some mathematicians of Plato's Academy. In particular, Cambiano suggests that it "was born probably with Eudoxus" (Cambiano 1967, 147). Anyhow, the material axiomatic method was first explicitly formulated by Aristotle.

Indeed, Aristotle states that, in order to justify and teach an already acquired proposition, we start from the principles proper to the subject matter of that proposition, and deduce the proposition from them, because "didactic arguments are those that deduce" propositions "from the principles proper to each subject matter" (Aristotle, *De Sophisticis Elenchis*, 2, 165 b 1–2). Thus, didactic arguments are the "demonstrative arguments" which are "treated in the *Analytics*" (ibid., 2, 165 b 9). Since didactic arguments are demonstrative arguments, the principles from which they deduce propositions must be true, because "a thing is demonstrated from what is true" (Aristotle, *Analytica Posteriora*, A 9, 75 b 39). Principles must be true in the sense that there must be things, specified in advance, for which the principles are true. Indeed, with respect to these things, "it is necessary to assume beforehand that they are" (ibid., A 1, 71 a 12). For, "it is not possible to have scientific knowledge of things that are not" (ibid., A 2, 71 b 25–26).

The purpose of the method thus formulated, namely the material axiomatic method, is not to acquire new knowledge, but only to justify and teach an already acquired proposition. Indeed, didactic arguments do not yield new knowledge, but "produce their teaching through what we already know" (ibid., A 2, 71 a 6–7). This follows from the fact that, generally, "all teaching and all intellectual learning come about from already existing knowledge. This is clear if we consider all the branches of intellectual learning. Indeed, the mathematical sciences are learned in this way and so is each of the other arts" (ibid., A 1, 71 a 1–4).

Then, for Aristotle, the material axiomatic method has quite a different purpose than the analytic-synthetic method. While the latter is a method to acquire knowledge, the material axiomatic method is a method to present, justify, or teach an already acquired proposition. That the analytic-synthetic method and the material axiomatic method have two different aims is underlined by Cicero by stating that "all methodical treatment of rational discourse involves two arts, one of discovering" and one of demonstrating or "judging; Aristotle came first in both," while, on the contrary, the Stoics "pursued the art of judging diligently" but "completely neglected" the "art of discovering" (Cicero, *Topica*, 6).

12.20 The Formal Axiomatic Method

In the material axiomatic method, the axioms must be true. Another form of the axiomatic method is the formal axiomatic method. The latter is like the material axiomatic method, except that the axioms are not required to be true, but only consistent.

There need not be a system of things, specified in advance, for which the axioms are true. There may be several systems of things for which the axioms may turn out to be true. In that case, all the propositions deduced from the axioms will be true for each of the systems of things for which the axioms are true.

Like the purpose of the material axiomatic method, the purpose of the formal axiomatic method is not to obtain new knowledge, but only to justify and teach an already acquired proposition, because deduction produces its teaching through what is already known.

12.21 Original Formulation of the Formal Axiomatic Method

In the second half of the nineteenth century, several people contributed to formulating the formal axiomatic method (see, for example, Cellucci 1998, Chap. 5). The formal axiomatic method, however, is especially known through Hilbert's formulation.

Hilbert states that "every theory is only a scaffolding or schema of concepts together with their necessary relations to one another," and "the basic elements can be thought of in any way one likes" (Hilbert 1980a, 40). Therefore, "any theory can always be applied to infinitely many systems of basic elements" (ibid., 40–41). For example, in the case of geometry, "if in speaking of my points I think of some system of things, e.g. the system: love, law, chimney-sweep … and then assume all my axioms as relations between these things, then my propositions, e.g. Pythagoras' theorem, are also valid for these things" (ibid., 40). Similarly, "all statements of the theory of electricity are of course also valid for any other system of things which is substituted for the concepts magnetism, electricity … provided only that the requisite axioms are satisfied" (ibid., 41).

The only requirement on the axioms is that one has "to prove that they are not contradictory, that is, that a finite number of logical steps based upon them can never lead to contradictory results" (Hilbert 2000, 250). This requirement is a necessary one because, "if contradictory attributes be assigned to a concept," then "mathematically the concept does not exist. So, for example, a real number whose square is –1 does not exist mathematically" (ibid., 251). Conversely, "if it can be proved that the attributes assigned to the concept can never lead to a contradiction by application of a finite number of logical processes, I say that the mathematical existence of the concept" is "thereby proved" (ibid.). In particular, "in the case before us, where we are concerned with the axioms of real numbers in arithmetic, the proof of the compatibility of the axioms is at the same time the proof of the mathematical existence" (ibid.).

The purpose of "the axiomatic exploration of a mathematical truth" is not to find "new or more general propositions connected to that truth, but to determine the position of that" truth "within the system of known truths" in "such a way that it can be clearly said which conditions are necessary and sufficient for giving a foundation of that truth" (Hilbert 1902–1903, 50). Thus, the purpose of the axiomatic method is "the final presentation and the complete logical grounding of our knowledge," an aim for which "the axiomatic method deserves the first rank" (Hilbert 1996a, 1093). Therefore, the purpose of the axiomatic method is not to obtain new knowledge, but only to present and justify an already acquired mathematical truth, and to teach it through textbooks such as Hilbert's *Grundlagen der Geometrie*.

12.22 Motivations of the Formal Axiomatic Method

According to a widespread opinion, the change from the material axiomatic method to the formal axiomatic method was due to the creation of non-Euclidean geometries and abstract algebras in the nineteenth century. This opinion, however, is unjustified, because the creators of non-Euclidean geometries and abstract algebras assumed their theories to be true of systems of things specified in advance. Thus, Lobachevsky assumed his geometry to be true of physical bodies, because "in nature there are neither straight nor curved lines, neither plane nor curved surfaces;

we find in it only bodies" (Lobachevsky 1898–1899, I, 82). And "we actually measure surfaces and lines by means of bodies" (ibid., 80). Boole assumed his algebra to be true of "those operations of the mind by which reasoning is performed" (Boole 1958, 1).

The change from the material axiomatic method to the formal axiomatic method was rather due to a somewhat late impact of Romanticism on mathematics. A basic characteristic of Romanticism was the claim of an absolutely free creativity of the mind and its power to solve all problems. The impact of Romanticism on mathematics is apparent from the assertion of several mathematicians in the second half of the nineteenth century, that mathematical concepts and axioms are absolutely free creations of the human mind, subject only to the requirement of consistency.

Thus Cantor states that "the essence of mathematics lies precisely in its freedom" (Cantor 1996, 896). Mathematics is "in its development entirely free and is only bound in the self-evident respect that its concepts must" be "consistent with each other" (ibid.).

Dedekind states that "numbers are free creations of the human mind" (Dedekind 1996a, 791). They are "something new" which "the spirit creates. We are of a divine race and, without any doubt, possess creative power not merely in material things (railroads, telegraphs) but especially in mental things" (Dedekind 1996b, 835). Our creative power is only subject to the condition that, after we have freely created a system of numbers through a system of axioms, the question arises: "Does such a system exist at all in the realm of our ideas? Without a logical proof of existence," namely, of consistency of the axioms, "it would always remain doubtful whether the notion of such a system might not perhaps contain internal contradictions. Hence the need for such proofs" (Dedekind 1967, 101).

Poincaré states that the facts with which the mathematician is concerned are not independent of him, but it is himself, one would almost say "his caprice, that creates these facts" (Poincaré 2013, 370). The latter are only subject to the condition that they must be consistent with each other, because "in mathematics the word exist can have only one meaning, it means free from contradiction" (ibid., 454).

Thus, in the second half of the nineteenth century, the Romantic view that mathematical concepts and axioms are absolutely free creations of the human mind, subject only to the requirement of consistency, gained acceptance. The Romantic view found a natural expression in the formal axiomatic method, which nurtured the Romantic dream that all mathematical problems could be solved through it.

This is apparent from Hilbert, who states that the mathematician must "be free to do as" he pleases "in giving characteristic marks" (Hilbert 1980a, 39). The formal axiomatic method "guarantees the maximum flexibility in research" (Hilbert 1996c, 1120). For, in the formal axiomatic method, the axioms can be taken quite arbitrarily, only subject to the condition that they must be consistent. Since the formal axiomatic method guarantees the maximum flexibility in research, we have "the firm conviction" that, "however unapproachable" mathematical "problems may seem to us and however helpless we stand before them," their "solution must follow by a finite number of logical processes" (Hilbert 2000, 248). So "for the mathematician there is no *ignorabimus*," nor "for any part of natural science" (Hilbert 1996f,

1165). Indeed, "there are absolutely no unsolvable problems. Instead of the foolish *ignorabimus*, our answer is on the contrary: We must know, we shall know" (ibid.).

The Romantic dream, however, was short-lived. In 1931 Gödel published his incompleteness results which showed the limitations of the formal axiomatic method. This was the end of the Romantic dream in mathematics.

12.23 The Axiomatic Method

The material axiomatic method and the formal axiomatic method are essentially the same method, except for the requirement on the axioms. In the material axiomatic method, the axioms are required to be true, in the sense that there must be a system of things, specified in advance, for which the principles are true. In the formal axiomatic method, the axioms are only required to be consistent.

However, being consistent may be considered to be a weak sense of being true. In fact, Hilbert states that "'consistent' is identical to 'true'" (Hilbert 2013, 987). From this point of view, it can be said that even in the formal axiomatic method the axiom are required to be true, although not in the strong sense that there must be a system of things, specified in advance, for which the axioms are true, but only in the weak sense that the axioms must be consistent.

On this basis, instead of speaking of the material axiomatic method and the formal axiomatic method, in what follows we will simply speak of the axiomatic method. We will assume that the axiomatic method is the method according to which, in order to justify and teach an already acquired proposition of a given field, one deduces the proposition from the axioms of that field, where the axioms must be true, either in the strong sense that there must be a system of things, specified in advance, for which the principles are true, or in the weak sense that they must be consistent. Which form of the axiomatic method is being referred to will be clear from the context.

References

Bacon, Francis. 1961–1986. *Works*. Stuttgart Bad Cannstatt: Frommann Holzboog.

Boole, George. 1958. *An investigation of the laws of thought on which are founded the mathematical theories of logic and probabilities*. Mineola: Dover.

Cambiano, Giuseppe. 1967. Il metodo ipotetico e le origini della sistemazione euclidea della geometria. *Rivista di Filosofia* 58: 115–149.

Cantor, Georg. 1996. Foundations of a general theory of manifolds: A mathematico-philosophical investigation into the theory of the infinite. In *From Kant to Hilbert: A source book in the foundations of mathematics*, ed. William Ewald, vol. 2, 881–920. Oxford: Oxford University Press.

Cellucci, Carlo. 1998. *Le ragioni della logica*. Rome: Laterza.

———. 2013a. *Rethinking logic: Logic in relation to mathematics, evolution, and method*. Dordrecht: Springer.

Dedekind, Richard. 1967. Letter to Keferstein, 27 February 1890. In *From Frege to Gödel: A source book in mathematical logic, 1879–1931*, ed. Jean van Heijenoort, 99–103. Cambridge: Harvard University Press.
———. 1996a. Was sind und was sollen die Zahlen? In *From Kant to Hilbert: A source book in the foundations of mathematics*, ed. William Ewald, vol. 2, 790–833. Oxford: Oxford University Press.
———. 1996b. Letter to Heinrich Weber, 24 January 1888. In *From Kant to Hilbert: A source book in the foundations of mathematics*, ed. William Ewald, vol. 2, 834–835. Oxford: Oxford University Press.
De Morgan, Augustus. 1835. Review of George Peacock, *A treatise on algebra*. *Quarterly Journal of Education* 9: 91–110, 293–311.
Descartes, René. 1996. *Oeuvres*. Paris: Vrin.
Doxiadis, Apostolos, and Michalis Sialaros. 2013. Sing, Muse, of the hypotenuse: Influences of poetry and rhetoric on the formation of Greek mathematics. In *Writing science: Medical and mathematical authorship in ancient Greece*, ed. Markus Asper, 367–409. Berlin: De Gruyter.
Doyle, Arthur Conan. 2003. *Sherlock Holmes: The complete novels and stories*. Volume I. New York: Bantham.
Dummett, Michael. 1991. *The logical basis of metaphysics*. Cambridge: Harvard University Press.
Frankfurt, Harry. 1958. Peirce's notion of abduction. *The Journal of Philosophy* 55: 593–597.
Frege, Gottlob. 1960. *The foundations of arithmetic: A logico-mathematical enquiry into the concept of number*. New York: Harper.
Gödel, Kurt. 1986–2002. *Collected works*. Oxford: Oxford University Press.
Hilbert, David. 1902–1903. Über den Satz von der Gleichheit der Basiswinkel im gleichschenkligen Dreieck. *Proceedings of the London Mathematical Society* 35: 50–68.
———. 1980a. Letter to Frege, 29 December 1899. In Gottlob Frege, *Philosophical and mathematical correspondence*, ed. Gottlob Frege, 38–41. Oxford: Blackwell.
———. 1996a. On the concept of number. In *From Kant to Hilbert: A source book in the foundations of mathematics, II*, ed. William Ewald, 1092–1095. Oxford: Oxford University Press.
———. 1996c. The new grounding of mathematics: First report. In *From Brouwer to Hilbert, II*, ed. William Ewald, 1117–1134. Oxford: Oxford University Press.
———. 1996f. Logic and the knowledge of nature. In *From Kant to Hilbert: A source book in the foundations of mathematics*, II, ed. William Ewald, 1157–1165. Oxford: Oxford University Press.
———. 2000. Mathematical problems. Appendix to Jeremy Gray, *The Hilbert challenge*, 240–282. Oxford: Oxford University Press.
———. 2013. Beweis des Tertium non datur. In David Hilbert, *Lectures on the foundations of arithmetic and logic 1917–1933*, ed. William Ewald, and Wilfried Sieg, 985–990. Berlin: Springer.
Hintikka, Jaakko, and Unto Remes. 1976. Ancient geometrical analysis and modern logic. In *Essays in memory of Imre Lakatos*, ed. Robert Cohen, Paul Feyerabend, and Max Wartofsky, 253–276. Dordrecht: Springer.
Hippocrates. 2005. *On ancient medicine*. Leiden: Brill.
Hoffmann, Michael. 1999. Problems with Peirce's concept of abduction. *Foundations of Science* 4: 271–305.
Husserl, Edmund. 1970. *The crisis of European sciences and transcendental phenomenology*. Evanston: Northwestern University Press.
Kant, Immanuel. 1998. *Critique of pure reason*. Cambridge: Cambridge University Press.
———. 2002. *Theoretical philosophy after 1781*. Cambridge: Cambridge University Press.
Lakatos, Imre. 1978. *Philosophical papers*. Cambridge: Cambridge University Press.
Lewis, David. 1996. Elusive knowledge. *Australasian Journal of Philosophy* 74: 549–567.
Lobachevsky, Nikolai. 1898–1899. *Zwei geometrische Abhandlungen*. Leipzig: Teubner.
Menn, Stephen. 2002. Plato and the method of analysis. *Phronesis* 47: 193–223.
Mill, John Stuart. 1963–1986. *Collected works*. Toronto: University of Toronto Press.

Patokorpi, Erkki. 2007. Logic of Sherlock Holmes in technology enhanced learning. *Educational Technology & Society* 10 (1): 171–185.

Peirce, Charles Sanders. 1931–1958. *Collected papers*. Cambridge: Harvard University Press.

Poincaré, Henri. 2013. *The foundations of science: Science and hypothesis – The value of science – Science and method*. Cambridge: Cambridge University Press.

Pólya, George. 1971. *How to solve it: A new aspect of mathematical method*. Princeton: Princeton University Press.

Proclus. 1992. *In primum Euclidis Elementorum librum commentarii*. Hildesheim: Olms.

Ribet, Kenneth A. 1990. From the Taniyama-Shimura conjecture to Fermat's last theorem. *Annales de la Faculté Mathematiques et des Sciences de Toulouse* 11: 116–139.

Suppes, Patrick. 1993. *Models and methods in the philosophy of science: Selected essays*. Dordrecht: Springer.

Wittgenstein, Ludwig. 2002. *Tractatus Logico-Philosophicus*. London: Routledge.

Chapter 13
Modelling Scientific Knowledge

Abstract The various methods to acquire knowledge are the basis of alternative models of science. With regard to science, one may speak of models in different senses, but the two main ones are models of science and models in science. Models of science are representations of how scientists build their theories. The chapter considers four models of science: the analytic-synthetic model, the hypothetico-deductive model, the semantic model, and the analytic model. Models in science are representations of empirical objects, phenomena, or processes of some area of science. The chapter discusses to what extent each of the above four models of science is capable of accounting for models in science.

13.1 Models of Science and Models in Science

The analytic method, the analytic-synthetic method, and the axiomatic method, considered in Chap. 12, are the basis of alternative models of science.

With regard to science, one may speak of models in different senses. The two main ones are models of science and models in science. A model of science is a representation of how scientists build their theories. A model in science is a representation of empirical objects, phenomena, or processes of some area of science.

This chapter considers four models of science: the analytic-synthetic model, the hypothetico-deductive model, the semantic model, and the analytic model. Then it briefly discusses to what extent each of these models of science is capable of accounting for models in science.

13.2 The Analytic-Synthetic Model

According to the analytic-synthetic model of science, a distinction must be made between science in the making, on the one hand, and finished science, that is, science in finished form, on the other hand. Science in the making is based on the analytic-synthetic method, finished science on the axiomatic method.

According to the analytic-synthetic model of science, there is a substantial difference between science in the making and finished science. Both science in the

© Springer International Publishing AG 2017

C. Cellucci, *Rethinking Knowledge*, European Studies in Philosophy of Science 4,
DOI 10.1007/978-3-319-53237-0_13

making and finished science involve principles which must be true. However, until one arrives at hypotheses which are principles, one uses hypotheses that are not known to be true but only plausible, so science in the making need not consist of truths only. Conversely, when deductions from principles have been found for all solution of problems of the science in question, the latter will consist entirely of truths. When this stage is reached, science becomes a finished science, ready to be presented in treatises such as Euclid's *Elements*.

13.3 The Analytic-Synthetic Model and Modern Science

The analytic-synthetic model is Aristotle's model of science, but is also followed by the creators of modern science.

It is widely believed that "modern science owes its origins and present flourishing state to a new scientific method which was fashioned almost entirely by Galileo Galilei" (Kline 1985, 284). In particular, "Galileo violates important rules of scientific method which were invented by Aristotle" (Feyerabend 1975, 297).

This belief, however, is unfounded, because the creators of modern science, from Galileo to Newton, did not fashion a new scientific method that violated important rules of scientific method which were invented by Aristotle. On the contrary, they followed Aristotle's analytic-synthetic model of science.

Indeed, Galileo states that the Aristotelians misunderstand Aristotle when they claim that he first laid the foundation of a conclusion "by means of natural, evident, and clear principles," hence by synthesis, and "afterwards he supported the same" conclusion "by the senses" (Galilei 1968, VII, 75). This, Galileo points out, "is the method by which he has written his doctrine," not "the one by which he investigated it" (ibid.). On the contrary, Aristotle "first procured, by way of senses, experiments and observations," hence by analysis, "to assure himself as much as possible of the conclusion," and only "afterwards he sought means to demonstrate it" (ibid.). Hence, to establish it by synthesis. Indeed, "when the conclusion is true, by making use of the resolutive method," namely analysis, "one may easily encounter some proposition which is already demonstrated, or arrive at some principle which is known by itself" (ibid.). From the certainty of these propositions, "the truth of our" conclusion "will acquire strength and certainty" (ibid., VII, 435). Galileo follows this method, that is, the method that "Aristotle teaches us in his Dialectic," and on this basis he may claim to be "truly Peripatetic, that is, an Aristotelian philosopher" (ibid., XVIII, 248).

Newton states that "as in mathematicks, so in natural philosophy, the inquiry of difficult things by the method of analysis, ought ever to precede the method" of synthesis, or "composition. This analysis consists in making experiments and observations, and in drawing general conclusions from them by induction" (Newton 1952, 404). By this way of analysis we may proceed "from effects to their causes, and from particular causes to more general ones, till the argument end in the most general. This is the method of analysis: and the synthesis consists in assuming the causes discovered, and established as principles, and by them explaining the phaenomena proceeding from them, and proving the explanations" (ibid., 404–405). Science in the making is based on this procedure. On the other hand, finished science is based on synthesis, because finished science is only concerned with showing how the causes, discovered and established as principles, "may be assumed in the method of composition," namely in synthesis, "for explaining the phaenomena arising from them" (ibid., 405).

From this it is clear that both Galileo and Newton followed Aristotle's analytic-synthetic model of science.

13.4 The Fading Out of Analysis

Despite the fact that the creators of modern science, from Galileo to Newton, followed Aristotle's analytic-synthetic model of science, in the period from the eighteenth century to the end of the nineteenth century the analytic part of the analytic-synthetic model fades out. At the origin of this there are at least four factors.

The first factor is Romanticism, which exalts intuition and genius. Thus Novalis states that scientific discoveries "are leaps – (intuitions, resolutions)" and products "of the genius – of the leaper *par excellence*" (Novalis 2007, 28). In human beings, there is "a powerful intuition of creative wilfulness, of boundlessness, of infinite diversity, of sacred originality and the omnipotence of inner humanity" (Novalis 1996, 74). This powerful intuition, first of all, manifests itself in mathematics, because the mathematician, when creating, "is *en état de Createur absolu*" (Novalis 2007, 140). In fact, "mathematics is genuine science – because it teaches one to be a genius" (ibid., 186). In mathematics, "a true method of progressing synthetically is the main thing," because this is the "method of the divinatory genius" (ibid., 100). Indeed, "genius is the synthesizing principle" (ibid., 215). The synthetic method gives "the regulation of genius" (ibid., 164). Admittedly, "the synthetic method" is "the freezing, wilting, crystallizing, structuring and successive method. The analytic method in contrast, is a warming, dissolving and liquefying method" (ibid., 175). Nevertheless, the synthetic method is the true method, because only it permits to build a system in an absolutely free way, so "the true philosopher has a synthetic method" (ibid., 73).

The second factor is the development, in the nineteenth century, of theories which employ unobservable entities and processes, and hence cannot be derived from observation. Thus Whewell states that "an art of discovery is not possible. At each step of the investigation are needed invention, sagacity, genius – elements which no art can give. We may hope in vain, as Bacon hoped, for an Organon which shall enable all men to construct scientific truths" (Whewell 1858, 5). Indeed, "scientific discovery must ever depend upon some happy thought, of which we cannot trace the origin – some fortunate cast of intellect, rising above all rules. No maxims can be given which inevitably lead to discovery. No precepts will elevate a man of ordinary endowments to the level a man of genius" (Whewell 1847, II, 20–21).

The third factor are the foundational problems of the infinitesimal calculus, which suggest that justification rather than discovery is the urgent question. Thus Frege states that "considerable, almost insuperable, difficulties stood in the way of any rigorous treatment" of the infinitesimal calculus, in particular "the concepts of function, of continuity, of limit and of infinity have been shown to stand in need of sharper definition" (Frege 1960, 1). So, today there is need for "rigour of proof, precise delimitation of extent of validity, and as a means to this, sharp definition of concepts" (ibid.). Therefore, logic must not concern itself with the question of discovery, but rather with the question of "how we can provide" a judgment "with the most secure foundation" (Frege 1967, 5).

The fourth factor is the opinion that a logic of discovery should provide an algorithmic method for problem solving. This factor has been already considered in Chap. 11.

The fading out of the analytic part of the analytic-synthetic method leads to a new model of science, the hypothetico-deductive model.

13.5 The Hypothetico-Deductive Model

According to the hypothetico-deductive model of science, science is based on the axiomatic method. Building a scientific theory is a matter of choosing certain hypotheses and deducing consequences from them. The hypotheses can be chosen arbitrarily, but are subject to the condition that they must be consistent with the total system of hypotheses. Moreover, they can and must be tested by comparing the consequences deduced from them with the observational and experimental data.

Thus Carnap states that building a scientific theory about a physical process is "a matter of deducing the concrete sentence which describes the process" from hypotheses consisting of "laws and other concrete sentences" (Carnap 2001, 320). There is "great freedom in the introduction" of hypotheses, or "primitive sentences" (ibid., 322). But "every hypothesis must be compatible with the total system of hypotheses" (ibid., 320). That is, every hypothesis must be consistent with the total system of hypotheses. Moreover, "the hypotheses can and must be tested by experience" (ibid.). That is, the hypotheses can and must be tested by comparing the sentence which describes the process with the observational and experimental data.

13.6 The Hypothetico-Deductive Model and Closed Systems

According to the hypothetico-deductive model, a scientific theory is a closed system. This means that the development of the theory remains completely internal to the theory, it involves no interaction with other theories, hence a scientific theory is a self-sufficient totality. The development of the theory entirely consists in deducing propositions from the hypotheses.

The propositions deduced from the hypotheses contain nothing essentially new with respect to the hypotheses, because deduction is non-ampliative, it simply makes explicit what is implicitly contained in the hypotheses. As Hempel states, deduction simply makes "explicit what is implicitly contained in a set of premisses. The conclusions to which" deduction leads "assert nothing that is theoretically new in the sense of not being contained in the content of the premisses" (Hempel 2001, 14). The conclusions can only be "psychologically new," in the sense that we were not aware of their being implicitly contained in the premisses, so we were not aware of "what we committed ourselves to in accepting a certain set of assumptions or assertions" (ibid.).

13.7 The Analytic-Synthetic Model, the Hypothetico-Deductive Model and Gödel's Incompleteness Theorems

The analytic-synthetic model and the hypothetico-deductive model are incompatible with Gödel's incompleteness theorems.

By Gödel's first incompleteness theorem, for any consistent, sufficiently strong, formal system, there is a sentence of the system which is true but cannot be deduced from the axioms of the system. The analytic-synthetic model is incompatible with Gödel's result because, according to it, all true sentences of a scientific theory must be deducible from the hypotheses of the theory. On the other hand, the hypothetico-deductive model is incompatible with Gödel's result because, according to it, the hypotheses of a scientific theory must solve all problems of the theory.

By Gödel's second incompleteness theorem, for any consistent, sufficiently strong, formal system, it is impossible to show, by absolutely reliable means, that the axioms of the system are consistent, let alone that they are true. The analytic-synthetic model is incompatible with Gödel's result because, according to it, the hypotheses of a scientific theory must be true. On the other hand, the hypothetico-deductive model is incompatible with Gödel's result because, according to it, the hypotheses of a scientific theory must be consistent.

13.8 An Alleged Way Out of Incompleteness

It might be objected that the hypothetico-deductive model may be viable if we assume that scientific knowledge is not represented by a single axiom system, but rather by a growing sequence of axiom systems.

Thus Curry states that Gödel's first incompleteness theorem only entails that "the concept of intuitively valid proof cannot be exhausted by any single formalization"

(Curry 1977, 15). But scientific knowledge can be represented by a growing sequence of formalizations. With such sequence, "proof is precisely that sort of growing thing which the intuitionists have postulated for certain infinite sets" (ibid.).

This argument, however, is not valid because the concept of proof as a growing thing is incompatible with the analytic-synthetic model and the hypothetico-deductive model. Indeed, in both such models, proof is a fixed thing. Each member of the growing sequence of formalizations is a step in such sequence and, as Gödel argues, "there cannot exist any formalism which would embrace all these steps" (Gödel 1986–2002, II, 151). But the existence of such a formalism is necessary if proof is to be a fixed thing.

13.9 Other Limitations of the Hypothetico-Deductive Model

In addition to being incompatible with Gödel's incompleteness theorems, the hypothetico-deductive model has other limitations. First, it leaves to one side the crucial issue of how to find hypotheses, it merely asserts that, to find them, "creative ingenuity is required" (Carnap 1966, 33). But this is a non-explanation, it completely evades the issue.

Moreover, it may happen that the observational and experimental data may confirm not only our hypotheses, but also other hypotheses which are incompatible with them. The hypothetico-deductive model has no criterion to assert that the test confirms our hypotheses in preference to the other hypotheses.

Furthermore, the hypothetico-deductive model is incapable of accounting for the process of theory change – the process through which a theory is replaced with another one. For, according to it, a theory has no rational connection with the preceding one, except that it agrees with more observational and experimental data. Thus, the hypothetico-deductive model leaves to one side not only the crucial issue of the discovery of hypotheses, but also the equally crucial issue of the process of theory change.

13.10 The Semantic Model

In the second half of the twentieth century, the support for the hypothetico-deductive model declines, and the hypothetico-deductive model is gradually replaced by the semantic model. This is the model of science underlying semantic structural realism, already discussed in Chap. 7.

According to the semantic model, to formulate a scientific theory about certain phenomena is to specify a family of models. The concept of model is supposed to be the same in mathematics and the empirical sciences. A model is a structure, consisting of a set along with a collection of operations and relations that are defined on it. A scientific theory is adequate if it has a model which is isomorphic to the phenomena that it is intended to theorize.

The semantic model, however, has some serious limitations. It leaves to one side the crucial issue of how to discover models. For, van Fraassen states that theories cannot be obtained by any process of systematic inference, "all those successes of science which so many people have thought must have been produced by induction or abduction were" in fact "initially good guesses under fortunate circumstances;" then they "were made effective by means of the precise formulation and disciplined teasing out of their implications through logic and mathematics" (van Fraassen 2000, 275). Thus, "if our pursuit of knowledge" is "to be successful, we must be lucky – we have no way to constrain such fortune" (ibid., 273).

In addition to leaving to one side the crucial issue of how to find models, the semantic model has all the other limitations of semantic structural realism that have been already pointed out in Chap. 7.

13.11 The Analytic Model

An alternative to the analytic-synthetic model, the hypothetico-deductive model, and the semantic model, is the analytic model. According to it, science is based on the analytic method.

Unlike the analytic-synthetic model, the analytic model assumes that analysis is not a finite process, terminating with principles, but a potentially infinite process, leading to ever more general hypotheses, so it is an unending quest. Therefore, unlike the analytic-synthetic model, the analytic model does not distinguish between science in the making and finished science. All science is always science in the making.

Moreover, unlike the hypothetico-deductive model and the semantic model, the analytic model deals with the issue of how to find hypotheses.

Furthermore, the analytic model establishes a rational connection between subsequent theories, because the hypotheses of the new theory are formulated through an analysis of the reasons why the hypotheses of the preceding theory are no longer plausible.

13.12 The Analytic Model and Open Systems

According to the analytic model, a scientific theory is an open system. This means that the development of the theory need not remain completely internal to the theory, but may involve interactions with other theories, so a scientific theory is not a self-sufficient totality.

No system of hypotheses may solve all the problems of a given field, any such system is inherently incomplete and must appeal to other systems to bridge its gaps.

Therefore, the hypotheses of the theory are not given once and for all, and developing the theory need not consist merely in deducing consequences from them. It may involve replacing the hypotheses with more general ones, obtained through interactions with other scientific theories, according to a potentially infinite process. (For more on these concepts of closed system and open system, see Cellucci 1998, Chaps. 6, 7, and 9).

13.13 The Analytic Model and Gödel's Incompleteness Theorems

While the analytic-synthetic model and the hypothetico-deductive model are incompatible with Gödel's incompleteness theorems, the analytic model is compatible with Gödel's results, and is even supported by them.

For, according to the analytic model, no system of hypotheses can solve all problems of a given field. The hypotheses are bound to be replaced sooner or later with other more general ones through a potentially infinite process, since every system of hypotheses is incomplete and needs to appeal to other systems to bridge its gaps. Thus, the analytic model is supported by Gödel's first incompleteness theorem.

Moreover, according to the analytic model, the hypotheses for the solution to a problem are not definitive, true and certain, but only provisional, plausible and uncertain. Thus, the analytic method is supported by Gödel's second incompleteness theorem.

13.14 An Example of Use of the Analytic Model

An example of use of the analytic model is Darwin's building of the theory of evolution by natural selection.

The problem Darwin wanted to solve by his theory was to explain the characteristics of existing living things, and how these characteristics came to be. His theory was based on two hypotheses: 1) Natural selection produced different species of animals and plants. 2) As more individuals of any species are produced than can possibly survive, there must be a struggle for existence.

Now, Darwin arrived at these two hypotheses through an analogy and an induction. On the one hand, Darwin arrived at hypothesis 1) through an analogy between artificial selection and natural selection. His starting point was that breeders used artificial selection to produce different breeds of animals and plants. From this, by analogy, Darwin inferred that nature used natural selection to produce different species of animals and plants. (An inference by analogy is one by which, if *a* is similar to *b* in certain respects, and *a* has a certain property, then *b* will also have that property). On the other hand, Darwin arrived at hypothesis 2) through an induction. His starting point was Malthus' observation that, as more human beings are produced than can possibly survive, there must be a struggle for existence. From this, by induction, Darwin inferred that, as more individuals of any species are produced than can possibly survive, there must be a struggle for existence. (An inference by induction is one by which, if a number of things of a certain kind have a certain property, all things of that kind will have that property).

In fact, Darwin states that he "came to the conclusion that selection was the principle of change from the study of domesticated productions" (Darwin 1903, I, 118). Namely, he came to that conclusion "from what artificial selection had done for domestic animals" (Darwin 2009b, II, 118). On the other hand, he came to the conclusion that, in any species, "as more individuals are produced than can possibly survive, there must in every case be a struggle for existence," from "the doctrine of Malthus applied with manifold force to the whole animal and vegetable kingdoms" (Darwin 2009a, 50).

Thus Darwin arrived at the hypotheses upon which his theory was based through an inference by analogy and an inference by induction.

13.15 The Neglect of the Analytic Model

Despite its advantages, in the past century the analytic model has been generally neglected. One of its few supporters has been Pólya, according to whom scientific hypotheses are found "by plausible reasoning" (Pólya 1954, I, v). Contrary to deductive reasoning from axioms, which is "safe, beyond controversy, and final," plausible reasoning is "hazardous, controversial, and provisional" (ibid.). On the other hand, deductive reasoning from axioms is "incapable of yielding essentially new knowledge about the world around us. Anything new that we learn about the world involves plausible reasoning" (ibid.). The latter is "the kind of reasoning on which" the mathematician's "creative work will depend," since to discover hypotheses one has "to combine observations and follow analogies" (ibid., I, vi). Thus Pólya shares the basic idea of the analytic model, that scientific hypotheses are obtained by logical procedures such as induction, analogy, metaphor, etc.

However, Pólya limits the scope of the analytic model, because he reduces plausibility to probability. Indeed, as we have seen in Chap. 9, he claims that we can use the calculus of probability to render more precise our views on plausible reasoning, since the calculus of plausibilities obeys the same rules as the calculus of probabilities. This claim is unjustified because, as already argued, plausibility is essentially different from probability. Pólya also limits the scope of the analytic model because, while stating that plausible reasoning is the kind of reasoning on which the mathematician's creative work will depend, he also states that deductive reasoning from axioms is the mathematician's "profession and the distinctive mark of his science" (ibid.). Moreover, Pólya claims that deductive reasoning from axioms is safe, beyond controversy, and final. This is incompatible with Gödel's second incompleteness theorem.

That, in the past century, the analytic model has been generally neglected does not mean, however, that it has not been tacitly used. As we have seen in Chap. 12, an example of this is the solution of Fermat's problem.

13.16 Models in Science

After considering models of science, we now briefly consider models in science. As already mentioned, a model in science is a representation of empirical objects, phenomena, or processes.

It is out of question that there is an optimal model in science. In the seventeenth century France, the minister Colbert charged the astronomer Gian Domenico Cassini to make an extremely detailed map of France. The map was the work of four different generations of the Cassini family, and is so detailed that, as Calvino says, "every forest in France is drawn tree by tree, every church has its bell-tower, every village is drawn roof by roof, so that one has the dizzying feeling that beneath one's eyes are all the trees and all the bell-towers and all the roofs of the Kingdom of France" (Calvino 2014, 23).

Perhaps this inspired Borges' story about an empire in which "the craft of cartography attained such perfection" that "the college of cartographers evolved a map of the empire that was of the same scale as the empire and that coincided with it point for point" (Borges 1975, 131).

A Lewis Carroll's story goes even further. One of the characters of the story says that in his kingdom they had "the grandest idea of all" about mapmaking, that is, to make "a map of the country, on the scale of a mile to the mile" (Carroll 1996, 556). But "the farmers objected: they said it would cover the whole country, and shut out the sunlight! So we now use the country itself, as its own map, and I assure you it does nearly as well" (ibid., 557).

Carroll's story suggests that there cannot be any optimal model in science, because the optimal model of reality is reality itself. However, contrary to what the story's character says, using reality itself as its own model would not do nearly as well. As Boltzmann states, "no theory can be objective, actually coinciding with nature," but "each theory is only a mental picture of phenomena" (Boltzmann 1974, 90–91).

In science there cannot be any optimal model, only models suited to particular purposes. Such are physical models, scale models, analogical models, mathematical models, just to name a few. It would be impossible to discuss all of them here. Instead, in what follows it is argued that not all models of science are equally capable of accounting for models in science.

13.17 The Analytic-Synthetic Model, the Hypothetico-Deductive Model and Models in Science

The analytic-synthetic model and the hypothetico-deductive model are incapable of accounting for models in science. For, according to them, a finished science consists of deductions from hypotheses, so use of models in science is inessential and can be eliminated. At best, it can only have a didactic or heuristic value in science in the making.

Thus, Carnap states that "it is important to realize that the discovery of a model has no more than an esthetic or didactic or at best a heuristic value, it is not at all essential for a successful application of the physical theory" (Carnap 1939, 68).

This conflicts with the fact that models have had a central role in the development of science. In particular, much of the current work of scientists consists in devising, testing, comparing and revising models.

The hypothetico-deductive model conflicts with mathematical practice also in another way. According to it, one must always be able to consider arbitrary models of axioms. However, in the actual practice of science, one does not consider arbitrary models of axioms, but only specific ones. For example, one never considers a model of geometry such as the system of love, law, chimney-sweep mentioned by Hilbert (see Chap. 12). In practice, there are always reasons for considering a model of the axioms rather than another one, and these reasons do not depend on the hypothetico-deductive model but are external to it. Therefore, the decision to consider a model of the axioms rather than another one cannot be justified in terms of the hypothetico-deductive model.

13.18 The Semantic Model and Models in Science

One would have thought that, unlike the analytic-synthetic model and the hypothetico-deductive model, instead the semantic model would be capable of accounting for models in science. For, according to it, formulating a scientific theory about certain phenomena means specifying a family of models.

But it is not so. From what has been said in Chap. 7, it is apparent that the semantic model is unable to account for the relation between a model, that is, a structure, and the phenomena. Moreover, the semantic model puts emphasis on the static aspect of physical systems, namely their structure, while physical systems have both structural and behavioural properties. Behaviour refers to state transitions and dynamic properties – operations and their relationships. Models should be able to express how and when changes occur to entities and relate with one another, but structures are unable to express that.

13.19 Scientific Realism and Models in Science

In particular, scientific realism is incapable of accounting for models in science. For, according to it, the aim of science is to have true theories about the world, where 'true' is understood in the classical correspondence sense. But, as argued in Chap. 8, the concept of truth as correspondence does not provide a criterion of truth. So, under scientific realism, the relation between theories and the world becomes a mysterious one, and models in science become useless, since it is generally impossible to ascertain that they are models of reality.

This reminds one of Calvino's story about the city of Eudoxia, where "a carpet is preserved in which you can observe the city's true form" (Calvino 1974, 96). Although "at first sight nothing seems to resemble Eudoxia less than the design of that carpet," nonetheless "each place in the carpet corresponds to a place in the city and all the things contained in the city are included in the design, arranged according to their true relationship, which escapes your eye" (ibid.). But the carpet and the city of Eudoxia are very dissimilar, therefore "an oracle was questioned about the mysterious bond between two objects so dissimilar" (ibid., 97). The oracle replied that "one of the two objects" has "the form the gods gave the starry sky and the orbits in which the worlds revolve; the other is an approximate reflection, like every human creation" (ibid.). The oracle's reply was interpreted as meaning that "the carpet's harmonious pattern" was a map of the universe, "but you could, similarly, come to the opposite conclusion: that the true map of the universe is the city of Eudoxia, just as it is" (ibid.).

The carpet and the city of Eudoxia are a metaphor of scientific theories and the universe, respectively. Scientific realism assumes that in scientific theories you can observe the universe's true form, and hence that scientific theories are true models of the universe. But, since there is no criterion of truth, this assumption is unfounded. Then, you could also come to the opposite conclusion: that, from the perspective of scientific realism, the true model of the universe is the universe itself, just as it is.

13.20 The Analytic Model and Models in Science

The analytic model is capable of accounting for models in science. According to it, solving a scientific problem involves formulating hypotheses. Now, while many hypotheses in science are expressed using sentences in language, many other hypotheses are expressed using models. Thus models are not ancillary to doing science, but central to the solution of scientific problems. A model is the hypothesis that certain properties of the world can be represented in a certain way for certain purposes.

While, according to the semantic model of science, models are structures, according to the analytic model of science they can be a wide range of things, including words, equations, diagrams, graphs, photographs, computer-generated images, dynamic entities, etc. Following Goodman, it could even be said that, according to the analytic model of science, a model can be "almost anything from a naked blonde to a quadratic equation" (Goodman 1976, 171). From the viewpoint of the analytic model of science, the question of isomorphism does not arise, because a model is only the hypothesis that certain properties of the world can be represented in a certain way for certain purposes. Therefore, the analytic model of science is capable of accounting for models in science.

That, unlike the hypothetico-deductive model of science and the semantic model of science, the analytic model of science is capable of accounting for models in science, justifies the claim that not all models of science are equally capable of accounting for models in science. Science is a more complex process than the hypothetico-deductive model or the semantic model suggest. To account for science it is necessary to account for theory formation and theory change.

Calvino states that "knowledge always proceeds via models, analogies, symbolic images, which help us to understand up to a certain point; then they are discarded, so we turn to other models," other analogies, other symbolic "images" (Calvino 2014, 119). Only the analytic model of science is capable of accounting for this dynamic character of scientific knowledge.

References

Boltzmann, Ludwig. 1974. *Theoretical physics and philosophical problems: Selected writings*. Dordrecht: Springer.

Borges, Jorge Luis. 1975. Of exactitude in science. In Jorge Luis Borges, *A universal history of infamy*, 131. London: Penguin Books.

Calvino, Italo. 1974. *Invisible cities*. New York: Harcourt Brace Jovanovich.

———. 2014. *Collection of sand*. New York: Houghton Mifflin Harcourt.

Carnap, Rudolf. 1939. *Foundations of logic and mathematics*. Chicago: Chicago University Press.

Carnap, Rudolph. 1966. *Philosophical foundations of physics: An introduction to the philosophy of science*. New York: Basic Books.

Carnap, Rudolf. 2001. *Logical syntax of language*. London: Routledge.

Carroll, Lewis. 1996. *Sylvie and Bruno concluded*. In *The complete illustrated Lewis Carroll*, 457–674. Ware: Wordsworth.

Cellucci, Carlo. 1998. *Le ragioni della logica*. Rome: Laterza.

Curry, Haskell Brooks. 1977. *Foundations of mathematical logic*. Mineola: Dover.

Darwin, Charles. 1903. *More letters of Charles Darwin*. London: Murray.

———. 2009a. *The origin of species by means of natural selection*. Cambridge: Cambridge University Press.

———. 2009b. *The life and letters of Charles Darwin, including an autobiographical chapter*. Cambridge: Cambridge University Press.

Feyerabend, Paul K. 1975. Machamer on Galileo. *Studies in History and Philosophy of Science* 5: 297–304.

Frege, Gottlob. 1960. *The foundations of arithmetic: A logico-mathematical enquiry into the concept of number*. New York: Harper.

———. 1967. *Begriffsschrift*, a formula language, modeled upon that of arithmetic, for pure thought. In *From Frege to Gödel: A source book in mathematical logic, 1879–1931*, ed. Jean van Heijenoort, 5–82. Cambridge: Harvard University Press.

Galilei, Galileo. 1968. *Opere*. Florence: Barbera.
Gödel, Kurt. 1986–2002. *Collected works*. Oxford: Oxford University Press.
Goodman, Nelson. 1976. *Languages of art: An approach to a theory of symbols*. Indianapolis: Hackett.
Hempel, Carl Gustav. 2001. *The philosophy of Carl G. Hempel*. Oxford: Oxford University Press.
Kline, Morris. 1985. *Mathematics for the nonmathematician*. Mineola: Dover.
Newton, Isaac. 1952. *Opticks, or a treatise of the reflections, refractions, inflections & colours of light*. Mineola: Dover.
Novalis. 1996. Christianity or Europe. A fragment. In *The early political writings of the German romantics*, ed. Frederick Beiser, 59–79. Cambridge: Cambridge University Press.
———. 2007. *Notes for a romantic encyclopedia: Das Allgemeine Brouillon*. Albany: State University of New York Press.
Pólya, George. 1954. *Mathematics and plausible reasoning*. Princeton: Princeton University Press.
van Fraassen, Bas. 2000. The false hopes of traditional epistemology. *Philosophy and Phenomenological Research* 60: 253–280.
Whewell, William. 1847. *The philosophy of the inductive sciences, founded upon their history*. London: Parker.
———. 1858. *Novum organon renovatum*. London: Parker.

Chapter 14
Knowledge as Problem Solving

Abstract Contrary to a philosophical tradition which maintains that sense perception is the starting point of knowledge, this chapter maintains that problems are the starting points of knowledge. This raises the questions: What are problems? How are problems posed? How are problems solved? The chapter gives an answer to these questions, arguing that problems are questions to be investigated in order to solve them; problems are posed by analysing the terms of the problem, namely the conditions that must be met to solve it; problems are solved discovering their solutions by the analytic method, therefore, knowledge is problem solving by the analytic method. The chapter also argues that the view that knowledge is problem solving by the analytic method permits to give an answer to the question: Is there any *a priori* knowledge?

14.1 Knowledge and Problems

The analytic method, on which the analytic model of science is based, when solving problems produces knowledge. Conversely, knowledge is produced solving problems by the analytic model. Therefore, knowledge is problem solving by the analytic method.

This implies that problems are the starting points of knowledge. Thus Russell states: "In all the creative work that I have done, what has come first is a problem, a puzzle involving discomfort. Then comes concentrated voluntary thought entailing great effort" (Russell 1949, 112).

Saying that problems are the starting point of knowledge contrasts with a philosophical tradition, going back to antiquity, according to which sense perception is the starting point of knowledge. Thus, Aristotle states that knowledge begins with "what we call sense perception," which "belongs to all animals" (Aristotle, *Analytica Posteriora*, B 19, 99 b 34–35). In less sophisticated animals "there is no knowledge outside the act of perceiving" (ibid., B 19, 99 b 38–39). For, they immediately forget what they have perceived through their senses. Conversely, more sophisticated animals "can retain something in the mind after perceiving it" (ibid., B 19, 99 b 39–100 a 1). Namely, they can have memory of what they have perceived. Such is the case of human beings. In them, "from sense perception there arises memory," and "from memory, when it occurs often in connection with the same thing, there arises

© Springer International Publishing AG 2017

C. Cellucci, *Rethinking Knowledge*, European Studies in Philosophy of Science 4,
DOI 10.1007/978-3-319-53237-0_14

experience; for, though there are many memories, they make up a single experience" (ibid., B 19, 100 a 3–6). And "from experience" there arises "the starting point of scientific knowledge, which "concerns how things are" (ibid., B 19, 100 a 6–9).

However, the assumption that sense perception is the starting point of knowledge is unjustified because, as it will be argued in Chap. 15, sense perception, in particular vision, is problem solving by the analytic method, and hence problems are also the starting points of sense perception.

Saying that problems are the starting points of knowledge raises the questions: What are problems? How are they posed? How are they solved? This chapter tries to give an answer to these questions.

14.2 The Nature of Problems

What are problems? The term 'problem' comes from the Greek *problema*, one of whose meanings is 'question proposed for solution', which refers to a question to be investigated in order to solve it. The claim that knowledge is problem solving by the analytic method refers to this meaning, because it involves that knowledge results from investigating and solving questions about the world, including ourselves. Another meaning of *problema* is 'obstacle', which refers to a condition that impedes or makes it difficult to achieve a desired aim, and needs to be solved to achieve it. The claim that knowledge is problem solving by the analytic method also refers to this meaning, because knowledge is pursued to achieve a desired aim, and an unsolved problem may impede or make it difficult to achieve that aim.

That knowledge is pursued to achieve a desired end follows from the fact that knowledge responds to certain needs. As argued in Chap. 6, contrary to Aristotle's claim that from knowledge nothing results apart from contemplation, all human beings desire knowledge because knowledge has a vital role, being a certain kind of response to the environment that is essential for survival. More generally, all human beings desire knowledge because knowledge responds to certain needs. As already stated, knowledge results from investigating and solving questions about the world. The questions about the world whose solution produces knowledge may arise from several kinds of needs. For example, they may arise from needs of social or economic life, such as commercial or financial transactions, the construction of edifices and churches, bridges and dams, ships and aircrafts. They may arise from the need to investigate the environment, or the physical world generally. They may even arise from the need for beauty, which often leads scientists or mathematicians to ask and seek answers to questions which give them great aesthetic satisfaction.

14.3 Problem Posing

How are problems posed? In order to pose a problem, one must analyse the terms of the problem, namely the conditions that must be met to solve it. Posing a problem is essential, because the terms of the problem provide guidance on what data to consider and what to disregard, what indications to use and what to reject, in what directions to search for a solution and in what directions not to search for it.

Generally, problems are posed through a long and complex process. The cases when a problem seems to have been posed in a flash have actually been preceded by a long preliminary process. Indeed often, when one succeeds in posing a problem, one is already well ahead in research. For this reason it is said that a well posed problem is already half solved. In fact, posing a problem suggests that there may be relations between the data, and solving it will establish that such relations really exist.

However, posing a problem does not guarantee that one will be able to solve it. Often a problem is posed when there are not yet the means to solve it, or when the position of the problem involves assumptions which are still not fully explicit. In these cases, finding a solution may be too difficult, and all efforts to find it may be in vain. But, even if posing a problem does not guarantee that one will be able to solve it, the trust in its solvability provides an emotional drive that brings one to feel intensively involved in it.

Indeed, one can only hope to solve problems whose solution is passionately desired. The search for a solution to a problem has a strong emotional component that leads one to engage intensively in it, even facing years of hard work and bitter disappointments. Thus Hadamard states: "That an affective element is an essential part in every discovery or invention is only too evident, and has been insisted upon by several thinkers; indeed, it is clear that no significant discovery or invention can take place without the will of finding" (Hadamard 1954, 31). Pólya states: "It would be a mistake to think that solving problems is a purely intellectual affair; determination and emotions play an important role," indeed, "to solve a serious scientific problem, will power is needed that can outlast years of toil and bitter disappointments" (Pólya 1971, 93). To solve a problem, one must be ready to face "the varying emotions of the struggle for the solution" (ibid., 94). Eliminating emotions would deprive knowledge of one of its most vital sustaining forces.

The strong emotional component that leads one to engage intensively in the search for a solution to a problem imposes an emotional strain. The discovery of a solution will release from it, giving a great satisfaction. An example of this is the story of Pythagoras who "offered a sacrifice of oxen on finding that in a right-angled triangle the square on the hypotenuse is equal to the squares on the sides containing the right angle" (Diogenes Laertius, *Vitae Philosophorum*, VIII, 12). Another example is the story of Archimedes who, when stepping into a bath, suddenly realized that the volume of the water displaced had to be equal to the volume of the part of his body he had submerged. This meant that he could solve the problem posed by the king of Syracuse on how to assess the purity of a golden crown. Therefore, he "lept out of the tub in joy and rushed home naked, crying with a loud voice that he had found what he was seeking, for he continued exclaiming *eureka*" (Vitruvius, *De Architectura*, 9.10).

The emotional drive has an essential role already in the choice of problems to investigate. Only a small part of the problems that can be posed is of interest to scientists and mathematicians, and intellectual tension serves as a guide to discriminate what is interesting from what is not, what is worth investigating from what is not. Research not driven by intellectual tension inevitably ends up degenerating and dispersing into triviality. Without a scale of interest in problems, one cannot find anything valuable. (For more on the role of emotion in knowledge, see Cellucci 2013a, Chap. 15).

14.4 Problem Solving

How are problems solved? That is, how are solutions to problems discovered? This question often receives incongruous answers, such as Pólya's answer that "the first rule of discovery is to have brains and good luck. The second rule of discovery is to sit tight and wait till you get a bright idea" (Pólya 1971, 172).

This answer is incongruous because it is of the same kind as that of Molière's Bachelierus: "Mihi a docto doctore domandatur causam et rationem quare opium facit dormire. A quoi respondeo, Quia est in eo virtus dormitiva, cujus est natura sensus assoupire [I am asked by a learned doctor for the cause and reason why opium makes one sleep. To which I reply, Because there is in it a dormitive virtue, whose nature is to make the senses drowsy]" (Molière, *The Imaginary Invalid*, Act III, Interlude III). Indeed, Pólya's answer amount to saying that the reason why the mind discovers solutions to problems is that there is in it a discoveritive virtue, whose nature is to make the mind inventive.

Pólya's answer is all the more incongruous as Pólya admits that there is a method of discovery. Indeed, he states that there are "procedures (mental operations, moves, steps) which are typically useful in solving problems. Such procedures are practiced by every sane person sufficiently interested in his problems" (Pólya 1971, 172). These procedures "may be less desirable than the philosophers' stone but can be provided" (ibid.). The best of them is "the method of analysis, or method of 'working backwards'" (ibid., 225).

In fact, a reasonable answer to the question, 'How can solutions to problems be discovered?' is that they can be discovered by the analytic method. From the antiquity, the latter has been recognized to be the main method for problem solving.

14.5 The Steps of Problem Solving

Specifically, the process by which problems are solved can be described as consisting in the following steps.

1) *We examine the problem in order to understand it.* Understanding the problem is essential to find a solution, because finding a solution requires realizing what task is to be carried out. Realizing this is a preliminary condition for thinking of ways and means to find a solution.

2) *In order to understand the problem we introduce a suitable way of expressing it, or draw a suitable diagram to represent it.* This will permit to consider the problem more easily or perspicuously, thus essentially contributing to its understanding.

3) *We consider the problem from different perspectives, looking for connections between the problem and the existing knowledge, in order to obtain some information useful to solve it.* This leads to a deeper understanding of the problem, because the latter is enriched with the questions connected to it. That a deeper understanding of a problem requires a comparison with the existing knowledge, provides evidence that problem solving involves interactions between several knowledge systems.

4) *We analyse and list general features that a solution of the problem should have.* These general features emerge from a deep understanding of the problem. While listing such features does not provide any hypothesis to solve the problem, it clarifies what conditions should be satisfied by a hypothesis. This is useful because, focusing on such conditions, may greatly reduce the search space where a hypothesis can be located.

5) *We investigate whether there is any solved problem to which our problem may be related.* For, if our problem is similar to the solved problem in all the relevant respects, then the procedure that was successful with the solved problem might be equally successful with our problem. Of course, given the differences between the two problems, some changes in the procedure may be required.

6) *We form a hypothesis on the basis of the general features that a solution to the problem should have.* The hypothesis is obtained by some non-deductive rule, where which rule is to be used depends on the problem. Indeed, as Aristotle states, "we must not seek a single universal method for all cases" (Aristotle, *Topica*, A 6, 102 b 35–36). Instead, we must seek a specific non-deductive rule for each case. Moreover, the same hypothesis may be formed by different non-deductive rules.

7) *If none of the hypotheses that we have formed permits us to solve the problem, we examine if they all depend on some common assumption.* If so, we take the negation of that assumption and consider what follows from it. For, it may happen that all the hypotheses that we have formed depend on some common assumption that prevent the solution, and taking the negation of that assumption may give indications for a more adequate hypothesis.

8) *If we arrive at a hypothesis that permits us to solve the problem, we examine if the hypothesis is plausible, namely, if the arguments for it are stronger than the arguments against it, on the basis of the existing knowledge.* This is necessary, because the hypothesis might permit us to solve the problem for the wrong reason, in particular because it contradicts the existing knowledge, and from a contradiction anything can be inferred. That examining if the hypothesis is plausible requires a comparison with the existing knowledge, provides further evidence that problem solving involves interactions between several knowledge systems.

9) *In order to check whether the hypothesis is plausible, we draw consequences from it.* The consequences will then be compared with the existing knowledge.

10) *If the hypothesis is not plausible, we investigate why it is not plausible.* From this investigation, we will draw suggestions as to how to modify the hypothesis and make it plausible. If that will turn out to be absolutely impossible, we will give up the hypothesis and look for a new one.

11) *Even when the hypothesis is plausible, this is inconclusive for its acceptance.* The latter also depends on other factors, first of all fruitfulness, namely the ability to open new lines of research. When the hypothesis is recognized to be not only plausible but also fruitful, it tends to consolidate and become stable.

12) *Even when the hypothesis consolidates and becomes stable, this is not the end of the investigation.* For, the hypothesis is in turn a problem that requires to be solved. In order to solve it we must go back to step 1), and so on *ad infinitum*.

Thus, solving a problem involves proceeding from the given problem to others of increasing depth, and is an infinitely proceeding process. This is a basic difference between the analytic method and the axiomatic method. The latter arbitrarily stops this infinitely proceeding process at a certain stage, considering the hypothesis reached at that stage as not requiring further justification. This is the basis of Plato's criticism of the axiomatic method (see Chap. 20).

14.6 *Meno*'s Paradox

At the beginning of this chapter it has been stated that knowledge is problem solving by the analytic method. Against the claim that knowledge is problem solving by the analytic method, it might be objected that it conflicts with *Meno*'s paradox: "It is impossible for a man to search either for what he knows or for what he does not know: for what he knows, because he already knows it and hence is in no need of searching for it; for what he does not know, because he does not even know what it is he is to search for" (Plato, *Meno*, 80 e 2–5).

This objection, however, is unjustified. On the one hand, what a man knows are solutions to problems that have already been found. Since he already knows them, he is in no need to search for them. Nevertheless, he is always in need to put them in question, because no solution to a problem by the analytic method is conclusive. Therefore, it is unjustified to say that it is impossible for a man to search for what he knows because he already knows it and hence is in no need to search for it. He never knows conclusively what he knows.

On the other hand, what a man does not know are the solutions to problems that have not been found yet. But the fact that he does not know them does not mean that he does not know what it is he is to search for. It only means that he has not found a solution yet, which does not exclude that he might find one at a later stage. Therefore, it is unjustified to say that it is impossible for a man to search for what he does not know because in that case he does not even know what it is he is to search for. He knows what he is searching for, simply, he has not found it yet.

14.7 Knowledge as Problem Solving and Certainty

Knowledge resulting from problem solving by the analytic method is not absolutely certain, because the hypotheses can only be plausible. This, however, is the best we can achieve. As Plato states, "certain knowledge is either impossible or extremely difficult to come by in this life," we can only "adopt the best and least refutable of human hypotheses, and embarking upon it as upon a raft, run the risk of sailing the sea of life" (Plato, *Phaedo*, 85 c 3–d 2).

It is often assumed that knowledge is about truth and certainty. But knowledge is about neither truth nor certainty, knowledge is about finding the most plausible hypotheses at that stage. Scientific hypotheses are credible not because they are true and certain, but because the arguments for them are stronger than the arguments against them, on the basis of the existing knowledge.

Here is a basic difference between science and religions. While religions – at least, monotheistic religions – claim to possess truth and certainty, science challenges anyone who claims to possess truth and certainty to answer the question: How can you say you know it to be truth, instead of merely saying that you believe it to be truth?

14.8 Remarks on a Different View of Knowledge as Problem Solving

The view that knowledge is problem solving by the analytic method, is related to Laudan's view that "science is essentially a problem-solving activity" (Laudan 1977, 11). According to Laudan, we "do not have any way of knowing for sure (or even with some confidence) that science is true," or "that it is getting closer to the truth" (ibid., 127). Therefore, we cannot say that the aim of science is truth or approximation to truth. Such aims "are utopian, in the literal sense that we can never know whether they are being achieved" (ibid.). Rather, "science fundamentally aims at the solution of problems" (ibid., 4–5). While a criterion of truth will never be found, "we can determine whether a given theory does or does not solve a particular problem" (ibid., 127).

So far so good. But, in addition to saying that science fundamentally aims at the solution of problems, Laudan also says that "the case has yet to be made that the rules governing the techniques whereby theories are invented (if such rules there be) are the sorts of things that philosophers should claim any interest in" (Laudan 1981, 191). Discovery cannot be a concern of philosophy, because "a theory is an artifact," and "the investigation of the mode of manufacture of artifacts" is "not normally viewed as philosophical activity. And quite rightly, for the techniques appropriate to such investigations are those of the empirical sciences, such as psychology, anthropology, and physiology" (ibid., 190–191).

Thus Laudan ends up agreeing with Frege's view that discovery cannot be an object of philosophy, and specifically of logic, but only of psychology (see Chap. 11). This view, however, is unjustified because, as already argued, since antiquity a method of discovery has been known and widely used, the analytic method, and there is nothing subjective and psychological about this method.

14.9 *A Priori* Knowledge

The view that knowledge is problem solving by the analytic method permits to give an answer to the question: Is there any *a priori* knowledge?

The terms *a priori* and *a posteriori* first occur in medieval philosophy with respect to demonstration (see Chap. 21). Subsequently, they undergo a number of changes, involving a shift from demonstration to knowledge. The most significant one is due to Kant.

A priori knowledge in Kant's sense has the following basic characters. It occurs "independently of all experience" (Kant 1998, A2). It has the character of "strict universality," namely, "no exception at all is allowed to be possible" (ibid., B4). It has "the character of inner necessity" (ibid., A2). And it is "certain" (ibid.).

In terms of *a priori* knowledge in Kant's sense, the answer to the question whether there is any *a priori* knowledge must be negative, because there can be no *a priori* knowledge in Kant's sense. Indeed, no knowledge occurs independently of all experience, since all knowledge, including mathematical knowledge, arises from interaction with the external world. No knowledge has the character of strict universality, since exceptions are always possible, knowledge being at most plausible. No knowledge has the character of inner necessity, since all knowledge, being the product of our limited experience and our fallible biology, is always contingent. No knowledge is certain, since all knowledge, being at most plausible, is potentially fallible.

Conversely, the answer to the question whether there is any *a priori* knowledge can be positive if we identify *a priori* knowledge with knowledge that is a result of solving problems by the analytic method, hence knowledge based on plausible hypotheses. Such knowledge goes essentially beyond experience, because in the analytic method hypotheses, being obtained by non-deductive rules, are not contained in the premisses, so even when the premisses depend on experience, the conclusion essentially goes beyond it.

Clearly, *a priori* knowledge in the sense considered here is not subject to the difficulties of *a priori* knowledge in Kant's sense. For, it does not occur independently of all experience, since it is obtained forming hypotheses by means of non-deductive inferences whose premisses may derive from experience, and whose conclusions draw their plausibility from experience. It has no character of strict universality, namely no exception at all is allowed to be possible, since the hypotheses on which it is based are only plausible, so exceptions are always possible. It has no character of inner necessity, since the hypotheses on which it is based are contingent, being possibly incompatible with future data. And it is not certain, since there is no guarantee that no counterexample to the hypotheses on which it is based will ever be found.

It might be objected that the view of *a priori* knowledge put forward here is incongruous, because it implies that *a priori* knowledge has none of the characters that motivate the introduction of the concept of *a priori* knowledge. This objection, however, is invalid because *a priori* knowledge in the sense considered here shares an important character with of *a priori* knowledge in Kant's sense. Kant maintains that *a priori* knowledge has the character of "indispensability for the possibility of experience itself" (Kant 1998, B5). Now, *a priori* knowledge in the sense considered here also has this character, because, according to it, any kind of knowledge, including perceptual knowledge, is possible only by forming plausible hypotheses. (As regards perceptual knowledge, see Chap. 15 below). Therefore, *a priori* knowledge in the sense considered here also has the character of indispensability for the possibility of experience itself. In fact, this character is the most important feature of *a priori* knowledge.

The crucial difference between *a priori* knowledge in the sense considered here and *a priori* knowledge in Kant's sense, lies elsewhere. Kant states that, "although all our cognition commences with experience, yet it does not on that account all arise from experience" (ibid., B1). That is, not all our cognition arises from experience. So far so good. But then he adds: "It is therefore at least a question requiring closer investigation" whether "there is any such cognition independent of all experience," where "one calls such cognitions *a priori*" (ibid., B2). His answer is that in fact there are such "*a priori* cognitions" (ibid., B3). That is, there are cognitions which are independent of all experience. Now, this is unjustified because, that not all our cognition arises from experience, does not imply that there is some cognition which is independent of all experience, it only implies that there is some cognition which does not reduce to experience. Such is *a priori* knowledge in the sense considered here, since in the analytic method hypotheses, while going essentially beyond experience, are not independent of all experience. For, they are obtained by means of non-deductive inferences whose premisses may derive from experience, and whose conclusions draw their plausibility from experience.

14.10 *A Priori* Knowledge, Individual, and Species

The view of *a priori* knowledge put forward here must not to be confused with other views to which it is apparently similar. In particular, it must not be confused with the view that the origin of the *a priori* is *a posteriori*. According to this view, the *a priori* is such for the individual, since it must be present if the individual's experience is to be possible, and is essential for the individual's survival and orientation in the external world. But it is *a posteriori* for the species, because it is based upon our ancestors' experience, and specifically, is the product of adaptation, the result of phylogenetic acquisition of *a posteriori* knowledge in the species. In particular, Kant's forms of intuition are *a priori* for the individual, but *a posteriori* for the species.

Thus Lorenz states that "the origin of the *a priori*" is "*a posteriori*" (Lorenz 2009, 233). The *a priori* is such for the individual, since it "exists *a priori* to the extent that it is present before the individual experiences anything, and must be present if" the individual's "experience it to be possible" (Lorenz 1977, 9). But it "is not something immutably determined by factors extraneous to nature" (Lorenz 2009, 232). Indeed, the *a priori* apparatus "that enables living things to survive and orient themselves in the outer world has evolved phylogenetically through confrontation with and adaptation to that form or reality which we experience as phenomenal space" (Lorenz 1977, 9). In particular, Kant's *a priori* forms of "intuition have to be understood just as any other organic adaptation" (Lorenz 2009, 239). They are *a priori* for the individual, since they are "fixed prior to individual experience," but are *a posteriori* for the species, because they "are adapted to the external world" (ibid., 233).

But it is unjustified to say that the origin of the *a priori* is *a posteriori*, because the *a priori* must be present if the individual's experience is to be possible. So, if the origin of the *a priori* is *a posteriori*, then experience would have been impossible for our remotest ancestors. And, if the *a priori* is essential for the individual's survival and orientation in the external world, then survival and orientation in the external world would have been impossible for our remotest ancestors. Therefore, for the individual, the *a priori* cannot be the result of phylogenetic acquisition of *a posteriori* knowledge in the species.

It is also unjustified to say that Kant's forms of intuition are *a priori* for the individual, but *a posteriori* for the species. This would imply that, while we are unable to derive the postulates of Euclid's geometry, in particular the parallel postulate, from experience, our remotest ancestors were able to do so. In this regard, Poincaré states: "It has often been said that if individual experience could not create geometry the same is not true of ancestral experience. But what does that mean? Is it meant that we could not experimentally demonstrate Euclid's postulate, but that our ancestors have been able to do it?" (Poincaré 2013, 91). This is absurd, because it would imply that our remotest ancestors had intellectual powers that we no longer have, and there is no evidence for this.

These problems do not arise with the view of *a priori* knowledge put forward here. According to it, our remotest ancestors formed their hypotheses about the external world by non-deductive inferences, so their hypotheses were not contained in the premisses of such inferences. Hence, even if the premisses were based on their experience, their hypotheses went beyond it, and indeed made their experience possible. Therefore, their hypotheses were *a priori* also for our remotest ancestors.

14.11 *A Priori* Knowledge, Trial and Error, and Innate Knowledge

The view of *a priori* knowledge put forward here must also not to be confused with the view that *a priori* knowledge is knowledge which we must possess before we can acquire observational knowledge, so it cannot be the result of induction – in fact there is no such thing as induction – but is obtained by the method of trial and error.

Moreover, *a priori* knowledge is innate knowledge, and forms 99 per cent of our knowledge.

Thus Popper states that *a priori* knowledge is knowledge "that we somehow must possess before we can acquire observational or *a posteriori* knowledge" (Popper 1999, 70). So *a priori* knowledge "cannot, in turn, be the result of observation" (ibid., 63). Hence, it cannot be the result of induction, in fact "there is no such thing as induction" (ibid., 54). Instead, *a priori* knowledge is obtained by the method of trial and error, since "we learn only through trial and error," where our trials are "our hypotheses" (ibid., 47). Such knowledge is "not valid *a priori*; not *a priori* necessary, not apodeictic," but only "genetically *a priori*" (ibid., 46). Namely, it is prior to sense experience. Moreover, "it is inborn knowledge" (ibid., 69). And it is very extensive, because "99 per cent of the knowledge taken by Kant to be *a posteriori* and to be the '*data*' that are 'given' to us through our senses is, in fact, not *a posteriori*, but *a priori*" (ibid., 70). Indeed, "99 per cent of" our knowledge "is inborn and incorporated in our biochemical constitution" (ibid., 70).

But it is unjustified to say that *a priori* knowledge cannot be the result of induction, and instead is obtained by the method of trial and error. For, Popper asserts that the success of the method of trial and error "depends very largely on the number and variety of the trials: the more we try, the more likely it is that one of our attempts will be successful" (Popper 1974, 312). So the success of the method of trial and error depends on induction. This contradicts the claim that there is no such thing as induction. Moreover, the number of the trials we can make is very small with respect to all possible ones, so the probability that a single trial succeeds is very low. This does not explain why our trials are often successful, namely, they lead to plausible hypotheses. This can be explained only if the search for hypotheses does not proceed blindly but is, to a certain extent, guided. Popper himself admits this when he states that by 'trial' he does "not mean a random trial" (Popper 2002, 47). However, Popper does not explain how we can make non-random trials, indeed he considers it inexplicable. For, he states that making successful trials belongs to successful thinking, and "the demand for a theory of successful thinking cannot be satisfied," since "success depends on many things – for example on luck" (ibid., 50).

It is also unjustified to say that *a priori* knowledge is innate knowledge and forms 99 per cent of our knowledge. For, innate knowledge is knowledge embodied in organisms as a result of biological evolution, and biological evolution is very slow, so innate knowledge is very limited. Such is the innate knowledge of space, number and physical objects mentioned in Chap. 9. On the contrary, *a priori* knowledge is very extensive, being the result not only of biological evolution but also of cultural evolution, which is relatively fast, so the *a priori* knowledge to which it leads is very extensive. Therefore, *a priori* knowledge cannot be all innate knowledge.

These problems do not arise with the view of *a priori* knowledge put forward here. According to it, *a priori* knowledge can be the result of induction. For, it is a result of solving problems by the analytic method, and in the analytic method hypotheses are obtained by non-deductive rules. The search for hypotheses does not proceed blindly but is, to a certain extent, guided by such rules. Moreover, only a tiny part of *a priori* knowledge is innate knowledge, because innate knowledge is very limited.

References

Cellucci, Carlo. 2013a. *Rethinking logic: Logic in relation to mathematics, evolution, and method.* Dordrecht: Springer.

Hadamard, Jacques. 1954. *The psychology of invention in the mathematical field.* Mineola: Dover.

Kant, Immanuel. 1998. *Critique of pure reason.* Cambridge: Cambridge University Press.

Laudan, Larry. 1977. *Progress and its problems.* Berkeley: University of California Press.

———. 1981. *Science and hypothesis: Historical essays on scientific methodology.* Dordrecht: Springer.

Lorenz, Konrad. 1977. *Behind the mirror.* London: Methuen.

———. 2009. Kant's doctrine of the *a priori* in the light of contemporary biology. In *Philosophy after Darwin: Classic and contemporary readings*, ed. Michael Ruse, 231–247. Princeton: Princeton University Press.

Poincaré, Henri. 2013. *The foundations of science: Science and hypothesis – The value of science – Science and method.* Cambridge: Cambridge University Press.

Pólya, George. 1971. *How to solve it: A new aspect of mathematical method.* Princeton: Princeton University Press.

Popper, Karl Raimund. 1974. *Conjectures and refutations: The growth of scientific knowledge.* London: Routledge.

———. 1999. *All life is problem solving.* London: Routledge.

———. 2002. *Unended quest: An intellectual autobiography.* London: Routledge.

Russell, Bertrand. 1949. Answer to questionnaire. In Eliot Dole Hutchinson, *How to think creatively*, 110–112. New York: Abingdon-Cokesbury Press.

Chapter 15
Perceptual Knowledge

Abstract According to an influential tradition, perception is a passive process, determined entirely by the features of the external world. On the contrary, this chapter maintains that perception is problem solving by the analytic method, hence it is an active process. Since, in the analytic method, hypotheses are obtained by non-deductive inferences, this means that perception is based on inference. The chapter discusses the evidence for this view, and the objections against it. It also maintains that perception may involve data from several sense organs. In particular, vision is based on the fact that we form hypotheses about objects of the external world from stimuli on the retina and from movements of the eye, head or of the whole body, by means of deductive inferences.

15.1 Philosophical and Psychological Theories of Perception

In Chap. 14 it has been maintained that knowledge is problem solving by the analytic method. This implies that knowledge is an active process. It might be objected that perceptual knowledge is a counterexample to this because, according to an influential tradition, perception is a passive process, determined entirely by the features of the external world.

In answer to this objection, in this chapter it is maintained that perceptual knowledge is problem solving by the analytic method, so perceptual knowledge is an active process. Since, in the analytic method, hypotheses are obtained by non-deductive inferences, this means that perception is based on inference.

In contemporary philosophy of perception, little attention is paid to the view of perception as inference. This contrasts with psychology of perception, where the inferential character of perception is duly recognized. The aim of this chapter is to present the view of perception as inference in a philosophical setting. Arguments for and against this view will be discussed.

© Springer International Publishing AG 2017

C. Cellucci, *Rethinking Knowledge*, European Studies in Philosophy of Science 4,
DOI 10.1007/978-3-319-53237-0_15

15.2 The View That Vision is a Passive Process

As mentioned above, according to an influential tradition, perception is a passive process, determined entirely by the features of the external world. For example, Newton states that the pictures projected on the retina, "propagated by motion along the fibres of the optick nerves into the brain, are the cause of vision," which depends only on the image on the retina and hence is entirely passive, because "accordingly as these pictures are perfect or imperfect, the object is seen perfectly or imperfectly" (Newton 1952, 15).

On this view, vision is supposed to be based on the fact that the eye is like a *camera obscura*, the retina is the screen, and the brain watches the screen. Indeed, Newton states that, as one may cast "objects from abroad upon a wall" in "a dark room," in "like manner when a man views any object" the "light which comes from the several points of the object is so refracted by the transparent skins and humours of the eye" as "to converge and meet again in so many points in the bottom of the eye, and there to paint the picture of the object upon that skin (called the tunica retina) with which the bottom of the eye is covered" (ibid.).

But considering the eye to be like a *camera obscura* is inadequate, because the eye has several limitations. For example, only a very restricted region of the retina, of a diameter of about half a millimetre, the fovea, has high visual acuity. Moreover, the focal length of the eye is different for red and blue light, so one of the two ends of the colour spectrum is always out of focus. To compensate for these limitations, we make very fast movements of the eye, called saccades, which direct the fovea to points of interest. These movements are essential to vision because, if they are prevented, the image fades and disappears.

This raises several problems. The stimuli on the retina are poor and unstable. Why, then, do we have a rich and stable visual experience? The image on the retina is two-dimensional. Why, then, do we have a three-dimensional impression? The image on the retina is upside down and reversed. Why, then, do we see things quite in order? The images of the same object on the retinas of our two eyes are different. Why, then, do we have a single impression?

15.3 Vision and Mental Images

From the limitations of the eye it is apparent that vision cannot be based on the fact that the eye is like a *camera obscura*. On the other hand, vision cannot be based either on the fact that we observe mental images resembling objects of the external world.

Mental images do not resemble objects of the external world. For example, the following two tables are equal, simply, they are oriented differently in the plane, and yet they appear different to us because their mental images are different. The mental images, then, do not resemble the two tables, since the latter are equal.

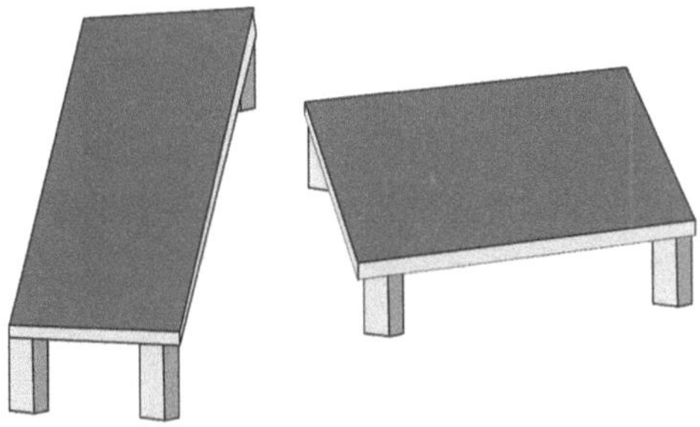

That mental images do not resemble objects of the external world is already pointed out by Descartes, who states that we do not judge shape "by the resemblance of the images in our eyes; for these images usually contain only ovals and rhombuses, while they make us see circles and squares" (Descartes 1996, VI, 140–141).

Moreover, if vision were based on the fact that we observe mental images resembling objects of the external world, there should be a homunculus within us capable of perceiving such resemblance, then there should be a homunculus inside such homunculus capable of perceiving the resemblance in question, and so on *ad infinitum*.

This also is already pointed out by Descartes, who states that "one must be careful not to suppose that in order to sense, the mind needs to contemplate certain images which are transmitted by objects to the brain" (ibid., VI, 112). Even when one of these images "still bears some resemblance to the objects from which it proceeds," one "must not think that it is by means of this resemblance that the image causes our sensory perception of these objects," since then it would be "as if there were yet other eyes within our brain," those of a homunculus, "with which we could perceive it" (ibid., VI, 130).

15.4 Vision as Problem Solving

But, if vision is not based on the fact that the eye is like a camera or that we observe mental images resembling objects of the external world, on what is it based? The answer is that it is based on inference. As Jaynes and Bretthorst state, "seeing is not a direct apprehension of reality," quite "the contrary: seeing is inference from incomplete information" (Jaynes and Bretthorst 2003, 133). Specifically, vision is problem solving by the analytic method. So perceptual knowledge is essentially of the same kind as all other knowledge, since, as argued in Chap. 14, knowledge is problem solving by the analytic method.

That vision is problem solving by the analytic method, means that vision is based on the fact that we form hypotheses about objects of the external world, from the stimuli on the retina and other data, by means of non-deductive inferences. From other data, because only a limited amount of visual data is normally derived from the stimuli on the retina, a great deal of visual data comes from the expectations of the visual system, gained via evolutionary or direct experience of the world. Thus, there is a sense in which what we see is affected by ancestral experience or our own experience of the world. Therefore, the hypotheses essentially go beyond the stimuli on the retina, and hence must be formed by means of non-deductive inferences.

The non-deductive inferences involved in vision are non-propositional, that is, they are not transformations of propositions, but rather transformations of data. It might be objected that, if perception is an inferential process "culminating in perceptual judgments pertaining to the qualities and kinds of distal objects, then perception too would appear to draw upon sentential structures to advance its inferential process," since "inferences must begin and end in sentences" (Maloney 1989, 18 and footnote 22). But this objection is based on the assumption that inference must be transformations of propositions. On the contrary, inferences can very well be transformations of data.

In addition to being non-propositional, the non-deductive inferences involved in vision are unconscious. Hanson states that Tycho and Kepler see different things at dawn, specifically, "Tycho sees the sun beginning its journey from horizon to horizon," namely, he sees it "circling our fixed earth," while Kepler sees "the horizon dipping, or turning away, from our fixed star" (Hanson 1965, 23). Such "difference is due neither to differing visual pictures, nor to any 'interpretation' superimposed on the sensation" (ibid.). It "depends on their knowledge, experience, and theories" (ibid., 18). Thus, according to Hanson, vision is based on processes depending on one's conscious knowledge, experience, and theories. This, however, conflicts with the fact that, for instance, although our conscious knowledge, experience, and theories tell us that the moon at the horizon has the same size as at its zenith, we see the moon at the horizon much larger than it does at its zenith. This means that vision involves unconscious inferences which overrule the conscious ones. As a result of biological evolution, the human brain embodies neural circuits that make inferences, through which the human brain categorizes data without our being aware of them.

That the inferences on which vision is based are non-deductive, non-propositional and unconscious was originally asserted by Descartes, saying that "size, distance and shape can be perceived one from the other by inference alone" (Descartes 1996, VII, 438). The inferences by means of which we perceive them are unconscious because, when we perceive objects, we "reason and judge those things at such high speed" that "we do not distinguish these operations from simple sense perception" (ibid.).

This view was further developed by von Helmholtz, saying that, although the inferences involved in perception "lack the purifying and scrutinizing work of conscious thinking," they "may be classed as inferences, inductive inferences unconsciously formed" (von Helmholtz 1867, 449). Thus, according to von Helmholtz, the inferences involved in perception are inductive and unconscious. Moreover, while "the 'conclusions' of logicians" are "capable of expression in words," the conclusions of such inferences "are not; because, instead of words, they only deal with sensations" (von Helmholtz 1995, 198). So the inferences involved in perception are non-propositional.

This requires an extension of logic. As Dilthey maintains, "logic is capable of complying with the demands of critical consciousness only by extending its province beyond the analysis of discursive thought" (Dilthey 1989, 166). The "achievement by which the perceptual process transcends what is given to it is an equivalent of discursive thought. The profound notion of unconscious inferences as developed by Helmholtz implies such an extension of logic" (ibid., 167).

However, von Helmholtz's assumption that the inferences involved in vision are inductive is somewhat limiting, because inductive inferences are only a special kind of non-deductive inferences. The above suggestion that vision is a problem which is solved by the analytic method removes this limitation. The inferences on which vision is based can be any kind of non-deductive inferences.

That the inferences on which vision is based are non-deductive, non-propositional and unconscious involves that, in the statement that knowledge is problem solving by the analytic method, knowledge must be meant in the widest sense. In particular, knowledge must not be restricted to propositional knowledge, because the inferences on which vision is based have no linguistic correlate. Moreover, knowledge must not be restricted to conscious knowledge, because visual knowledge is the result of processes which occur too fast and at too low a level in the mind to be accessible to direct inspection.

15.5 Vision and the Limitations of the Eye

Stating that vision is problem solving by the analytic method, we may give an answer to the questions raised by the limitations of the eye.

Thus we may explain why we have a rich and stable impression, although the stimuli on the retina are poor and unstable. This occurs because what we see is not the image on the retina, but rather the result of forming hypotheses which go beyond the stimuli on the retina. For example, when we see an object moving fast, the eye moves with the object, so the image of the object is rather stable on the retina, while the background with respect to which the object moves, which we perceive as steady, actually moves fast on the retina. Thus what we see is based on a hypothesis by which we distinguish the stability of the object on the retina from the movement of the object in the world. This is the result of an adaptation, because we would be unable to take appropriate behaviour with respect to the environment if we were unable to make such distinction.

Also, we may explain why we have a three-dimensional impression, although the image on the retina is two-dimensional. This occurs because what we see is not the image on the retina, but rather the result of forming the hypothesis that the object in front of us is three-dimensional. This is the result of an adaptation, because we would be unable to take appropriate behaviour with respect to the environment if we saw all objects as two-dimensional.

Moreover, we may explain why we see things quite in order, although the image on the retina is upside down. Once again, this occurs because what we see is not the image on the retina, but rather the result of forming hypotheses which go beyond the stimuli on the retina. For example, if we wear special glasses with prism lenses that show an upside down image of the world, at first we can hardly do anything, in particular we are unable to walk. This is due to the fact that, turning the image of the world upside down, upsets the hypotheses on the base of which we put the stimuli on the retina in relation with the objects of your experience. We have to form new hypotheses and, until we will have formed them, our visual experience is upset. This is the result of an adaptation, because we would be unable to take appropriate behaviour with respect to the environment if we were unable to perceive the orientation of objects properly.

Furthermore, we may explain why we have a single impression, although the images of the same object on the retinas of our two eyes are different. This occurs because what we see is not the image on the retina, but rather the result of forming the hypothesis that the stimuli on the two retinas are produced by the same object. This is the result of an adaptation, because we would be unable to take appropriate behaviour with respect to the environment if we saw the same object as two different objects.

On the view that vision is problem solving by the analytic method, vision is active. So such view is part of a philosophical tradition going back to Plato, who is "the ancestor of the idea of perception as a force that is not merely receptive, but an active power of grasping the world" (Remes 2014, 10). Of course, a link with Plato arises also because, as stated in Chap. 12, Plato was the first to give an explicit formulation of the analytic method.

On the view that vision is problem solving by the analytic method, we perceive a scene not all at once, in a flash. Vision is a dynamic process during which the eyes continually sample the environment, and we form hypotheses about the environment by non-deductive inferences from the data coming from the eyes. As an old Chinese proverb says, two-thirds of what we see is behind our eyes.

15.6 Evidence for Vision as Problem Solving

Several arguments can be given in favour of the view that vision is problem solving by the analytic method. Here are some of them.

(a) *Size Constancy*

Although the size of the image of an object on the retina changes as the distance from the object to the observer changes – the greater the distance, the smaller the image on the retina – in familiar situations they appear to us as being of the same size. For instance, the woman in the background, although smaller, looks normal in size when compared to the nun in the foreground. But, if the woman in the background is brought to the same elevation as the nun in the foreground, the woman no longer looks normal in size.

This depends on the fact that we form the hypothesis that the woman in the background is normal in size on the basis of our ordinary experience, that the size of an object remains constant as the object gets closer or moves away from us. On this basis, we infer that the woman in the background, though smaller, is normal in size. On the same basis, we infer that the woman in the background, brought to the same elevation as the nun in the foreground, is not normal in size since, being in the foreground, she must be smaller and hence looks smaller. This is the result of an adaptation, because we would be unable to take appropriate behaviour with respect to the environment, if the size of an object changed when the object gets closer or moves away from us. This is apparent from the fact that size constancy occurs only in familiar situations. For example, buildings, cars or persons, seen from an airplane – a situation unfamiliar from our evolutionary past – do not look normal in size.

(b) *Shape Constancy*

Although two equal objects, differently oriented in space, project different images on the retina, in familiar situations they look like they have the same shape. Such is the case of two equal plates differently oriented in space.

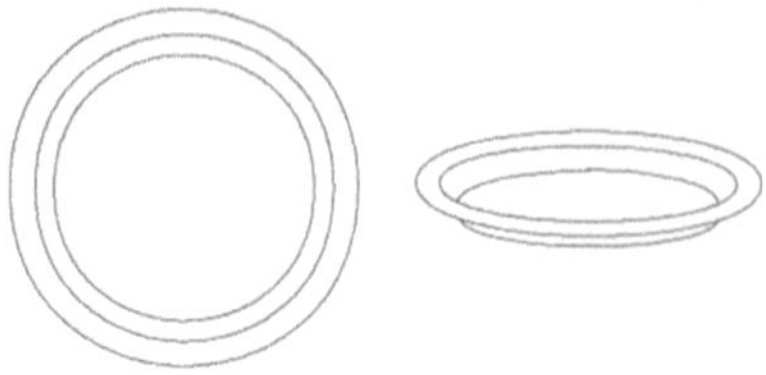

This is the phenomenon to which Descartes refers when he says that we do not judge shape by the resemblance of the images in our eyes (see above).

The phenomenon in question depends on the fact that we form the hypothesis that the two plates have the same shape on the basis of our ordinary experience, that the shape of an object remains constant as the object is differently oriented in space. On such basis, we infer that the two plates have the same shape. This is the result of an adaptation, because we would be unable to take appropriate behaviour with respect to the environment if the shape of an object changed when the object is differently oriented. This is apparent from the fact that shape constancy occurs only in familiar situations. For example, a circle and an ellipsis look to have different shapes.

(c) *Colour Constancy*

Although two objects of the same colour, when illuminated by lights of different colour, project different images on the retina, in familiar situations they look like they have the same colour. For example, two white sheets of paper look white when one of them is illuminated by the white sunlight of midday and the other is illuminated by the red light of sunset.

This depends on the fact that we form the hypothesis that the two sheets of paper are white on the basis of our ordinary experience, that the colour of an object remains constant independently of its illumination. On such basis, we infer that the two sheets of paper have the same colour, even if they look to have different colours. This is the result of an adaptation, because we would be unable to take appropriate behaviour with respect to the environment if the colour of an object changed with the illumination. This is apparent from the fact that shape constancy occurs only in familiar situations. For example, if the two sheets of white paper are illuminated by non-natural light, say, one of them by monochromatic blue light and the other one by monochromatic red light, they will appear blue and red, respectively. Colour cinematography is based on this fact, otherwise, projecting a movie on a white screen, the screen would remain white.

(d) *Illusory Contours*

If we are shown the following picture and are asked what it is, we will answer that it is a tiger, not five anatomical sections of a tiger, even if this answer would be compatible with the picture.

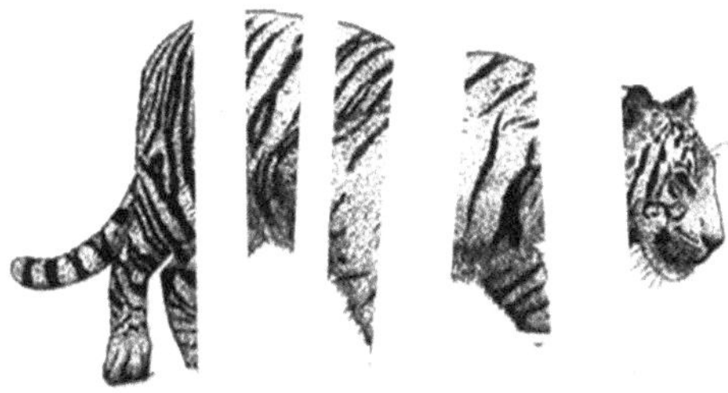

This depends on the fact that we form the hypothesis that it is a tiger partially hidden by four vertical objects, say, four tree trunks. We form such hypothesis on the basis of our ordinary experience of the world, that this would be our actual visual experience if a tiger were partially hidden by four tree trunks. This is the result of an adaptation, because we would be unable to take appropriate behaviour with respect to the environment if this situation did not produce such visual experience. Indeed, if it produced the visual experience of five anatomical sections of a tiger, we would have little chance of surviving an encounter with the tiger.

(e) *Born Blind People Who Recover Sight*

Born blind people who, through surgery, recover sensitivity to visual stimuli, do not acquire the capacity of seeing immediately but only after a long exercise. In order to see, they must first learn to connect visual stimuli to objects, making hypotheses on the objects which produce those visual stimuli. Until then, they are unable to adequately interpret stimuli.

Thus, after a lady underwent surgery, "her mother explicitly taught her objects around the house," she "learned to recognize her siblings and parents six months after surgery, and after a year could name objects around the house purely by sight" (Ostrovsky et al. 2006, 1010).

(f) *Touch as a Surrogate for Vision*

Descartes discusses the experience of walking at night over rough ground without a light, using a stick to feel the various objects situated around. He says that "it is true that this kind of sensation is somewhat confused and obscure in those who do not have long practice with it; but consider it in those born blind, who have made use of it in all their lives, and you will find that, with them it is so perfect and so exact that one might almost say that they see with their hands" (Descartes 1996, VI, 84). This depends on the fact that those born blind have learned to form hypotheses about the objects around them from the tactile stimuli obtained through their stick.

A more elaborate version of this argument can be given in terms of the tactile vision substitution systems, which consist in a camera that records an image in real-time, and sends signals to an array of vibrating elements located on the subjects' thorax, abdomen or back. At first, users report experiences in terms of the sensations on the area of skin which is receiving the stimuli, but afterwards their reports are in terms of objects localized externally in space in front of them. Granted enough experience with the system, users "begin to recognize shapes of simple, familiar objects via the tactile display," then "users' awareness of the proximal tactile sensation fades" and "users begin to 'perceive' stable three-dimensional objects in space," where this progression occurs "without conscious processing" (Visell 2009, 40). Here again, this depends on the fact that they have learned to form hypotheses about the objects around them from the tactile stimuli produced by the array of vibrating elements.

(g) *Movement as a Trigger for Vision*

At least fifty percent of people can see the movement of their own hand even in total darkness, so, "in total darkness, self-generated body movements are sufficient to evoke normally concomitant visual perceptions" (Dieter et al. 2013, 1). Such movements transmit sensory signals that can give rise to real visual perceptions even in the complete absence of external visual input. This means that people who can see the movement of their own hand even in total darkness combine information from different senses to make hypotheses about the objects in front of them. Generally, it means that what we normally perceive of as sight is really as much a function of our brain making hypotheses as our eyes.

15.7 Objections to Vision as Problem Solving

Notwithstanding the above arguments in favour of the view that vision is problem solving by the analytic method, several objections have been raised against this view. Here are some of them.

(a) *Propositional Inferences*

One must not be misled "by the incoherent assertion that perceptions are conclusions of unconscious inferences. One may form hypotheses about what one sees, but to see is not to form a hypothesis" (Bennett and Hacker 2003, 137). For, "inferences are transformations of propositions in accordance with a rule," while "perceiving something does not involve transformations of propositions by a perceiver (or his brain)" (ibid.).

This objection, however, is unjustified, because it is based on the assumption that inferences are transformations of propositions in accordance with a rule, from which it follows that perceptions, not being transformations of propositions in accordance with a rule, are not conclusions of unconscious inferences. This assumption is unwarranted, because inferences need not be transformations of propositions, they can be transformations of data in accordance with a rule.

(b) *Answerability*

Even admitting that vision is based on forming hypotheses, "an hypothesis is something answerable to evidence, to data. To what data could the perceptual hypotheses" be "answerable, but to perceptions?" (Anscombe 1981, II, 65).

This objection, however, is unjustified, because the hypotheses involved in vision are answerable to other sense stimuli, or rather, to hypotheses formed from other sense stimuli. For example, we observe a stick in water and it appears crooked, so we form the hypothesis that it is crooked, then we verify the hypothesis, for example, by touching the stick, and we realize that it is not crooked. Thus the hypothesis that the stick is crooked, formulated from the stimuli on the retina, is checked by tactile stimuli, or rather, by another hypothesis formed from tactile stimuli.

(c) *Constraints*

One need not assume that vision involves the formation of hypotheses, because vision is based only on the fact that our visual system embodies "(without explicitly representing and drawing inferences from) certain very general constraints on the interpretations that it is allowed to make" (Pylyshyn 2003, 96). These constraints "produce the correct interpretation under specified conditions," so they yield vision "without 'unconscious inference'" (ibid.). They "derive from principles of optics and projective geometry" (ibid., 120).

This objection, however, is unjustified because, saying that our perceptual system embodies certain very general constraints which produce the correct interpretation of stimuli, amounts to saying that the human brain embodies specialized circuits that make inferences by which the human brain categorizes data.

(d) *Invariants*

It is unwarranted to say that the stimuli on the retina are poor, and hence vision needs inference. In vision, "information consists of invariants underlying change. It does not consist of stimuli, nor of patterns of stimuli, nor of sequences of stimuli. A perceptual system does not respond to stimuli" but "extracts invariants" (Gibson 1976, 236). For example, while the size of objects decreases in proportion to the square of the distance, two objects of the same size, at different distances from the observer, will appear of the same size to the observer. This occurs because the relation between the size of the first object and its distance from the horizon is equal to the relation between the size of the second object and its distance from the horizon. Namely, the relation is invariant with respect to all distances from the observer.

This objection is unjustified, because saying that the relation between the size of an object and its distance from the horizon is invariant with respect to all distances from the observer, amounts to formulating an inference rule. Then extracting invariants is an active process which involves applying inference rules. Two objects of the same size, at different distances from the observer, will appear of the same size to the observer as a result of an inference.

15.8 Vision and Movement

It has been stated above that vision is based on the fact that we form hypotheses about objects of the external world, from stimuli on the retina and other data, by means of non-deductive inferences. This statement, however, must be integrated, taking into account the relation between vision and movement.

It has been pointed out above that vision requires very fast movements of the eye. But vision may also require movements of the head or the whole body. For example, if we observe a cube from the front, we see a square. To realize that it is a cube, we must move the head or the whole body. So, to find out what are the objects of the external world, we also use the movements of the head or the entire body. Such movements produce changes in the way these objects appear to us, and permit us to make appropriate inferences about their spatial properties.

Then, the above statement must be integrated by saying that vision is based on the fact that we form hypotheses about objects of the external world, from stimuli on the retina and other data, as well as from movements of the eye, head or of the whole body, by means of deductive inferences.

15.9 Vision and Touch

The most immediate interpretation of the above example of the cube is that we perceive spatial properties of objects by sight. Some people, however, maintain that we perceive spatial properties of objects by touch, which acquires spatial content through movement.

Thus Berkeley maintains that, "whenever we say an object is at a distance, whenever we say it draws near, or goes farther off, we must always mean it of the latter sort, which properly belong to the touch, and are not so truly perceived as suggested by the eye" (Berkeley 1948–1957, I, 189–190). Touch acquires spatial content through movement because, whenever someone says that "he sees this or that thing at a distance," what he actually sees "only suggests to his understanding that after having passed a certain distance, to be measured by the motion of his body, which is perceivable by touch, he shall come to perceive such and such tangible ideas which have been usually connected with such and such visible ideas" (ibid., I, 188). Thus, according to Berkeley, the perception of distance ultimately belongs to touch.

However, saying that the perception of distance ultimately belongs to touch is unjustified, because sight does not need touch to perceive distance, but can perceive it directly through movement. Objects in vision are not given at once in their totality but are given in stages, through the movements of the eye, the head or the whole body. Producing changes in the way objects appear to us, these movements permit us to make appropriate inferences about the spatial properties of the objects which yield hypotheses on them. Therefore, sight acquires spatial content through movement without passing through touch.

Moreover, sight acquires spatial content differently from touch. While we perceive that an object is a cube by sight depending on how it changes when we move the eye, the head or the body, we perceive that the object is a cube by touch depending on how it guides or resists the movements of our hands on it, which suggests how its faces and vertices are structured.

Of course, that sight acquires spatial content through movement without passing through touch, does not mean that tactile stimuli cannot be used to form hypotheses about spatial properties of objects. It only means that the stimuli on the retina are the primary basis for forming them.

References

Anscombe, Elisabeth. 1981. *The collected philosophical papers*. Oxford: Blackwell.
Bennett, Maxwell Richard, and Peter Michael Stephan Hacker. 2003. *Philosophical foundations of neuroscience*. Oxford: Blackwell.
Berkeley, George. 1948–1957. *Works*. London: Nelson.
Descartes, René. 1996. *Oeuvres*. Paris: Vrin.
Dieter, Kevin, Bo Hu, David Knill, Randolph Bike, and Duje Tadin. 2013. Kinestesis can make an invisible hand visible. *Psychological Science* 25 : 66–75.
Dilthey, Wilhelm. 1989. *Introduction to the human sciences*. Princeton: Princeton University Press.
Gibson, James. 1976. The myth of passive perception: A reply to Richards. *Philosophy and Phenomenological Research* 37 : 234–238.
Hanson, Norwood Russell. 1965. *Patterns of discovery*. Cambridge: Cambridge University Press.
Jaynes, Edwin, and George Bretthorst. 2003. *Probability theory*. Cambridge: Cambridge University Press.
Maloney, Christopher. 1989. *The mundane matter of the mental language*. Cambridge: Cambridge University Press.
Newton, Isaac. 1952. *Opticks, or a treatise of the reflections, refractions, inflections & colours of light*. Mineola: Dover.
Ostrovsky, Yuri, Aaaron Andalman, and Pawan Sinha. 2006. Vision following extended congenital blindness. *Psychological Science* 17 : 1009–1014.
Pylyshyn, Zenon. 2003. *Seeing and visualizing: It's not what you think*. Cambridge: The MIT Press.
Remes, Pauliina. 2014. Plato: Interaction between the external body and the perceiver in the *Timaeus*. In *Active perception in the history of philosophy, from Plato to modern philosophy*, ed. José Filipe Silva, and Mikko Yrjönsuuri, 9–30. Cham: Springer.
Visell, Yvon. 2009. Tactile sensory substitution: Models for enaction in HCI. *Interacting with Computers* 21 : 38–53.
von Helmholtz, Hermann. 1867. *Handbuch der Physiologischen Optik*. Leipzig: Voss.
———. 1995. *Science and culture: Popular and philosophical essays*. Chicago: The University of Chicago Press.

Chapter 16
Knowledge and Error

Abstract In the analytic method, knowledge is obtained by use of deductive and non-deductive rules. Since such rules are not plausibility preserving, they may give rise to errors. On this basis, contrary to a tradition according to which error is heterogeneous to knowledge, this chapter maintains that error is homogeneous to knowledge and inherent to it. Indeed, error is even homogeneous and inherent to logical and mathematical knowledge. The chapter also maintains that, though error may be a serious problem for logic, mathematics and science, it may also be fruitful. For, an analysis of the causes of error may provide useful indications for forming new hypotheses, and may also help to formulate new concepts.

16.1 The Heterogeneity View

That, as argued in Chap. 12, the deductive and non-deductive rules through which knowledge is obtained are not plausibility preserving, implies that they may give rise to errors. So error is not heterogeneous to knowledge, but homogeneous and inherent to it.

This conflicts with the fact that, since antiquity, error has been considered heterogeneous to knowledge. It has been considered heterogeneous to knowledge in two different senses.

On the one hand, error has been considered to be non-knowledge, lack of knowledge, ignorance. Either there is knowledge or there is error, so, if there is knowledge, then there can be no error. Therefore, knowledge is always true.

Thus, Aristotle states that to know a thing is to know its essence, and one knows the essence of a thing when one grasps it by intuition and states it, since a proposition which "states what a thing is according to its essence is true" (Aristotle, *De Anima*, Γ 6, 430 b 28). Now, about essences "it is not possible to be mistaken" (Aristotle, *Metaphysica*, Θ 10, 1051 b 31). For, "not grasping them is not knowing them" (ibid., Θ 10, 1051 b 25). Thus, "it is not possible to be mistaken about the essence except accidentally" (ibid., Θ 10, 1051 b 25–26). On the contrary, we may be mistaken about attributes which are incidental to objects, for example "we may be mistaken" with respect to the question "whether the white object is this thing or another" (Aristotle, *De Anima*, Γ 3, 428 b 21–22).

© Springer International Publishing AG 2017

C. Cellucci, *Rethinking Knowledge*, European Studies in Philosophy of Science 4,
DOI 10.1007/978-3-319-53237-0_16

On the other hand, error has been considered to be the result of a bad application of the rules of knowledge, and specifically, of forming a judgment for which we do not have enough cognition. Therefore, error hinders the entry of truth.

Thus Kant states that "an error is a *vitium* of judgment" (Kant 1992, 272). It consists in considering true a judgment which is false. Therefore, "a false cognition and an error are distinct. If I propound and examine a false judgment, there is not yet any error: error is the holding-to-be-true of falsehood" (ibid., 288). But then "no error is unavoidable in itself, because one simply need not judge about things of which one understands nothing" (ibid.). Ignorance and error are distinct, because ignorance "does not always rest on our will, but often on the restrictions of our nature," so "ignorance can well be unavoidable" (ibid.). On the contrary, with error "we are ourselves always culpable, in that we are not cautious enough in venturing a judgment, for which we do not have enough cognition" (ibid.). The "fate of someone who errs is worse, then, than that of someone who is ignorant, for error hinders the entry of truth" (ibid., 275). (On Kant's conception of error, see Capozzi 2013, Chaps. 12 and 14).

16.2 Limitations of the Heterogeneity View

It has been stated above that, since antiquity, error has been considered heterogeneous to knowledge in two different senses. This view, however, is unjustified in both senses.

On the one hand, it is unjustified to consider error as non-knowledge, lack of knowledge, ignorance. This applies to Aristotelian essentialist science, in which to know a thing is to know the essence of that thing, with respect to which one cannot be mistaken, error is simply lack of knowledge. But it does not apply to modern science, in which to know a thing is to know some phenomenal properties of it, mathematical in kind. Indeed, if knowledge is knowledge of such properties, then error cannot be considered to be lack of knowledge, ignorance, because error tells us that the hypothesis is inadequate to solve the problem. So it tells us something about the relation between the hypothesis and the problem, which can help us to formulate a new hypothesis.

On the other hand, it is unjustified to consider error as the result of a bad application of the rules of knowledge, and specifically, of forming a judgment for which we do not have enough cognition, therefore, as something which hinders the entry of truth. This is unjustified in three respects.

To consider error as the result of a bad application of the rules of knowledge is to overlook that knowledge and error have the same origin. They both arise from formulating hypotheses to solve a problem by a non-deductive rule. That the hypothesis yields knowledge or error does not depend on the non-deductive rule, but on the plausibility of the hypothesis, which in turn depends on compatibility with the existing knowledge. Error can be distinguished from knowledge only through a comparison with the existing knowledge. If the hypothesis is compatible with it there is knowledge, otherwise there is error. The difference between knowledge and error, then, is not intrinsic but extrinsic, depending on a comparison with the existing knowledge. As Mach says, "knowledge and error flow from the same mental sources; only success can tell the one from the other" (Mach 1976, 84).

Moreover, to consider error as the result of forming a judgment for which we do not have enough cognition, is to overlook that hypotheses are always the result of forming a judgment for which we do not have enough cognition. Hypotheses serve to acquire knowledge which goes beyond the existing data, and hence, by their very nature, they are the result of forming a judgment for which we do not have enough cognition.

Furthermore, to consider error as something which hinders the entry of truth, is to overlook that error can produce knowledge. Indeed, an analysis of the causes of error can provide useful indications for forming new hypotheses, which can give new knowledge. For example, as Rota states, "an important step in solving a mathematical problem, perhaps the most important step, consists of analyzing other attempts" that have been previously "carried out, with a view to discovering how such previous approaches failed" (Rota 1997, 99). Of course, although one can err in many ways, one can learn only from some of them. Even Kant admits that error is not something completely negative, since he states that, "because the understanding is in fact active in every error, men always bring out truth when they judge at the risk of error" (Kant 1992, 282). Indeed, "in all our judgments there is always something true. A man can never err completely and utterly. For in accordance with what he perhaps presupposes, he can always have some truth, even if only partial. A total error would be a complete opposition to the laws of the understanding" (ibid.). But, stating that error hinders the entry of truth, Kant contradicts his own statement that men always bring out truth when they judge at the risk of error.

In fact, error is not heterogeneous to knowledge but homogeneous to it since, as already mentioned, knowledge and error have the same origin. They both arise from formulating hypotheses to solve problems.

16.3 Logic and Error

That error is not heterogeneous to knowledge but homogeneous to it, also applies to logic. Since, as it has been argued in Chap. 12, both deductive rules and non-deductive rules are not plausibility preserving, even when the premisses are plausible, the conclusion need not be plausible.

This contrasts with the widespread opinion that logic is a canon of the understanding and reason, so in logic there can be no error.

Thus Kant states that logic "is a canon of the understanding and reason" (Kant 1998, A53/B77). It "contains the absolutely necessary rules of thinking, without which no use of the understanding takes place" (ibid., A52/B76). In logic, "truth is agreement of cognition with the laws of the understanding" (Kant 1992, 281). Then, the question if error is possible is the question "to what extent the form of thought contrary to the understanding is possible" (ibid.). But a power cannot "deviate from its own laws, since it acts only according to certain laws. If these laws are essential," like the laws of logic, "then the power cannot deviate from them" (ibid.). So in logic there can be no error, since the understanding "can produce nothing that conflicts with its nature" (ibid., 282–283). Indeed, "if we had no other power of cognition but the understanding, we would never err" (ibid., 560).

Similarly, Gödel states that in logic there can be no error, since "every error is due to extraneous factors; reason itself does not commit mistakes" (Wang 1996, 291).

This opinion is based on the assumption that the aim of science is truth, and on the fact that deductive rules are truth preserving. However, as argued in Chap. 9, the aim of science is not truth but plausibility, so the relevant fact to be taken into account here is not that deductive rules are truth preserving, but instead that both deductive rules and non-deductive rules are not plausibility preserving. Moreover, by the strong incompleteness theorem for second-order logic, there is no consistent set of rules capable of deducing all second-order logical consequences of any given set of formulas, so logic cannot be restricted to deductive rules.

16.4 Mathematics and Error

That error is not heterogeneous to knowledge but homogeneous to it, also applies to mathematics. As a matter of fact, mathematicians often make errors. Lecat lists "about five hundred errors, made by some 330 mathematicians, many of them famous" (Lecat 1935, vii–viii). Davis states that "a mathematical error of international significance may occur every twenty years or so," where, by 'a mathematical error of international significance', he means the "conjunction of a mathematician of great reputation and a problem of great notoriety" (Davis 1972, 262).

This contrasts with the widespread opinion that in mathematics there can be no error. Such opinion is based on the following two arguments.

1) *Intuition.* In mathematics there can be no error, because therein everything can be set forth through intuition.

Thus Kant maintains that "one believes mathematicians because it is not possible that they can err, since they would hit upon false consequences at once" (Kant 1992, 469). Indeed, "mathematics involves infallible reason. Hence as soon as a proposition is maintained by mathematics it is infallible" (ibid., 342). This depends on the fact that in mathematics "everything can be set forth through intuition" (ibid., 479). Mathematical concepts "are confirmed by themselves through intuition" (ibid.).

But appealing to intuition does non guarantee against the possibility of error in mathematics. As we have seen in Chap. 3, Frege's, Russell's, and Hilbert's attempts to base the certainty of mathematics on different kinds of intuition collapsed. Thus the aim of appealing to intuition to justify the certainty of mathematics fails. It makes way for the task of accounting for mathematics as it is, dubious, uncertain and fallible as any other human activity, and yet so rich and often also so useful and fruitful.

2) *Formalization*. In mathematics there can be no error, because therein everything can be concretely exhibited through formalization. In particular, demonstrations can be concretely exhibited through formal demonstrations, whose correctness is beyond doubt since it can be mechanically verified by computer.

Thus Hilbert maintains that, through formalization, we can make "every mathematical statement into a formula that can be concretely exhibited and rigorously derived, and thereby bring mathematical concept-formations and inferences into such a form that they are irrefutable" (Hilbert 1996e, 1152). Indeed, formal demonstrations deduce theorems from axioms by means of mechanical rules, "hence on the basis of a pure formula game, without extraneous considerations being adduced" (Hilbert 1967b, 475). So the correctness of formal demonstrations can be mechanically verified, without "need to fall back upon intuition or meaning" (ibid.).

Wiedijk even maintains that, since the correctness of formal demonstrations, being mechanically verifiable by computer, is beyond doubt, "all mathematicians will start using formalization for their proofs," and "referees will insist on getting a formalized version before they want to look at a paper" (Wiedijk 2008, 1414).

But appealing to formalization does not guarantee against the possibility of error. For, mechanically verifying the correctness of formal demonstrations presupposes mechanically verifying the correctness of the computer program for mechanically verifying the correctness of formal demonstrations. Now, by Rice's theorem, there is no algorithm for deciding with generality non-trivial questions on computer programs, such as their correctness. So, as Hersh points out, "it just is not the case that a doubtful proof would become certain by being formalized. On the contrary, the doubtfulness of the proof would then be replaced by the doubtfulness of the coding and programming" (Hersh 1979, 43).

Moreover, even if the correctness of formal demonstrations, being mechanically verifiable by computer, were beyond doubt, an important problem would remain. In order to assert that axiomatic demonstration can attain complete reliability through formalization, it would be necessary to demonstrate the consistency of the axioms on which formal demonstrations are based, and to demonstrate it by absolutely reliable means. But, by Gödel's second incompleteness theorem, this is impossible.

16.5 Demonstration and Error

That in mathematics there can be errors, means that we must safeguard ourselves against them, and specifically we must safeguard ourselves against errors in mathematical demonstrations.

The ancients were so well aware of this problem that Euclid devoted a whole book, the *Pseudaria*, to it. Euclid's book has been lost, but Proclus informs us that, "since many matters seem to adhere to truth and to follow from scientific principles, but really lead away from them and deceive the unexperienced students," in the *Pseudaria* Euclid gives us "methods by which a perspicuous mind may detect such errors; and, possessing these methods, we may instruct beginners in this study in discovering fallacies and not being deceived" (Proclus 1992, 70.1–9).

However, this kind of exercise may not be enough. As Hume points out, even if every time a mathematician "runs over his proofs, his confidence increases," and it "is rais'd to its utmost perfection by the universal assent and applauses of the learned world," nevertheless, "this gradual increase of assurance is nothing but the addition of new probability" (Hume 1978, 180). Indeed, "in every judgment, which we can form concerning probability" of demonstrations, "we ought always to correct the first judgment" ineludibly "by another judgment" (ibid., 181–182). But the new judgment will be itself only probable, so, "in every probability, beside the original uncertainty," there arises "a new uncertainty deriv'd from the weakness of that faculty, which judges," and "so on *in infinitum*" (ibid., 182).

In fact, despite all precautions, errors in mathematical demonstrations may remain undetected even by the most attentive eye. They can be so subtle that it can take years to notice them. This is often overlooked or even denied. Indeed, it is widely believed that no error in a mathematical demonstration can remain undetected for a long time.

Thus Thom states that, in the history of mathematics "never has a significant error slipped into a conclusion without almost immediately being discovered" (Thom 1998, 72).

But this contrasts with several historical cases, which show that errors in mathematical demonstrations may remain undetected for a long time. For example, the errors in Jordan's demonstration of the Jordan curve theorem were detected twenty years after its publication, and the errors in Kempe's demonstration of the four colour theorem were detected eleven years after its publication.

A further problem is raised by computer-aided demonstrations. For example, the computer-aided demonstration, by Lam and collaborators, that there are no finite projective planes of order ten, involves the examination of 10^{14} cases, and could contain errors due either to programming mistakes or to undetected hardware failures. Therefore, Lam states that, "as physicists have learned to live with uncertainty, so we" mathematicians "should learn to live with an 'uncertain proof'" (Lam 1990, 12).

This shows that errors in mathematical demonstrations are not only a serious problem for mathematics, but a problem which cannot be easily solved.

16.6 Fruitfulness of Error

That error is a serious problem for mathematics does not mean, however, that error cannot be fruitful. Indeed, as Lecat points out, "generally, the greater the mathematician, the more fruitful and serious his type of error" (Lecat 1935, viii).

For example, Shimura says that Taniyama "was gifted with the special capability of making many mistakes, mostly in the right direction. I envied him for this and tried in vain to imitate him, but found it quite difficult to make good mistakes" (Singh 1998, 174).

Making good mistakes is difficult because, when one makes a good mistake, one is already on the good way to find a solution. Good mistakes are fruitful because they may lead to new knowledge, since an analysis of their causes may provide useful indications for forming new hypotheses. For example, an analysis of the cause of Russell's paradox led Zermelo to formulate his axioms for set theory.

Not only an analysis of the causes of error may provide useful indications for forming new hypotheses, but may also help to formulate new concepts. For example, an analysis of the implications of his axioms for set theory helped Zermelo to formulate the so-called 'iterative concept of set', according to which "a set is something obtainable form the integers" by "iterated application of the operation 'set of'" (Gödel 1986–2002, II, 259). More specifically, a set is a member of the iterative hierarchy of sets which is formed in successive stages, starting from the integers, by transfinite iteration of the power-set operation.

16.7 Error and the Rationality of Hypothesis Formation

Of course, an analysis of the causes of error may provide useful indications for formulating new hypotheses only if the process through which hypotheses are formulated is a rational one.

Popper claims that there is no rational means for formulating new hypotheses, the latter are "the result of an almost poetic intuition" (Popper 1974, 192). Therefore, from Popper's viewpoint, an analysis of the causes of error cannot provide useful indications for formulating new hypotheses.

It could be objected that it is unfair to attribute this position to Popper, because he maintains that "we can learn from our mistakes," so controlling our hypotheses by criticism "is of decisive importance: by bringing out our mistakes it makes us understand the difficulties of the problem which we are trying to solve. This is how we become better acquainted with our problem, and able to propose more natural solutions" (ibid., vii).

However, what Popper means by this is not that an analysis of the causes of error may provide useful indications for formulating new hypotheses, but only that the refutation of a hypothesis, leading to its elimination, is a progress in itself. Indeed, Popper states that "the very refutation" of a hypothesis – "that is, of any serious tentative solution to our problem – is always a step forward that takes us nearer to the truth. And this is how we can learn from our mistakes" (ibid.). Thus the step forward consists in the refutation itself.

According to Popper, what is indispensable to formulate new hypotheses is not an analysis of the causes of error, but "the creative ability to produce new guesses, and more new guesses" (Popper 1972, 260). Now, producing new guesses is not a rational process, it requires originality, and "originality is a gift of the gods" (Popper 2002, 68). Therefore, according to Popper, formulating new hypotheses is a gift of the gods.

Contrary to Popper's view, formulating new hypotheses is not a gift of the gods. As argued in Chap. 14, formulating new hypotheses and finding solutions to problems is a rational process. Only by admitting that the process through which hypotheses are formulated is a rational process we can assert that error can lead to new knowledge. For, only so it is justified to say that an analysis of the causes of error can provide useful indications for formulating new hypotheses, and that error can be fruitful.

References

Capozzi, Mirella. 2013. *Kant e la logica I*. Naples: Bibliopolis.

Davis, Philip. 1972. Fidelity in mathematical discourse: Is one and one really two? *The American Mathematical Monthly* 79 : 252–263.

Gödel, Kurt. 1986–2002. *Collected works*. Oxford: Oxford University Press.

Hersh, Reuben. 1979. Some proposals for reviving the philosophy of mathematics. *Advances in Mathematics* 31 : 31–50.

Hilbert, David. 1967b. The foundations of mathematics. In *From Frege to Gödel: A source book in mathematical logic, 1879–1931*, ed. Jean van Heijenoort, 464–479. Cambridge: Harvard University Press.

———. 1996e. The grounding of elementary number theory. In *From Kant to Hilbert: A source book in the foundations of mathematics*, II, ed. William Ewald, 1149–1157. Oxford: Oxford University Press.

Hume, David. 1978. *A treatise of human nature*. Oxford: Oxford University Press.

Kant, Immanuel. 1992. *Lectures on logic*. Cambridge: Cambridge University Press.

———. 1998. *Critique of pure reason*. Cambridge: Cambridge University Press.

Lam, Clement. 1990. How reliable is a computer-based proof? *The Mathematical Intelligencer* 12 (1): 8–12.

Lecat, Maurice. 1935. *Erreurs de mathématiciens des origines à nos jours*. Bruxelles: Castaigne.

Mach, Ernst. 1976. *Knowledge and error: Sketches on the psychology of enquiry*. Dordrecht: Springer.

Popper, Karl Raimund. 1972. *Objective knowledge*. Oxford: Oxford University Press.

———. 1974. *Conjectures and refutations: The growth of scientific knowledge*. London: Routledge.

———. 2002. *Unended quest: An intellectual autobiography*. London: Routledge.

Proclus. 1992. *In primum Euclidis Elementorum librum commentarii*. Hildesheim: Olms.

Rota, Gian-Carlo. 1997. *Indiscrete thoughts*. Boston: Birkhäuser.

Singh, Simon. 1998. *Fermat's last theorem*. New York: Anchor Books.

Thom, René. 1998. 'Modern' mathematics: An educational and philosophical error? In *New directions in the philosophy of mathematics*, ed. Thomas Tymoczko, 67–78. Princeton: Princeton University Press.

Wang, Hao. 1996. *A logical journey: From Gödel to philosophy*. Cambridge: The MIT Press.

Wiedijk, Freek. 2008. Formal proof – Getting started. *Notices of the American Mathematical Society* 55 : 1408–1414.

Chapter 17
Knowledge and Mind

Abstract This chapter discusses the relation of knowledge to mind. It distinguishes between two views of knowledge, the view of disembodied knowledge and the view of embodied knowledge. According to the view of disembodied knowledge, the mind is separate and independent of the body, and knowledge belongs to the mind alone, it is entirely based on ideas or representations of the mind, and its object are ideas or representations of the mind. Conversely, according to the view of embodied knowledge, the mind consists of certain capacities of the body, and knowledge is a natural process based on these capacities, which are implemented through processes external to the mind. The chapter maintains that the view of disembodied knowledge is untenable and only the view of embodied knowledge is defensible.

17.1 The View of Disembodied Knowledge

What has been said in Chap. 6 about the nature of knowledge, must be completed with a discussion of the relation of knowledge to mind. In this regard, a distinction must be made between two views of knowledge, the view of disembodied knowledge and the view of embodied knowledge.

According to the view of disembodied knowledge, the mind is separate and independent of the body. Knowledge belongs to the mind alone, it is entirely based on ideas or representations of the mind, and its object are such ideas or representations.

This view of knowledge underlies representative realism (see Chap. 7), and has had large following in the modern and contemporary period.

Thus Descartes states that the mind is separate and independent of the body, because "the mind can be understood as a subsisting thing" and "nothing belonging to the body is attributed to it" (Descartes 1996, VII, 226). Knowledge "belongs to the mind alone, and not to the composite of mind and body" (ibid., VII, 83). Indeed, knowledge is based only on ideas, because "we cannot have any knowledge of things except by the ideas we conceive of them" (ibid., III, 476). Knowledge does not arise because external objects "conveyed the ideas into our mind through sense organs," but rather because they "conveyed something which gave the mind occasion to form the ideas" (ibid., VIII-2, 359). Not only knowledge is based only on ideas, but it has ideas as its object, because ideas are "all that is immediately

© Springer International Publishing AG 2017
C. Cellucci, *Rethinking Knowledge*, European Studies in Philosophy of Science 4,
DOI 10.1007/978-3-319-53237-0_17

perceived by the mind" (ibid., VII, 181). In knowledge, the mind "turns in some way toward itself, and regards one of the ideas that are within itself" (ibid., VII, 73).

Fodor states that the mind is separate and independent of the body, because "there can be no serious hope that" a being "whose psychology is effectively identical to that of some organism can be described by physical natural kind predicates" (Fodor 1974, 105). Knowledge belongs to the mind alone, because it is entirely based on "mental representations (often called 'ideas' in the older literature)" (Fodor 1981, 26). For, "we have access to the world only via the ways in which we represent it" (ibid., 241). That knowledge is entirely based on mental representations is "a Good Old Theory – one to which both Locke and Descartes (among many others) would certainly have subscribed" (ibid., 26). Not only knowledge is entirely based on mental representations, but it has mental representations as its object, because the mind is "an organ whose function is the manipulation of representations and these, in turn, provide the domain of mental processes and the (immediate) objects of mental states" (ibid., 203).

Since, for the view of disembodied knowledge, the body plays no essential role in knowledge, as Auden says, for such view, "Self was the one city, | The cell where each must find his comfort and his pain, | The body nothing but a useful favourite machine | To go upon errands of love and to run the house, | While the mind in its study spoke with its private God" (Auden 1950, 290).

17.2 Shortcomings of the View of Disembodied Knowledge

Despite its large following in the modern and contemporary period, the view of disembodied knowledge has several shortcomings.

1) The assumption that knowledge belongs to the mind alone, conflicts with the fact that many features of our knowledge depend on the kind of body we have, including our sensorimotor capacities. For example, our knowledge would be completely different if our eyes perceived electromagnetic radiations of wavelengths other than those which they actually perceive, or if our body was much larger in size that it actually is.

2) The assumption that knowledge belongs to the mind alone, conflicts with the fact that knowledge is present in all organisms, even the most elementary ones, such as the prokaryotes. The view of disembodied knowledge denies this, in particular it assumes that the prokaryotes are like thermostats in that they respond only to nomic properties, namely, properties which can be completely explained in terms of physical laws. Thus Fodor states that the prokaryotes are like thermostats because they "respond selectively" to nomic properties such as "temperature and light intensity" and do not "respond selectively to clear cases of nonnomic properties" (Fodor 1986, 12). According to Fodor, this sharply distinguishes the prokaryotes from human beings, who respond selectively to nonnomic properties, and excludes the idea that the prokaryotes have knowledge, because there is a strict connection "between mental representation and the capacity to respond selectively to nonnomic properties"

(ibid., 13). While human beings have mental representations, "it would be preposterous to attribute mental representations" to the prokaryotes, for "where would they keep them?" (ibid., 3). But, contrary to Fodor's claims, the prokaryotes do have knowledge. As stated in Chap. 6, they acquire knowledge about the environment by interpreting the data received from their sensors, on the basis of the information encoded in their DNA.

3) The assumption that knowledge belongs to the mind alone, conflicts with the fact that, as argued in Chaps. 2 and 14, emotions play an important role in knowledge, including scientific knowledge. They play such role both in the choice of problems and in the choice of hypotheses to solve them. Now, as Aristotle states, all emotions "involve a body: anger, gentleness, fear, pity, courage, joy, loving, and hating; in all these there is a concurrent affection of the body" (Aristotle, *De Anima*, A 1, 403 a 16–19). Therefore, if knowledge belongs to the mind alone, it is impossible to account for the role of emotions in knowledge.

4) The assumption that knowledge is entirely based on representations of the mind, and its object are ideas or representations of the mind, leads to an infinite regress. It implies that inside us there must be a homunculus capable of interpreting the representations, because nothing is a representation in itself but only with respect to an interpreter. But then, for the same reason, inside this homunculus there must be a homunculus capable of interpreting the representations. And so on, *ad infinitum*.

17.3 The View of Embodied Knowledge

Because of its shortcomings, the view of disembodied knowledge is untenable. An alternative to it is the view of embodied knowledge.

According to the view of embodied knowledge, the mind consists of certain capacities of the body, and knowledge is a natural process based on these capacities. The latter include sensorimotor capacities, which have an important role in knowledge since, as it will be argued below, in knowledge we make an essential use of physical or biological processes external to the mind which involve sensorimotor capacities.

That the capacities of the body on which knowledge is based include sensorimotor capacities has some important implications.

1) Since sensorimotor capacities involve processes which are mostly unconscious, not all knowledge is conscious. In particular, as already argued in Chap. 15, vision is based on inferences which are unconscious. That not all knowledge is conscious is opposed to Descartes' view that "there can be nothing in the mind, in so far as it is a thinking thing, of which it is not aware," otherwise "it would not belong to the mind *qua* thinking thing; and we cannot have any thought of which we are not aware at the very moment when it is in us" (Descartes 1996, VII, 246).

2) Since sensorimotor capacities are present not only in more complex organisms but also in simple ones, knowledge is something that human beings have in common

with non-human organisms. This is opposed to Descartes' view that animals cannot "have any knowledge of things" (ibid., X, 415). For "animals do not see as we do when we are aware that we see; but only as we do when our mind is elsewhere" (ibid., I, 413).

3) Since sensorimotor capacities are vehicles of emotions, knowledge depends on emotions. This is opposed to Descartes' view that knowledge does not depend on emotions, since emotions "are all ordained by nature to relate to the body," and "belong to the mind only in so far as it is joined with the body" (ibid., XI, 430).

17.4 Objections and Replies to the View of Embodied Knowledge

It has been pointed out above that the view of embodied knowledge implies that not all knowledge is conscious, because sensorimotor capacities involve processes which are mostly unconscious. It might be objected that knowledge requires consciousness, which only human beings have, therefore only human beings can have knowledge. But this objection is unjustified, because consciousness is neither necessary nor sufficient for knowledge. It is not necessary for knowledge, because a significant part of human knowledge is arrived at through processes which are unconscious. Such is the case of perceptual knowledge. It is not sufficient for knowledge, because beliefs arrived at consciously need not be knowledge. For example, as already pointed out in Chap. 6, Frege arrived at his Basic Law V consciously, but this did not prevent the Basic Law V from leading to a contradiction. Thus, Frege's claim that the Basic Law V was true was not knowledge, even if Frege thought so.

It has also been stated above that the view of embodied knowledge implies that knowledge is something that human beings have in common with non-human organisms. It might be objected that "animals have no knowledge," since "there is no knowledge without rational criticism" (Popper 1996, 21). But this objection is unjustified because, as stated above, not all knowledge is conscious, and knowledge which is not conscious cannot be based on rational criticism. Moreover, the objection assumes that knowledge is entirely based on representations of the mind and its object are representations of the mind, while the body plays no essential role in it. But this objection is unjustified, because knowledge is based on certain capacities of the body. We do not merely have a body, we are a body. The mind does not really exist as an independent entity, it consists of certain capacities of the body. The capacities of the body which make up the mind and on which knowledge is based, are present to a certain extent in all organisms. Therefore, it is not odd to say that knowledge is something that human beings have in common with non-human organisms.

It has also been stated above that that the view of embodied knowledge implies that knowledge depends on emotions. It might be objected that emotions are a hindrance to knowledge. But this objection is unjustified because, as already argued in Chaps. 2 and 14, emotions play an important role in knowledge, including scientific knowledge.

17.5 Processes Internal and Processes External to the Mind

As already stated, according to the view of embodied knowledge, the capacities of the body on which knowledge is based include sensorimotor capacities. Now, the latter are implemented through processes external to the mind, which greatly enhance the capacities of the mind. Therefore, knowledge involves not only processes internal to the mind, but also processes external to it.

There is large evidence that knowledge involves not only processes internal to the mind, but also processes external to it. For example, the Neanderthals and the archaic *Homo sapiens* had a mental activity that, with respect to the present human beings, was almost exclusively based on their brain. As a result, the capacity of their mental apparatus was significantly lower than that of the present human beings.

But, about 60,000 years ago, by the use of processes external to the mind, human beings could overcome the limitations of their brain, starting a cooperation between processes internal to the mind and processes external to it which led to the formation of an integrated cognitive system. This permitted human beings to pass from having thoughts about basic necessities of life, such as getting food, to having thoughts on issues such as the origin of the universe or the nature of the mind.

17.6 Examples of Processes External to the Mind

The processes external to the mind, which cooperate with those internal to the mind thus forming an integrated cognitive system, can be of many kinds. As an example, we consider four of them: writing, diagrams, symbols, deductions. All these processes involve the sight of eyes, hand movements and motor coordination between the eyes and the hands, so they are essentially based on our sensorimotor capacities.

(a) *Writing*

In order to acquire much knowledge, it is necessary to make use of writing. This has sometimes been denied.

Thus, Plato challenges the importance of writing for knowledge by narrating the myth of Theuth, the god inventor of writing, who went to Thamous, the king of Egypt, and offered him his invention, saying: "O king, this invention will make the Egyptians wiser and more capable of memory, because it has been invented as an elixir of memory and wisdom" (Plato, *Phaedrus*, 274 e 4–7). But Thamous declined the offer, replying: "What you, as the father of writing, now say out of affection for it, is the opposite of what writing is capable of making. In fact, it will produce forgetfulness in the minds of those who will have learned it, because they will not practice their memory. For, putting their trust in writing, they will depend for their memory on external characters, instead of using their own memory within them" (ibid., 274 e 9–275 a 5). So "you offer your pupils the appearance of wisdom, not true wisdom" (ibid., 275 a 6–7). Only "the discourse that is written down, with knowledge, in the mind of the learner" is "the living, breathing discourse of the man who knows, of which the written discourse can be properly said to be only a semblance" (ibid., 276 a 5–6, 8–9).

Contrary to Plato's claims, without the help of writing, no elaborate discourse could be written down, with knowledge, in the mind of the learner. Without this help, the mind would not have gone very far. Only the invention of writing made possible abstract thought and the rise of philosophy and science. Such invention must not have been an easy one if the first forms of writing known to us are not older than 3200 BCE.

(b) *Diagrams*

In order to acquire geometrical knowledge, it is necessary to make use of diagrams. For example, let us consider Euclid's demonstration of the first proposition of his *Elements*, Proposition I.1: For any finite straight line, there exists an equilateral triangle having that finite straight line as one of its sides. Euclid's demonstration runs as follows. Let *AB* be a given finite straight line. With center *A* and radius *AB*, let the circle *BCD* be described (Postulate 3: For any two distinct points *A* and *B* there exists a circle with center *A* and radius *AB*). With center *B* and radius *BA*, let the circle *ACE* be described (Postulate 3).

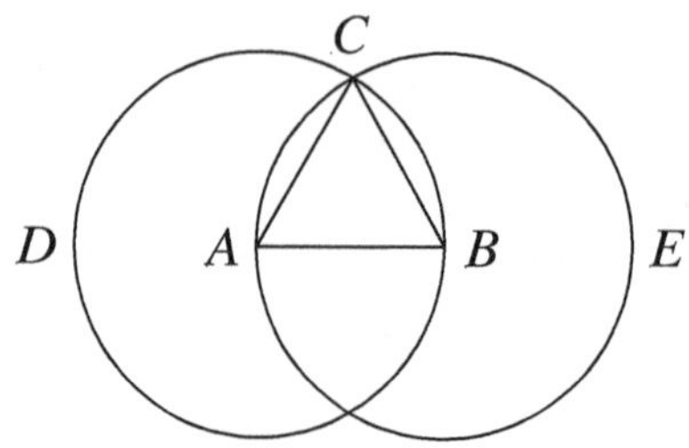

From a point *C* in which the circles meet, draw the straight lines *CA* and *CB* (Postulate 1: For any two distinct points *A* and *B* there exists a unique straight line passing through *A* and *B*). Since *A* is the center of the circle *BCD*, *AC* is equal to *AB* (Definition 15: A circle is a plane figure contained by one line such that all the straight lines falling upon it from one point among those lying within the figure equal one another). Since *B* is the center of the circle *ACE*, *BC* is equal to *AB* (Definition 15). Since *AC* and *BC* are both equal to *AB*, they are equal to one another (Common notion 1: Things which are equal to the same thing are also equal one another). Therefore, *ABC* is an equilateral triangle having the given finite straight line *AB* as one of its sides (Definition 20: An equilateral triangle is that which has its three sides equal).

Euclid's demonstration is not a deduction, because in it Euclid makes use of a fact which he does not demonstrate but tacitly assumes on the basis of the diagram, namely, that the two circles *BCD* and *ACE* meet at a point *C*.

(c) *Symbols*

Diophantus' sepulchral epitaph stated: "God granted him the sixth part of his life for his boyhood. After a twelfth more his cheeks acquired a beard. After an additional seventh, he kindled the light of marriage, and in the fifth year there came a son" (*Anthologia Graeca*, XIV, 126). Then, "when the dear but unfortunate child had reached the measure of half of his father's life, the chill grave took him. After consoling his grief for four years by this science of numbers," Diophantos "reached the end of his life" (ibid.).

The epitaph raises the problem: Determine the number of years Diophantus lived. To solve this problem, let x be this number of years. The epitaph suggests the equation $x = x/6 + x/12 + x/7 + 5 + x/2 + 4$. Reducing both sides of the equation to a common denominator, we obtain $84x = 14x + 7x + 12x + 420 + 42x + 336$, so $9x = 756$, hence $x=84$.

Equations play a role in algebra similar to the role of diagrams in geometry. As Peirce states, "the very idea" of algebra "is that it presents formulae which can be manipulated, and that by observing the effects of such manipulation we find properties not to be otherwise discerned," and such formulae "are the icons par excellence of algebra" (Peirce 1931–1958, 3.363).

(d) *Deductions*

Carroll proposes the problem: Given the following ten propositions as premisses, "to ascertain what conclusion, if any, is consequent from them" (Carroll 1977, 135):

(1) The only animals in this house are cats;
(2) Every animal is suitable for a pet, that loves to gaze at the moon;
(3) When I detest an animal, I avoid it;
(4) No animals are carnivorous, unless they prowl at night;
(5) No cat fails to kill mice;
(6) No animals ever take to me, except what are in this house;
(7) Kangaroos are not suitable for pets;
(8) None but carnivora kill mice;
(9) I detest animals that do not take to me;
(10) Animals, that prowl at night, always love to gaze at the moon.

His answer to the problem is that the conclusion which is consequent from them is: "I always avoid a kangaroo" (ibid., 187). In order to obtain this conclusion, order the premisses as follows: (1), (5), (8), (4), (6), (10), (2), (7), (9), (3). By means of the following Euler diagram,

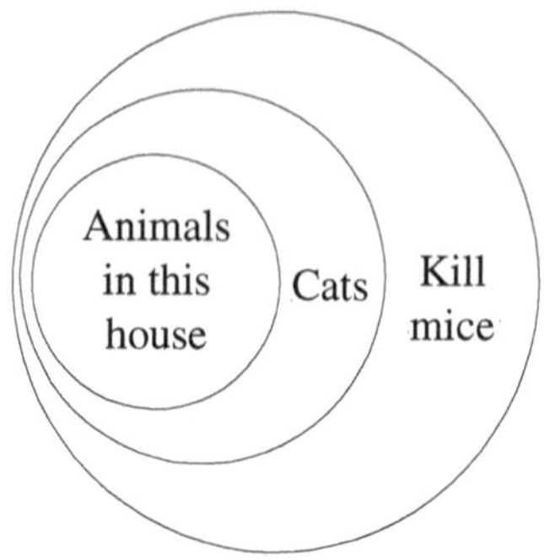

from (1) and (5), we obtain: All animals in this house kill mice. Similarly, from the latter and (8) we obtain: All animals in this house are carnivorous. From the latter and (4) we obtain: All animals in this house prowl at night. From the latter and (6) we obtain: All animals which take to me prowl at night. From the latter and (10) we obtain: All animals which take to me love to gaze at the moon. From the latter and (2) we obtain: All animals which take to me are suitable for a pet. From the latter and (7) we obtain: Kangaroos do not take to me. From the latter and (9) we obtain: I detest kangaroos. Finally, from the latter and (3) we obtain: I always avoid a kangaroo.

17.7 Strengthening of the Mind with External Processes

There is clear evidence that the use of processes external to the mind greatly enhances the capacities of the mind.

For example, preverbal babies are able to do some simple arithmetic, because there is "a number module" which "is built into all our brains when we are born" (Butterworth 1999, 9). But the capacities of the number module are limited. In order to extend them, we use a number of tools external to the mind, resulting from cultural evolution. This means that our brain "contains these two elements: a number module and our ability to use the mathematical tools supplied by our culture" (ibid., 7).

The Neanderthals and the archaic *Homo sapiens* must have had a number module, but they could not use external tools supplied by culture. Therefore, their mental capacities were confined to what their brain was able to do on its own, without the aid of such tools, which was very little.

Even Frege acknowledges that the use of processes external to the mind greatly enhances the capacities of the mind. Indeed, he maintains that "in order to think we must use sense symbols" (Frege 1964, 155). Sense "symbols hold the selfsame significance for thinking as did the discovery of using the wind to sail cross-wind for navigation. Let no one be contemptuous of symbols! A good deal depends upon a practical selection of them" (ibid., 156). Without sense symbols, "we would further hardly raise ourselves to the level of conceptual thought" (ibid.). Indeed, "the concept itself is first gained by our symbolising it, for, since the concept is of itself imperceptible to the senses, it requires a perceptible representative in order to appear to us" (ibid.).

These statements are rather surprising, since Frege also maintains that "it is of the essence of a thought to be non-temporal and non-spatial" (Frege 1979, 135). This raises the problem of explaining how, for Frege, sense symbols could open us the doors of a non-sensible world of non-temporal and non-spatial thoughts, a problem for which there is unlikely to be a solution.

17.8 External Processes and Brain Plasticity

The use of processes external to the mind required new capacities of the human brain, not available to the Neanderthals and the archaic *Homo sapiens*. Such new capacities cannot be attributed to biological evolution, because the use of processes external to the mind is too recent to have exerted any evolutionary pressure on brain evolution. They must instead be attributed to brain plasticity, the ability of the human brain to partially rewire itself as a result of experience, through interactions with external processes. Thus "cultural inventions invade evolutionary older brain circuits" (Dehaene and Cohen 2007, 384).

The architecture of the human brain is laid down under tight genetic constraints but has a certain variability range, so the nervous system can rewire itself to a certain extent as the result of experiences that occur during an individual lifetime. Specifically, when the human brain is faced with tasks for which it has not been prepared by biological evolution, it partially reconverts some cerebral circuits, initially selected to support evolutionary relevant functions, but sufficiently plastic to acquire new functions. Thus cognitive abilities, originally evolved for evolutionary purposes, are co-opted for new tasks.

That the human brain partially reconverts a limited number of cerebral circuits, originally selected to support evolutionary relevant functions, means that the human brain is not a blank slate but is highly structured. Biological evolution has built up specialized cerebral circuits that carry out specific mental functions. But the functions of the human brain are sufficiently modifiable through experience, especially in young children, so as to make it capable of functions different from those for which it was prepared by biological evolution. As a result, although we are born with the same brain as early human beings, our mind is not the same as their mind.

On the other hand, the modifications of the functions of the human brain can never overturn the architecture of the human brain resulting from biological evolution. The reconversion of a limited number of cerebral circuits is possible only insofar as it remains within a certain variability range. Thus, the new capacities reflect the intrinsic constraints of the underlying architecture of the human brain. The cultural inventions which invade evolutionary older brain circuits, inherit many of their structural constraints. The modifications in question are not genetically but only culturally heritable, so they need to be reiterated by each child since birth. Each child can participate into the phylogenetic acquisition of culture only through that personal (ontogenetic) acquisition of skills and experience.

17.9 The Mind as an Incomplete Cognitive System

The need to help the mind with external processes in order to enhance its capacities, arises from the fact that, in a basic sense, the mind is an incomplete cognitive system, and hence, in order to operate efficiently, must rely on external processes. The mind solves problems, not by a self-sustaining force, but by cooperating with these processes.

It is part of our deepest human nature to annex, exploit, and incorporate, processes external to our mind into our mental profiles. As Clark says, we are essentially "hybrid beings, joint products of our biological nature and multilayered linguistic, cultural, and technological webs" (Clark 2003, 195). It is therefore necessary to abandon the idea that the mind is a sort of ethereal substance, radically distinct from physical media. We are the conjunction of biological processes and physical processes external to the mind, and are inseparable from them. Many of those that are commonly considered as our mental capacities, are capacities of a cognitive system of which the mind is just a component.

These mental capacities are a product of evolution. The latter has shaped the mind to enter into a relationship with external processes. It is owing to this that human beings are capable of abstract thought. Processes external to the mind, such as writing, diagrams, symbols, deductions, etc., have permitted the human mind to develop abstract thought, arriving where no other animal thought had been able to arrive.

Thinking to oneself gives the impression that abstract thought is an entirely internal to the mind, but it is not so. Abstract thought would be impossible without the ability of the mind to enter into relations with external processes. Thinking to himself is only the internalization of such relations.

The propensity of the human mind to expand and become integrated with external processes, explains why human beings are such special beings, although, from a biological point of view, they are not much different from other animals, with whom they share many genes. The biological differences between human beings and other animals, that have made it possible for the human mind to integrate with external processes, are perhaps not very great, but their effects are substantial.

17.10 Distributed Character of Knowledge

The use of processes external to the mind results in a change in the subject of knowledge. Objects in the environment come to play a role in knowledge similar to that of neural states, so the subject of knowledge no longer consists in the mind alone, but rather in the mind plus certain external processes. The latter are not merely perceptual inputs or stimuli for the mind, but rather an essential component of the cognitive system. As the stick of the blind is an integral part of the way the blind perceives the world, these external processes are an integral part of the way human beings know the world. Therefore, the external processes are not simply cultural artefacts, but rather essential tools for the biological endowment of the mind.

That the subject of knowledge no longer consists in the mind alone implies that the cognitive system cannot be identified with the mind, but must be identified with the system formed by the mind plus external processes. The dividing line between the mind and external processes is not very sharp, because the higher the integration between the mind and the external processes, the less such processes are external. They are increasingly an integral part of the cognitive system.

Therefore, with respect to knowledge, it is of little importance where the cognitive processes are located, whether in the mind or outside it, and whether the processes in question are implemented by biological or physical means. It is more important to distinguish, within the cognitive system, which parts are used for the general livelihood of the system, and which parts play an essential and specific role in the acquisition of knowledge. But one should not too hastily take for granted that a certain part of the system plays an essential and specific role in such acquisition. Thus, in the case of writing, the use of pen and paper as a writing medium is inessential, one may use a computer. What is essential is only the use of some physical external medium.

Insofar as knowledge involves both a component internal to the mind and a component external to it, knowledge has a distributed character. In order to produce knowledge, what is essential is the coordination between these two components.

17.11 Knowledge and Other Minds

Of course, the external processes which greatly enhance the capacities of the mind include not only processes of a physical kind, such as writing, diagrams, symbols, deductions, etc., but also processes of a biological kind, namely, other minds. This gives rise to cognitive systems of an essentially different kind, because they consist of the mind plus external processes of a physical and a biological kind.

This does not mean that the knowing subject consists of a single collective mind. The distributed character of knowledge does not involve the problematic idea of a single collective mind, it is perfectly compatible with that of a plurality of interacting individual minds. Nevertheless, it involves abandoning the traditional division between inside and outside, recognizing that knowledge transcends the boundaries of the individual mind. Rather than analysing knowledge in terms of processes that operate on internal representations of an individual mind, we must analyse it in terms of interactions between an individual mind and external processes of a physical and a biological kind.

References

Auden, Wystan. 1950. *Collected shorter poems 1930–1944*. London: Faber & Faber.

Butterworth, Brian. 1999. *The mathematical brain*. London: Macmillan.

Carroll, Lewis. 1977. *Symbolic logic*. New York: Clarkson N. Potter.

Clark, Andy. 2003. *Natural-born cyborgs: Minds, technologies, and the future of human intelligence*. Oxford: Oxford University Press.

Dehaene, Stanislas, and Laurent Cohen. 2007. Cultural recycling of cortical maps. *Neuron* 56: 384–398.

Descartes, René. 1996. *Oeuvres*. Paris: Vrin.

Fodor, Jerry. 1974. Special sciences (or: the disunity of science as a working hypothesis). *Synthese* 28: 97–115.

———. 1981. *Representation: Philosophical essays on the foundations of cognitive science*. Brighton: Harvester Press.

———. 1986. Why paramecia don't have mental representations. *Midwest Studies in Philosophy* 10: 3–23.

Frege, Gottlob. 1964. On the scientific justification of a concept-script. *Mind* 73: 155–160.

———. 1979. *Posthumous writings*. Oxford: Blackwell.

Peirce, Charles Sanders. 1931–1958. *Collected papers*. Cambridge: Harvard University Press.

Popper, Karl Raimund. 1996. *In search of a better world*. London: Routledge.

Part IV
The Nature of Mathematical Knowledge

Chapter 18
Mathematics as Problem Solving

Abstract Contrary to the belief of several mathematicians that philosophy is irrelevant to mathematics, this chapter argues that philosophy is relevant to it, because it may expose the inadequacy of some basic mathematical concepts, it may provide an analysis of some basic mathematical concepts, and may help to formulate new rules of discovery. What is relevant to mathematics, however, is not classical philosophy of mathematics, which has basic limitations, but an alternative approach to the philosophy of mathematics according to which, like all other knowledge, mathematics is problem solving by the analytic method. In this perspective, the chapter argues that the aim of mathematics is not truth but plausibility, and that intuition has no role in mathematics.

18.1 A Global Approach to the Nature of Mathematical Knowledge

After examining the methods to acquire knowledge, this part of the book examines the nature of mathematical knowledge.

In Chap. 2 it has been maintained that philosophy is not limited to sectorial questions, it gives a global view. So, there cannot be a philosophy of mathematics alone, or of physics alone, or of biology alone, etc. For, the question of the nature of mathematics, or physics, or biology, etc., cannot be adequately approached locally, that is, in isolation from all other knowledge, but only globally, as part of a general approach to knowledge.

Accordingly, this part of the book carries out a philosophical examination of the nature of mathematical knowledge, not in isolation from all other knowledge, but as part of the general approach to knowledge outlined in the previous parts of the book.

The choice to focus on the nature of mathematics, rather than on that of physics, or biology, etc., is motivated by the fact that, with respect to natural sciences, mathematics raises special problems.

© Springer International Publishing AG 2017
C. Cellucci, *Rethinking Knowledge*, European Studies in Philosophy of Science 4,
DOI 10.1007/978-3-319-53237-0_18

18.2 The Relevance of Philosophy to Mathematics

Contrary to a philosophical examination of the nature of mathematical knowledge as part of a general approach to knowledge, several mathematicians claim that philosophy is irrelevant to mathematics.

Thus Gowers states: "Suppose a paper were published tomorrow that gave a new and very compelling argument for some position in the philosophy of mathematics" (Gowers 2006, 198). Then, "what would be the effect on mathematics? I contend that there would be almost none," because "the questions considered fundamental by philosophers are the strange, external ones that seem to make no difference to the real, internal business of doing mathematics" (ibid.).

This claim, however, is unjustified, because philosophy is relevant to mathematics in several respects.

1) *Philosophy may expose the inadequacy of some basic mathematical concepts.* For example, Berkeley exposed the inadequacy of the basic concepts of Newton's and Leibniz's infinitesimal calculus, in which, "if we remove the veil and look underneath," we "shall discover much emptiness, darkness, and confusion," or rather, "direct impossibilities and contradictions" (Berkeley 1948–1957, IV, 69). If, despite the inadequacy of its basic concepts, the infinitesimal calculus leads to correct conclusions, it is only because "two errors being equal and contrary destroy each other; the first error of defect being corrected by a second error of excess" (ibid., IV, 78). Berkeley's attack had positive effects on the development of the infinitesimal calculus. As Grabiner states, it "pointed out real deficiencies," it "served to keep the question of foundations alive and under discussion," and "pointed to the questions which had to be answered if a successful foundation were to be given" (Grabiner 2005, 27).

2) *Philosophy may provide an analysis of some basic mathematical concepts.* For example, Gödel thought that "there was no good definition of effective calculability" (Church 2011, 226). Then Turing provided such definition, through a philosophical analysis of the actions of a human being who is making calculations according to a fixed routine. On the basis of this analysis, Turing concluded: "We may now construct a machine to do the work of this" human "computer" (Turing 2001, 251). Indeed, the Turing machine is modelled on Turing's analysis of the actions of a human being who is making calculations according to a fixed routine. Eventually, Gödel declared: "That this really is the correct definition of mechanical computability was established beyond any doubt by Turing" (Gödel 1986–2002, III, 168). With "the concept of general recursiveness (or Turing's computability)" one "has for the first time succeeded in giving an absolute definition of an interesting epistemological notion" (ibid., II, 150).

3) *Philosophy may help to formulate new rules of discovery.* For example, Aristotle formulated induction as a means of scientific discovery, by stating that "people for the most part obtain universal premisses by induction" (Aristotle, *Topica*, Θ 8, 160 a 38–40). Aristotle was not the first to use induction. He himself acknowledges that "two things may be fairly ascribed to Socrates, inductive

reasoning and general definition, both of which are concerned with the starting point of scientific knowledge" (Aristotle, *Metaphysica*, M 4, 1078 b 27–30). But Aristotle was the first to formulate induction as a means of scientific discovery.

18.3 Philosophy of Mathematics and Philosophy of Mathematicians

The main reason why several mathematicians would claim that philosophy is irrelevant to mathematics, is that they believe that only mathematicians can say what mathematics is.

This belief, however, is unjustified, because we gain perspective on any human activity, including mathematics, by standing outside it. Although mathematicians are experienced in doing mathematics, this does not automatically mean that they are experienced in reflecting on what mathematics is.

Barrow even states that, "if you want to know what mathematics is, a mathematician is probably the last person to ask. Historians know what history is, sociologists know what sociology is, but most mathematicians neither know nor care what mathematics is" (Barrow 2000, 83).

The belief that only mathematicians can say what mathematics is underlies the attitude of most mathematicians towards philosophers of mathematics, which can be summarized as follows: Who are these outsiders to tell us what mathematics is? Indeed, most mathematicians assume that the philosophy of mathematics is the philosophy of mathematicians.

Thus Hersh declares: "By 'philosophy of mathematics' I mean the working philosophy of the professional mathematician, the philosophical attitude toward his work that is assumed by the researcher, teacher, or user of mathematics" (Hersh 1979, 31).

If the philosophy of mathematics is the philosophy of mathematicians, then it appears natural to believe that only mathematicians can say what mathematics is.

However, the assumption that the philosophy of mathematics is the philosophy of mathematicians is unjustified, because the philosophy of mathematicians changes from period to period, within the same period it changes from school to school, and within the same school it changes from mathematician to mathematician. In fact, no two mathematicians seem to have the same philosophy because, for each mathematician, mathematics is what he does, the way he does it.

Thus, the assumption that the philosophy of mathematics is the philosophy of mathematicians, reduces the philosophy of mathematics either to the history of mathematics, or to the sociology of mathematics, or to the psychology of mathematics. A philosophy of mathematics so intended could only give an answer to the question: How did a particular school of mathematicians at a particular age, or particular mathematicians within that school, view mathematics?

On the contrary, a genuine philosophy of mathematics should give an answer to questions such as: What is the nature of mathematical knowledge? How is mathematical knowledge acquired? How is the question of mathematical knowledge related to the general question of human knowledge? What is the role of mathematical knowledge in human life? These are typically questions for philosophy.

18.4 Basic Limitations of Classical Philosophy of Mathematics

That such questions are typically questions for philosophy does not mean, however, that they are the questions that classical philosophy of mathematics addresses.

Here, 'classical philosophy of mathematics' refers to the philosophical tradition started by Frege, which developed through the three main foundational schools of the first half of the twentieth century, namely logicism (Frege, Russell), formalism (Hilbert), and intuitionism (Brouwer, Heyting), and continued in the second half of the twentieth century through their offsprings, namely neo-logicism (Wright, Hale), neo-formalism (Bourbaki, Curry, Mac Lane), and neo-intuitionism (Bishop, Dummett).

All classical philosophy of mathematics is part of the foundationalist view of knowledge. This is at the origin of the fact that classical philosophy of mathematics has powerful limitations. Here are some of them.

1) Classical philosophy of mathematics assumes that the main philosophical questions about mathematics are the ontological question and the epistemological question, where the ontological question is whether "there are" mathematical "entities and if so what their nature is," and the epistemological question is "how mathematical beliefs come to be completely justified" (Lehman 1979, 1). This assumption, however, is problematic. On the one hand, the ontological question is unimportant because, as Thomas points out, "since we must be able to reason as dependably about what does not exist – even in a mathematical sense – as about what does, for instance in *reductio* proofs, whether some things exist or not is not of any practical importance" (Thomas 2014, 248). On the other hand, the epistemological question is empty because, by Gödel's second incompleteness theorem, mathematical propositions cannot come to be completely justified. The assumption that the main philosophical questions about mathematics are the ontological and the epistemological question, has made classical philosophy of mathematics a subject devoted to the study of questions which seem irrelevant to mathematicians. Such is Frege's question of "what the number one is" (Frege 1960, xiii). Frege states that it is "a scandal that our science should be so unclear about the first and foremost among its objects" (ibid., xiv). But virtually no mathematician would agree that this is really an urgent question.

2) Classical philosophy of mathematics assumes that the philosophy of mathematics must focus on finished mathematics, namely, mathematics as it appears in

books and journal articles, ignoring mathematics in the making. Classical philosophy of mathematics justifies this assumption with the argument that mathematics in the making cannot be accounted for rationally, because it is based on intuition. Thus Dieudonné states that "what goes on in a creative mind never has a rational 'explanation', in mathematics any more than elsewhere. All that we know is that" it involves "sudden 'illuminations', and a 'formalizing' of what these have revealed" (Dieudonné 2013, 27). Feferman states that "the mathematician at work relies on surprisingly vague intuitions and proceeds by fumbling fits and starts with all too frequent reversals. In this picture the actual historical and individual processes of mathematical discovery appear haphazard and illogical" (Feferman 1998, 77). But the argument that mathematics in the making cannot be accounted for rationally is unfounded. Since antiquity, it has been recognized that mathematical discovery is a rational process, because there is a method for solving mathematical problems, namely, the analytic method. The latter gave ancient mathematicians great heuristic power, and had a decisive role in the new developments of mathematics and physics at the beginning of the modern period. Within the analytic method, logic plays an essential role, both in the discovery of hypotheses, which are obtained by non-deductive rules, and in their justification by the plausibility test procedure described in Chap. 12, which involves deducing conclusions from hypotheses. Since mathematical discovery is a rational process, an examination of the nature of mathematics must focus on this process, further developing the analytic method.

3) Classical philosophy of mathematics assumes that the philosophy of mathematics does not contribute to the advancement of mathematics. Thus Körner states that, "as the philosophy of law does not legislate, or the philosophy of science devise or test scientific hypotheses, so – we must realize from the outset – the philosophy of mathematics does not add to the number of mathematical theorems and theories" (Körner 1986, 9). This is just a special case of the view of classical analytic philosophy that philosophy does not contribute to the advancement of knowledge, it only clarifies what we already know. But saying that the philosophy of mathematics does not contribute to the advancement of mathematics is unjustified because, as argued above, philosophy may contribute to the advancement of mathematics in several ways. In particular, it may contribute to the advancement of mathematics by further developing the analytic method. Even Frege acknowledges that "a development of method, too, furthers science. Bacon, after all, thought it better to invent a means by which everything could easily be discovered than to discover particular truths, and all steps of scientific progress in recent times have had their origin in an improvement of method" (Frege 1967, 6).

4) Classical philosophy of mathematics assumes that the question of the nature of mathematics can be adequately approached locally, namely, in isolation from all other knowledge, rather than globally, that is, as part of a general approach to knowledge. Thus George and Velleman state that "mathematics is the purest product of conceptual thought," which "sets it apart from all else" (George and Velleman 2002, 1). The effect of this is that, as Hersh points out, "the philosophy of mathematics as practiced in many articles and books is a thing unto itself, hardly connected either to living mathematics or to general philosophy. But how can it be

claimed that the nature of mathematics is unrelated to the general question of human knowledge?" (Hersh 2014, 68).

5) Classical philosophy of mathematics assumes that the question of the nature of mathematics can be approached taking no account of what cognitive architecture makes mathematical thought possible. Thus George and Velleman state that, in order to understand the nature of mathematics, we need not ask "such questions as 'What brain, or neural activity, or cognitive architecture makes mathematical thought possible?' or 'What kind of environment is needed to facilitate the development of the capacity for such thought?'" (George and Velleman 2002, 2). These questions "focus on phenomena that are really extraneous to the nature of mathematical thought itself," namely, "the neural states that somehow carry thought", while "philosophers, by contrast, are interested in the nature of those thoughts themselves, in the content carried by the neural vehicles" (ibid.). But this is unjustified, because mathematics is a human product, and the only mathematics human beings can produce is what their brain, neural activity and cognitive architecture enable them to produce. Thus the nature of mathematical thought essentially depends on the biological makeup of human beings.

This last part of the book aims at outlining an approach to mathematics, hopefully not subject to the limitations of classical philosophy of mathematics.

18.5 Natural Mathematics

First, a distinction must be made between natural mathematics and artificial mathematics. Natural mathematics is the mathematics embodied in organisms as a result of biological evolution, artificial mathematics is mathematics as a discipline.

In Chap. 9 it has been mentioned that experiments have revealed that human beings have some innate knowledge of space and number. Specifically, human beings have "a set of cognitive systems that are phylogenetically ancient, innate, and universal across humans: systems of core knowledge" about space and number, which capture "the primary information in the system of Euclidean plane geometry" and "the primary information in the system of positive integers" (Spelke 2011, 287). These systems of core knowledge are a result of biological evolution and are essential for survival, so they have a biological role. They form what may be called 'natural mathematics' – in particular 'natural geometry' and 'natural arithmetic' – to stress that they are a result of biological evolution. Natural geometry and natural arithmetic were the first forms of mathematics to develop.

That human beings have some innate knowledge of space and number contrasts with the empiricist view that the mind is originally "white paper, void of all characters, without any ideas" (Locke 1975, 104). According to this view, children are born without knowledge, and gradually learn to deal with the world as they are confronted with it, since the responses that a brain produces are shaped by experience alone. In particular, there is no natural mathematics, all of mathematics is a cultural product.

The empiricist view has been very influential until recently. For example, Kitcher states that "children come to learn the meanings of 'set', 'number', 'addition' and to accept basic truths of arithmetic by engaging in activities of collecting and segregating," such as collecting and segregating "blocks on the floor" (Kitcher 1983, 107–108). So they learn that, "if one performs the collective operation called 'making two', then performs on different objects the collective operation called 'making three', then performs the collective operation of combining, the total operation is an operation of 'making five'" (ibid., 108). Since children are incapable of making these collective operations before two or three years of age, this means that, according to Kitcher, children cannot come to learn the meanings of 'set', 'number', 'addition' and to accept basic truths of arithmetic before two or three years of age.

But the empiricist view seems untenable on account of the fact that experiments have shown that human beings have some innate knowledge of space and number. Therefore, the empiricist view seems to be no longer viable or sustainable.

18.6 Natural Mathematics and Evolution

Natural mathematics, being the mathematics embodied in organisms as a result of biological evolution, provides some basic tools by which human beings solve their problems, starting from that of survival.

Human beings live in a world which changes continually and irregularly, so they are confronted all the time with new problems, some of which are essential for their survival. Some of these problems are mathematical in kind. For example, in order to survive, our remotest ancestors had to face, on the one hand, the problem of how to capture animals for food, on the other hand, the problem of how to avoid becoming food for the great predators. In order to solve these problems, they had to solve other ones, such as the problem of how to recognize the shapes of bodies, which was necessary to classify bodies as good, bad, or indifferent; the problem of how to localize the position of bodies, which was necessary to make pursuit or escape movements; and the problem of how to evaluate the number of bodies, which was necessary to determine the most appropriate movements.

These problems were mathematical in kind. Indeed, solving the problem to recognize the shapes of bodies or the problem to localize their position, required making geometrical hypotheses; solving the problem to evaluate the number of bodies, required making arithmetical hypotheses. Human beings obtained these hypotheses by the analytic method, on the basis of which they formed them by non-deductive inferences from experience, and established their plausibility through a comparison with experience. The hypotheses which did not survive such comparison were eliminated. This was an adaptation to the environment, and in this sense, like all knowledge, mathematical knowledge is part of a natural process of adaptation to the environment. Natural mathematics, specifically natural geometry or natural arithmetic, has a main place in this process, because it contributes to satisfy needs that are essential for survival, so, in a very basic sense, natural mathematics is important in human life.

Hart claims that, if "mathematics has attracted philosophical reflection since Plato, and most of the great philosophers since then have had at least something to say about it," it "is not because of the importance of mathematics in human life. Agriculture is at least as important, but there is no philosophy of farming" (Hart 1996, 1). But this claim is unjustified. The importance of agriculture in human life is obvious and requires no elucidation. On the contrary, the importance of mathematics in human life is not so clear and requires elaboration. Therefore, it is only too natural that mathematics should have attracted philosophical reflection since Plato.

18.7 Artificial Mathematics

Natural mathematics, however, has a restricted scope. Each of the systems of core knowledge mentioned in the previous section "is limited in its domain of application" and "in the information that it makes available," in particular, none of them has the power of Euclidean geometry or of the system of integers, so the "geometrical and numerical concepts that we possess as adults may not be given to us as infants" (Spelke 2011, 313). Such concepts are a result of cultural evolution. They do not belong to natural mathematics but to what may be called 'artificial mathematics', or mathematics as a discipline – mathematics as a organized field of study.

Unlike natural mathematics, which is innate, artificial mathematics is acquired. The name 'artificial mathematics' is intended to stress that it is a human product rather than a natural product, not being a result of biological evolution. Calling it 'artificial mathematics' may appear awkward, but the first dictionary meaning of 'artificial' is, 'produced by human art or effort rather than originating naturally', which is exactly what is meant here. Devlin designates artificial mathematics with the name "abstract mathematics" (Devlin 2005, 249). But calling it 'abstract mathematics' is somewhat limiting, because the term 'abstract mathematics' is usually used to designate only a part of mathematics as a discipline, so the term 'artificial mathematics' seems more suitable,

Cooper claims that artificial "mathematics must itself be evolutionarily reducible" (Cooper 2001, 135). But this is unjustified, because artificial mathematics has arisen too recently to be a direct product of biological evolution, it must have been a product of cultural evolution. This required new capacities. Human beings could acquire them only by an ever more extensive use of processes external to the mind, from material to symbolic ones, such as pictures, models, maps, and symbol systems. As Grosholz states, artificial "mathematics cannot be carried out without good notation and diagrams" (Grosholz 2007, xiii). Such processes external to the mind formed an integrated whole with those internal to it.

Artificial mathematics is obtained with the help of processes external to the mind. It goes beyond the systems of core knowledge of natural mathematics, because it makes use of cultural artefacts, such as pictures, models, maps, and symbol systems. Therefore, natural mathematics and artificial mathematics are distinct. On the other hand, they are not opposed, because artificial mathematics is the product of brains that are a result of biological evolution, and is based on them. Rather, between natural mathematics and artificial mathematics there is continuity, not in the sense that artificial mathematics is reducible to natural mathematics, but in the sense that natural mathematics and artificial mathematics are based on the very same capacities, which are a result of biological evolution.

Moreover, natural mathematics and artificial mathematics are based on the same method, namely, the analytic method. As already stated in Chap. 6, our remotest ancestors solved their survival problem making hypotheses about the location of predators or prey, starting from the clues they found in the environment, such as footprints, crushed or bent grass and vegetation, bent or broken branches or twigs, mud displaced from streams, excrements, and so on. Much in the same way, scientists solve problems about the world, making hypotheses by the analytic method, starting from the clues they find in nature.

18.8 Mathematics and Truth

That natural mathematics and artificial mathematics are based on the same method, namely the analytic method, means that the only mathematical knowledge we can have is plausible knowledge, and that the aim of mathematics is plausibility. This contrasts with the view of many mathematicians, that the aim of mathematics is truth, hence to obtain absolute certainty.

For example, Byers states that "mathematics is about truth: discovering the truth, knowing the truth, and communicating the truth to others" (Byers 2007, 327). Mathematics "is a way of using the mind with the goal of knowing the truth, that is, of obtaining certainty" (ibid.). For, "truth is normally seen as knowledge that is certain" (ibid., 330). The certainty of mathematics is one about which one cannot "have the slightest doubt" (ibid., 328). It is absolute certainty. Indeed, "mathematical truth has this kind of certainty, this quality of inexorability. This is its essence" (ibid.).

However, the belief that the aim of mathematics is truth, hence to obtain absolute certainty, is unjustified because, by Gödel's second incompleteness theorem, mathematical knowledge cannot be said to be absolutely certain. This implies that, if the aim of mathematics is truth, then this aim will be generally unachievable.

It could be objected that this implication depends on the claim that, by Gödel's second incompleteness theorem, mathematical knowledge cannot be said to be absolutely certain, and this claim is self-defeating. For, if mathematical knowledge cannot be said to be absolutely certain, then Gödel's second incompleteness theorem, being a mathematical result, cannot be said to be absolutely certain. But the conclusion that mathematical knowledge cannot be said to be absolutely certain is based on Gödel's second incompleteness theorem. Therefore, this conclusion cannot be said to be absolutely certain.

This objection, however, is unjustified because the claim that, by Gödel's second incompleteness theorem, mathematical knowledge cannot be said to be absolutely certain, does not depend on the assumption that Gödel's second incompleteness theorem can be said to be absolutely certain. It is a reduction to the impossible, since it is of the following kind. Let us suppose, for argument's sake, that mathematical knowledge can be said to be absolutely certain. Then Gödel's second incompleteness theorem, being a mathematical result, can be said to be absolutely certain. But, by Gödel's second incompleteness theorem, mathematical knowledge cannot be said to be absolutely certain. Hence mathematical knowledge cannot be said to be absolutely certain. Contradiction. Therefore, mathematical knowledge cannot be said to be absolutely certain. This conclusion does not depend on the assumption that Gödel's second incompleteness theorem can be said to be absolutely certain.

Since, if the aim of mathematics is truth, then this aim will be generally unachievable, we may conclude that the aim of mathematics cannot be truth. As Kline states, "mathematics is a body of knowledge. But it contains no truths. The contrary belief, namely, that mathematics is an unassailable collection of truths," is "a popular fallacy" (Kline 1964, 9).

18.9 Mathematics and Plausibility

An alternative to the view that the aim of mathematics is truth is that the aim of mathematics is plausibility.

That the aim of mathematics is plausibility is clear from the fact that both natural and artificial mathematics are based on the analytic method. In Chap. 14 it has been argued that knowledge is problem solving by the analytic method. This includes mathematics. Like all knowledge, mathematics is problem solving by the analytic method. This explains why the aim of mathematics is not truth but plausibility.

That mathematics is problem solving by the analytic method does not mean that mathematics is only problem solving, if only because problems, in order to be solved, must first be posed (on problem posing, see Chap. 14). Nevertheless, problem solving is the core of mathematical activity, so it seems justified to say that problem solving is the essential character of mathematics.

That mathematics is problem solving by the analytic method implies that no solution to a mathematical problem is final. As Poincaré states, there are not "solved problems and others which are not; there are only problems more or less solved," where "it often happens however that an imperfect solution guides us toward a better one" (Poincaré 2013, 377–378).

The view that mathematics is problem solving by the analytic method provides a sense in which there are not solved problems and others which are not, there are only problems more or less solved. No hypothesis is absolutely justified, any hypothesis which provides a solution to a problem is liable to be replaced with another one when new data emerge, so every solution is always provisional. The view that mathematics is problem solving by the analytic method also provides a sense in which it often happens that an imperfect solution guides us toward a better one. An analysis of the reasons why a hypothesis gives an only imperfect solution to a problem, may provide useful indications for forming a better one.

That mathematics is problem solving by the analytic method shows the limitations of the widespread belief that "the two main pillars of mathematics are deductive reasoning and abstraction," and that the Greeks "essentially created mathematics, as we know it today, based on these two pillars" (Ó Cairbre 2009, 42). In the analytic method, the hypotheses for the solution of a problem are obtained by non-deductive rules and, as it is apparent from Hippocrates of Chios, the Greeks essentially created mathematics, as we know it today, based on the analytic method. As to abstraction, its limits are discussed in Chap. 24.

18.10 Origin of the View That Mathematics Is Problem Solving

The view that mathematics is problem solving by the analytic method can be traced back to Plato, who states that we will approach geometry "by working through problems" (Plato, *Respublica*, VII 530 b 6). And we will solve problems "making use of a hypothesis," meaning by this "what geometers often do" (Plato, *Meno*, 86 e 3–5). That is, we will solve problems by the analytic method.

According to a widespread view, Plato contrasts the method of philosophy with the method of mathematics. Thus, Bostock states that "Plato thinks of the method of mathematics as one that starts by assuming some hypotheses and then goes 'downwards' from them (i.e. by deduction), whereas the method of philosophy (i.e. dialectic) is to go 'upwards' from the initial hypotheses, finding reasons for them (when they are true), until eventually they are shown to follow from an 'unhypothetical first principle'" (Bostock 2009, 13–14).

But it is not so. Plato does not contrast the method of philosophy with the method of mathematics, on the contrary, as argued in Chap. 12, he takes the method of philosophy to be the same as the method of mathematics. Indeed, he bases philosophy on the method that Hippocrates of Chios used to solve mathematical problems, such as the problem of doubling the cube or the problem of the quadrature of certain lunules, namely the analytic method. As Knorr states, Plato "introduces a method of reasoning 'from hypothesis' which, but for its name, is identical" to that "used, for instance, by Hippocrates in his attack on the cube duplication" (Knorr 1993, 71). Thus, Plato's method of reasoning from hypothesis is identical to the analytic method. Therefore, both the method of philosophy and the method of mathematics are 'upward' methods.

18.11 An Objection to the View that Mathematics is Problem Solving

The view that mathematics is problem solving by the analytic method focuses on plausibility instead of truth. This means that no solution to a mathematical problem is final, every solution is revisable. Against this, it could be objected that the things of mathematics are determinate and unchanging as the things of nature are not, and their justification is final and not revisable.

Thus Grosholz states: "I do not think that the naturalist view that we should focus on plausibility instead of truth, and construe justification as always somewhat empirical and revisable, captures the peculiar nature of mathematical reasoning" (Grosholz 2015, 139). For, "the things of mathematics are determinate and unchanging as the things of nature are not: why would we ever revise our belief that the sum of the squares of the two legs of a Euclidean right triangle is equal to the square of its hypotenuse, or that 2 +3 = 5?" (ibid.).

However, the claim that the things of mathematics are determinate and unchanging as the things of nature are not is problematic because, as Rodin points out, virtually all "mathematical concepts change through time: some of them change their content preserving the name (and something like 'general idea'), some new mathematical concepts regularly emerge and some old concepts stop growing and die off" (Rodin 2014, 215). In fact, "the phenomenon of conceptual change in mathematics is so evident that nobody can deny its very existence" (ibid.).

This depends on the fact that mathematics is a means by which we make the world understandable to ourselves. Now, the world poses ever newer problems and challenges, with respect to which the existing hypotheses and concepts often turn out to be inadequate, so they must be modified or replaced with new ones.

In particular, if we would not revise our belief that the sum of the squares of the two legs of a Euclidean right-angled triangle is equal to the square of its hypotenuse, or that 2 + 3 = 5, this is not because the things of mathematics are determinate and unchanging as the things of nature are not. It is rather because we have evolved in a world in which, at our scale, the sum of the squares of the two legs of a Euclidean right-angled triangle is equal to the square of its hypotenuse, and 2 + 3 = 5. The belief in question is based on natural mathematics, specifically on natural geometry and natural arithmetic, and natural mathematics is essential for survival. Natural mathematics would have been different if we had evolved in a different world. But, if natural mathematics had been different, it is dubious that we would have been able to survive in this world.

This is already pointed out by Poincaré, who states that, if our natural geometry "has three dimensions, this is because it has adapted itself to a world having certain properties," and "it is in order to be able to live in this world that this" natural geometry "has been established" (Poincaré 2013, 427). Although "we could conceive, living in our world, thinking beings whose" natural geometry "would be four-dimensional and who consequently would think in hyperspace," it is dubious "that such beings" could live "and defend themselves against the thousand dangers by which they would be assailed" (ibid., 427–428).

Natural geometry and natural arithmetic were the first forms of mathematics to arise because, as stated above, the capacity to recognize the shapes of bodies, or to localize their position, or to evaluate their number, was essential for survival. A reflection of the fact that natural geometry and natural arithmetic were the first forms of mathematics to arise, is that the Rhind Mathematical Papyrus, one of the oldest mathematical documents, dating to around 1650 BC, is a kind of instruction manual in geometry and arithmetic.

Since natural mathematics is mathematics which is embodied in organisms as a result of biological evolution, the things of natural mathematics are not determinate and unchanging as the things of nature are not, and their justification is not final and unrevisable. So are also the things of artificial mathematics and their justification, because all propositions of artificial mathematics are based on hypotheses which, by Gödel's second incompleteness theorem, cannot be said to be true but only plausible.

As Goodman states, "the traditional philosophy holds that mathematical theorems, once generally accepted, are accepted forever, whereas scientific laws are often overthrown," but in fact "the results of mathematics are no more certain or everlasting than the results of any other science, even though, for sociological reasons, our histories of mathematics tend to disguise that fact" (Goodman 1991, 126). For example, when the theory of infinite series of eighteenth century mathematicians "was replaced by the modern one at the beginning of the nineteenth century, essentially by the work of Cauchy, many of Euler's results were simply discarded," and "almost all of his arguments" were "replaced by new argument" (ibid., 125–126). But our histories of mathematics "do not describe this as a scientific revolution," instead "they say that Euler made some mistakes which were later corrected by Cauchy" (ibid., 126). This is a misrepresentation of historical reality.

18.12 Mathematics and Intuition

The view that mathematics is problem solving by the analytic method implies that intuition has no role in mathematics. This is a special case of the fact that, as argued in Chap. 12, intuition has no role in solving problems by the analytic method.

On the contrary, several mathematicians and philosophers claim that intuition has a central role in mathematics, because intuition is the faculty by which we grasp mathematical objects.

Thus Poincaré states that, "to make arithmetic, as to make geometry, or to make any science, something else than pure logic is necessary. To designate this something else we have no word other than intuition" (Poincaré 2013, 215). Intuition is "a sort of sudden illumination, after an unconscious working somewhat prolonged" (ibid., 391). Such "a sudden illumination seizes upon the mind of the mathematician" (ibid., 392). It makes the mathematician "divine hidden harmonies and relations" (ibid., 385).

Bourbaki states that we "can not over-emphasize the fundamental role played in" the mathematician's "research by a special intuition, which is not the popular sense-intuition, but rather a kind of direct divination (ahead of all reasoning) of the normal behavior" of "mathematical beings, with whom a long acquaintance has made him as familiar as with the beings of the real world" (Bourbaki 1950, 227). For the mathematician, this special intuition "is like a sudden modulation which orients at one stroke in an unexpected direction the intuitive course of his thought, and which illuminates with a new light the mathematical landscape in which he is moving about" (ibid.).

Davis and Hersh state that "those ideas whose properties are reproducible are called mathematical objects, and the study of mental objects with reproducible properties is called mathematics. Intuition is the faculty by which we can consider or examine these (internal mental) objects" (Davis and Hersh 1981, 399).

But the claim that intuition has a central role in mathematics, because intuition is the faculty by which we grasp mathematical objects, is problematic. As Dieudonné points out, it is doubtful that anyone seriously has an intuition of an integer greater than ten (see Chap. 10). It is even more doubtful that anyone seriously has an intuition, say, of large cardinals.

Admittedly, some cognitive neuroscientists claim that mathematical intuition is a valid concept, because the systems of core mathematical knowledge are sources of our geometrical and numerical intuitions.

Thus Dehaene claims that "all great mathematicians appeal to their 'intuition'," and "cognitive neuroscience research shows that mathematical intuition is a valid concept" which "relates to the availability of 'core knowledge' associated with evolutionarily ancient and specialized cerebral subsystems" (Dehaene 2009, 232).

But a large part of the systems of core knowledge of which natural mathematics consists is embodied in our perceptual system and, as argued in Chap. 15, perception is based on inference, so core knowledge involves inference. Therefore, intuition cannot relate to the availability of core knowledge.

References

Barrow, John David. 2000. *Between inner space and outer space: Essays on science, art, and philosophy*. Oxford: Oxford University Press.

Berkeley, George. 1948–1957. *Works*. London: Nelson.

Bostock, David. 2009. *Philosophies of mathematics: An introduction*. Oxford: Wiley-Blackwell.

Bourbaki, Nicholas. 1950. The architecture of mathematics. *The American Mathematical Monthly* 57 : 221–232.

Byers, William. 2007. *How mathematicians think*. Princeton: Princeton University Press.

Church, Alonzo. 2011. Letter to Stephen Kleene, 29 November 1935. In Appendix B to Rod Adams, *An early history of recursive functions and computability: From Gödel to Turing*, 223–227. Boston: Docent Press.

Cooper, William. 2001. *The evolution of reason: Logic as a branch of biology*. Cambridge: Cambridge University Press.
Davis, Philip J., and Reuben Hersh. 1981. *The mathematical experience*. Boston: Birkhäuser.
Dehaene, Stanislas. 2009. Origins of mathematical intuitions: The case of arithmetic. *Annals of the New York Academy of Sciences* 1156 : 232–259.
Devlin, Keith. 2005. *The math instinct: Why you're a mathematical genius (along with lobsters, birds, cats, and dogs)*. New York: Thunder's Mouth Press.
Dieudonné, Jean. 2013. *Mathematics – the music of reason*. Berlin: Springer.
Feferman, Solomon. 1998. *In the light of logic*. Oxford: Oxford University Press.
Frege, Gottlob. 1960. *The foundations of arithmetic: A logico-mathematical enquiry into the concept of number*. New York: Harper.
———. 1967. *Begriffsschrift*, a formula language, modeled upon that of arithmetic, for pure thought. In *From Frege to Gödel: A source book in mathematical logic, 1879–1931*, ed. Jean van Heijenoort, 5–82. Cambridge: Harvard University Press.
George, Alexander, and Daniel J. Velleman. 2002. *Philosophies of mathematics*. Oxford: Blackwell.
Gödel, Kurt. 1986–2002. *Collected works*. Oxford: Oxford University Press.
Goodman, Nicolas D. 1991. Modernizing the philosophy of mathematics. *Synthese* 88 : 119–126.
Gowers, Timothy. 2006. Does mathematics need a philosophy? In *18 unconventional essays on the nature of mathematics*, ed. Reuben Hersh, 182–200. New York: Springer.
Grabiner, Judith. 2005. *The origins of Cauchy's rigorous calculus*. Mineola: Dover.
Grosholz, Emily. 2007. *Representation and productive ambiguity in mathematics and the sciences*. Oxford: Oxford University Press.
———. 2015. Review of Carlo Cellucci, *Rethinking logic: Logic in relation to mathematics, evolution and method. Philosophia Mathematica* 23 : 136–140.
Hart, Wilbur Dyre. 1996. Introduction. In *The philosophy of mathematics*, ed. Wilbur Dyre Hart, 1–13. Oxford: Oxford University Press.
Hersh, Reuben. 1979. Some proposals for reviving the philosophy of mathematics. *Advances in Mathematics* 31 : 31–50.
———. 2014. *Experiencing mathematics: What do we do, when we do mathematics?* Providence: American Mathematical Society.
Kitcher, Philip. 1983. *The nature of mathematical knowledge*. Oxford: Oxford University.
Kline, Morris. 1964. *Mathematics in Western culture*. Oxford: Oxford University Press.
Knorr, Wilbur Richard. 1993. *The ancient tradition of geometric problems*. Mineola: Dover.
Körner, Stephan. 1986. *The philosophy of mathematics: An introductory essay*. Mineola: Dover.
Lehman, Hugh. 1979. *Introduction to the philosophy of mathematics*. Oxford: Blackwell.
Locke, John. 1975. *An essay concerning human understanding*. Oxford: Oxford University Press.
Ó Cairbre, Fiacre. 2009. The importance of being beautiful in mathematics. *IMTA Newsletter* 109: 29–45.
Poincaré, Henri. 2013. *The foundations of science: Science and hypothesis – The value of science – Science and method*. Cambridge: Cambridge University Press.
Rodin, Andrei. 2014. *Axiomatic method and category theory*. Berlin: Springer.
Spelke, Elizabeth. 2011. Natural number and natural geometry. In *Space, time and number in the brain*, ed. Stanislas Dehaene and Elizabeth Merrit Brannon, 287–317. London: Elsevier.
Thomas, Robert. 2014. Reflections on the objectivity of mathematics. In *From a heuristic point of view*, ed. Cesare Cozzo, and Emiliano Ippoliti, 241–256. Newcastle upon Tyne: Cambridge Scholars Publishing.
Turing, Alan Mathison. 2001. *Collected works: Mathematical logic*. Amsterdam: North-Holland.

Chapter 19
Mathematical Objects, Definitions, Diagrams

Abstract The aim of this chapter is to give answer to the question: What is mathematics about? The chapter maintains that, being problem solving by the analytic method, mathematics is about objects which are hypotheses human beings make to solve mathematical problems. Therefore, mathematical objects exist only in the minds of the mathematicians who hypothesize them, and in the minds of the people who make use of them. Like mathematical objects, the mathematical definitions by which mathematical objects are introduced are hypotheses human beings make to solve mathematical problems by the analytic method. Mathematical diagrams also fit in the analytic method, since they are important tools to solve problems by that method.

19.1 What Mathematics Is About

In Chap. 18 it has been argued that mathematics is problem solving by the analytic method. This answers the question: What is the method of mathematics? But another important question about mathematics is: What is mathematics about?

Proclus tells us that "geometry was first discovered by the Egyptians and originated in the measurement of lands; this was necessary for them because of the inundations of the Nile, which washed away the boundaries between their properties. It is not surprising that the discovery of this as well as other sciences has sprung from practical needs" (Proclus 1992, 64.18–24).

In that period, mathematics was about concrete, physical realities, such as rectangular or triangular pieces of land. But, passing from Egypt to Greece, mathematics became an abstract discipline, detached from its empirical origin. Specifically, "Pythagoras transformed the study of geometry into a liberal discipline, tracing its first principles and studying its problems from a purely abstract and theoretical viewpoint" (ibid., 65.15–19).

Detaching mathematics from its empirical origins led the Greeks to assume that mathematics is about non-physical things, existing independently of the human mind, and knowable only by intuition.

Thus Plato states that, although mathematicians "use visible figures and make their claims about them," they "are not thinking of them but of those ideas of which they are likeness. They pursue their inquiry for the sake of the square itself and the

© Springer International Publishing AG 2017
C. Cellucci, *Rethinking Knowledge*, European Studies in Philosophy of Science 4,
DOI 10.1007/978-3-319-53237-0_19

diagonal itself, and not for the sake of the diagonal they draw, and similarly with the others" (Plato, *Respublica*, VI, 510 d 5–e 1). They use the visible figures they draw "as only images, but what they really seek is to get sight of those ideas, which can be seen only by the mind" (ibid., VI, 510 e 3–511 a 1). In fact, "there really exist these ideas in themselves, that we cannot perceive with the senses but only with intuition" (Plato, *Timaeus*, 51 d 4–5). Then, "things being so, we must admit that there exists a kind of reality which is unchanging, ungenerated, and imperishable," which is "invisible and imperceptible with other senses," and which "only intuition has been granted to contemplate" (ibid., 51 e 6–52 a 4).

19.2 Mathematical Platonism

The assumption that mathematics is about non-physical things, existing independently of the human mind, is at the core of the contemporary view of mathematics called 'mathematical platonism'. The most articulated formulation of mathematical platonism is Gödel's. His formulation can be presented as follows.

1) Mathematics is about "a non-sensual reality, which exists independently both of the acts and of the dispositions of the human mind and is only perceived, and probably perceived very incompletely, by the human mind" (Gödel 1986–2002, III, 323). So "mathematical objects and facts (or at least something in them) exist objectively and independently of our mental acts and decisions" (ibid., III, 311). The "objects and theorems of mathematics are as objective and independent of our free choice and our creative acts as is the physical world" (ibid., III, 312, footnote 17). In particular, "the set-theoretical concepts and theorems describe some well-determined reality, in which Cantor's conjecture must be either true or false" (ibid., II, 260). This well-determined reality is a non-sensual one, because "the objects of transfinite set theory" do "not belong to the physical world, and even their indirect connection with physical experience is very loose" (ibid., II, 267).

2) But, "despite their remoteness from sense experience, we do have something like a perception also of the objects of set theory," and this kind of perception is "mathematical intuition" (ibid., II, 268). The latter is a superior faculty, which is not "derived from subconscious induction or Darwinian adaptation" (ibid., III, 354).

3) Mathematical intuition "is sufficiently clear to produce the axioms of set theory and an open series of extensions of them" (ibid., II, 268). It gives us insights both "for obtaining unambiguous answers to the questions of transfinite set theory," such as Cantor's continuum hypothesis, and "for the solution of the problems of finitary number theory (of the type of Godlbach's conjecture)" (ibid., II, 269). Thus, mathematical intuition gives us "knowledge of the abstract concepts" of transfinite set theory, and permits us "to gain insights into the solvability, and the actual methods for the solution, of all meaningful mathematical problems" (ibid., III, 383).

4) We can "extend our knowledge of these abstract concepts, i.e., to make these concepts themselves precise and to gain comprehensive and secure insight into the fundamental relations that subsist among them, i.e., into the axioms that hold for

them," by cultivating (deepening) knowledge of the abstract concepts themselves" (ibid.). The procedure will consist "in focusing more sharply on the concepts concerned by directing our attention in a certain way, namely, onto our own acts in the use of these concepts, onto our powers of carrying out our acts, etc." (ibid.). This will "produce in us a new state of consciousness in which we describe in detail the basic concepts we use in our thought, or grasp other basic concepts hitherto unknown to us" (ibid.). So we will arrive at an "intuitive grasping of ever newer axioms that are logically independent from the earlier ones, which is necessary for the solvability of all problems even within a very limited domain" (ibid., III, 385).

5) Mathematical intuition, however, "need not be conceived as a faculty giving an immediate knowledge of the objects concerned. Rather," as "in the case of physical experience, we form our ideas also of those objects on the basis of something else which is immediately given" (ibid., II, 268). This something else here is the concept of set. In fact, "there is a close relationship between the concept of set" – given by the iterative concept of set described in Chap. 16 – "and the categories of pure understanding, in Kant's sense. Namely, the function of both is 'synthesis', i.e., the generation of unities out of manifolds (e.g., in Kant, of the idea of one object out of its various aspects)" (ibid., II, 268, footnote 40).

6) Besides mathematical intuition, there exists also "another (though only probable) criterion of truth of mathematical axioms, namely their fruitfulness in mathematics and, one may add, possibly also in physics" (ibid., II, 269). Fruitfulness means "fruitfulness in consequences, in particular in 'verifiable' consequences, i.e., consequences demonstrable without the new axiom" (ibid., II, 261). The "simplest case of an application of the criterion under discussion arises when some set-theoretical axiom has number-theoretical consequences verifiable by computation up to any given integer" (ibid.).

19.3 Limitations of Mathematical Platonism

Mathematical platonism has had widespread following among mathematicians. Nevertheless, it has some serious limitations which make it untenable. With reference to Gödel's formulation, the main limitations are as follows.

1) Gödel claims that mathematics is about a non-sensual reality, which exists independently both of the acts and of the dispositions of the human mind. But then, why is mathematics useful in physics?

Gödel's answer is that mathematics "might help us considerably in knowing" the physical reality "if in some respect, or as to certain parts, it happened to be similar or isomorphic to it. In fact this would correspond closely to the manner in which mathematics is applied in theoretical physics" (Gödel 1986–2002, III, 353–354, footnote 44). Thus, according to Gödel, mathematics can be useful in physics if it is isomorphic to the physical reality.

But this conflicts with the fact that different mathematical theories are used to deal with the same physical phenomena. For example, on the one hand, Weyl states that group theory is exactly what is "necessary for an adequate description of the quantum mechanical relations" (Weyl 1950, xxi). On the other hand, von Neumann states that Hilbert space theory is exactly what is necessary for an adequate description of the quantum mechanical relations, since it permits to present "quantum mechanics in a unified representation" which "is mathematically rigorous" (von Neumann 1955, vii).

2) Gödel claims that, despite their remoteness from sense experience, we do have something like a perception also of the objects of set theory, and this kind of perception is mathematical intuition.

But this is faced with the problem that, as Kant states, sensation is "the effect of an object on the capacity for representation, insofar as we are affected by it" (Kant 1998, A20/B34). Similarly, intuition can take "place only insofar as the object is given to us; but this in turn, is possible only if it affects the mind in a certain way" (ibid., A19/B33). However, while physical objects can affect us in sensation through our sense organs, how could the objects of set theory affect us in mathematical intuition?

Gödel's answer is that "some physical organ is necessary to make the handling of abstract impressions (as opposed to sense impressions) possible," and "such a sensory organ must be closely related to the neural center for language" (Wang 1996, 233).

But there is no evidence that such a sensory organ actually exists, and indeed its existence makes no sense in evolutionary terms. For, the highly abstract objects of set theory were not known until the nineteenth century, and it is highly implausible that the sensory organ in question could have evolved only then, particularly when most members of the species appear not to possess it. As to the claim that this sensory organ must be closely related to the neural center for language, experiments have shown that "mathematics and language are functionally and neuroanatomically independent" (Brannon 2005, 3178).

3) Gödel claims that mathematical intuition is sufficiently clear to produce the axioms of set theory and an open series of extensions of them. Mathematical intuition gives us insights both for obtaining unambiguous answers to the questions of transfinite set theory, such as Cantor's continuum hypothesis, and for the solution of the problems of finitary number theory, of the type of Godlbach's conjecture.

But, at least since Kant, intuition has been conceived of as singular. Indeed, Kant states that "an intuition is a singular representation," so it differs from a concept, which is "a universal representation, or a representation of what is common to several objects, hence a representation insofar as it can be contained in various ones" (Kant 1992, 589). Then, how can mathematical intuition, being a singular representation, produce the axioms of set theory, which are universal?

Moreover, intuition has not given insights to Gödel, or anybody else, for obtaining unambiguous answers to questions of transfinite set theory, such as Cantor's continuum hypothesis, or for the solution of the problems of finitary number theory, of the type of Godlbach's conjecture. These problems remain unsolved.

4) Gödel claims that we can extend our knowledge of the abstract concepts of transfinite set theory by focusing more sharply on the concepts concerned. So we

will arrive at an intuitive grasping of ever newer axioms, which is necessary for the solvability of all problems.

This, however, is problematic. Suppose that, by focusing more sharply on the concept of set Σ, we get an intuition of that concept. Let S be a formal system for set theory, whose axioms this intuition ensures us to be true of Σ. So Σ is a model of S, hence S is consistent. Then, by Gödel's first incompleteness theorem, there is a sentence A of S which is true of Σ but is unprovable in S. Since A is unprovable in S, the formal system $S' = S \cup \{\neg A\}$ is consistent, and hence has a model, say Σ'. Then $\neg A$ is true of Σ' and hence A is false of Σ'. Now, Σ and Σ' are both models of S, but A is true of Σ and false of Σ', so Σ and Σ' are not equivalent. Suppose next that, by focusing more sharply on the concept of set Σ', we get an intuition of this concept. Then we have two different intuitions, one ensuring us that Σ is the concept of set, and the other ensuring us that Σ' is the concept of set, where the sentence A is true of Σ and false of Σ'. This raises the question: Which of Σ and Σ' is the genuine concept of set? Gödel's procedure gives no answer to this question.

5) Gödel claims that mathematical intuition need not be conceived as a faculty giving an immediate knowledge of the objects concerned. Rather, we form our ideas of mathematical objects on the basis of the iterative concept of set. The latter is closely related to Kant's categories of pure understanding, since the function of both is synthesis, namely the generation of unities out of manifolds.

Here Gödel refers to the fact that, according to Kant, only through categories "can any object of experience be thought at all" (Kant 1998, A93/B126). Similarly, according to Gödel, only through the iterative concept of set can any set be thought at all. Gödel also refers to the fact that, according to Kant, "the functions of composing (of synthesis) come before, but they do not have any object yet; they get it through schematizing, namely through a priori intuitions to which they can be applied. This produces knowledge of things as phenomena" (Kant 1900, XIII, 468). Similarly, according to Gödel, the functions of composing objects into a set come before, but they do not have any object yet; they get it through schematizing, namely through a priori intuitions to which they can be applied, and this produces knowledge of sets.

But, according to Kant, schematizing a mathematical concept, say the concept of triangle, consists in "exhibiting an object corresponding to this concept, either through mere imagination, in pure intuition, or on paper, in empirical intuition, but in both cases completely a priori, without having had to borrow the pattern for it from any experience" (Kant 1998, A713/B741). Now, while it is clear how Kant could think of exhibiting an object corresponding to the concept of triangle in pure or empirical intuition, since triangles are spatio-temporal objects, it is completely unclear how Gödel could think of exhibiting an object corresponding to the iterative concept of set in pure or empirical intuition, since sets are not spatio-temporal objects.

6) Gödel claims that, besides mathematical intuition, there exists another, though only probable, criterion of truth of mathematical axioms, namely their fruitfulness in consequences, in particular in verifiable consequences, i.e., consequences demonstrable without the new axiom. The simplest case of application of this criterion of

truth is when some set-theoretical axiom has number-theoretical consequences verifiable by computation up to any given integer.

But, as Gödel himself admits, this criterion of truth "cannot yet be applied to the specifically set-theoretic axioms (such as those referring to great cardinal numbers), because very little is known about their consequences in other fields" (Gödel 1986–2002, II, 269). In particular, as concerns number-theoretical consequences verifiable by computation up to any given integer, "on the basis of what is known today" it "is not possible to make the truth of any set-theoretical axiom reasonably probable in this way" (ibid.).

Moreover, this criterion of truth fails because from a false axiom one can deduce true consequences. In order to assert that an axiom is true, one ought to be able to show that all of its logical consequences are true, but this is generally unfeasible. (For more on this, see Chap. 21).

19.4 Early Modern Philosophers and Mathematical Objects

In addition to mathematical platonism, other approaches have been developed as to what mathematics is about. They too, however, are unsatisfactory. All of them consider the ontological question – the question whether mathematical objects exist and, if so, what their nature is – to be central to mathematics. This contrasts with the seventeenth and eighteenth centuries, when Descartes, Locke, Hume, and Kant considered the ontological question to be irrelevant to mathematics.

Thus Descartes states that "arithmetic, geometry and other subjects of this kind" really "care little whether" the things with which they deal "exist in nature or not" (Descartes 1996, VII, 20).

Locke states that "all the discourses of the mathematicians about the squaring of a circle, conick sections, or any other part of mathematicks, concern not the existence of any of those figures: but their demonstrations, which depend on their ideas, are the same, there be any square or circle existing in the world, or no" (Locke 1975, 566).

Hume states that, "though there never were a circle or triangle in nature, the truths demonstrated by Euclid would for ever retain their certainty and evidence" (Hume 1975, 25).

Kant states that "in mathematical problems the question is not" about "existence as such at all, but about the properties of the objects in themselves, solely insofar as these are combined with the concept of them" (Kant 1998, A719/B747).

19.5 Mathematical Objects as Hypotheses

Those seventeenth and eighteenth century philosophers were right, the ontological question is irrelevant to mathematics. Indeed, there is no more to the existence of mathematical objects than that they are hypotheses human beings make to solve mathematical problems by the analytic method. Their status is similar to that of the theoretical terms of scientific theories – expressions that refer to nonobservational entities – which are hypotheses human beings make to solve physical problems by the analytic method.

Mathematical objects enter into mathematics through a process in which three phases may be distinguished:

1) Mathematical objects are introduced as hypotheses to solve mathematical problems.

2) They are shown to lead to the solution of certain mathematical problems.

3) They become objects of study themselves.

For example, Menaechmus introduced conic sections as hypotheses to solve the problem of doubling the cube. As we have seen in Chap. 12, Hippocrates of Chios had shown that a sufficient condition for solving that problem was the hypothesis that, given any two straight lines, a and b, we can always find two other straight lines, x and y, which are the mean proportionals in continued proportion between a and b. This hypothesis, however, was in turn a problem that had to be solved. In order to solve it, Menaechmus introduced conic sections, specifically, parabolas and hyperbolas, and showed that they led to the solution of the problem of doubling the cube (see Cellucci 2013a, Chap. 4). Then conic sections became objects of study themselves, leading to Apollonius of Perga's *Conics*.

Of course, in addition to leading to the solution of mathematical problems, the hypotheses that introduce mathematical objects must be plausible. Only then mathematical objects can be said to 'exist'. But this does not mean that they exist literally, in the physical or in a superphysical world. It was their literal existence that Descartes, Locke, Hume and Kant objected to. That mathematical objects exist simply means that the hypotheses which introduce them lead to the solution of mathematical problems and are plausible. This remains the status of mathematical objects even when they become objects of study themselves.

19.6 The Mental-Cultural Reality of Mathematical Objects

Being hypotheses human beings make to solve mathematical problems by the analytic method, mathematical objects exist only in the minds of the mathematicians who hypothesize them, and in the minds of the people who make use of them. As Hersh says, while mathematical "platonism mistakenly locates" mathematical objects "'out there', in an external unspecified realm of non-human, non-physical reality," mathematical objects "are right here, in our individual minds, shared also with many other individual minds. Their reality is both psychological and social, it is mental-cultural" (Hersh 2014, 90–91).

Mathematics is a human product, it is part of being human. Natural mathematics, which is a result of biological evolution, is a set of capacities that are essential for the survival of human beings. Artificial mathematics, which is a result of cultural evolution, is a set of concepts and procedures that allow human beings to deal with the world and make it understandable to themselves. The purpose of mathematics is to develop means to see and think about the world. Mathematical objects are among such means. They are made by human beings and allow them to see and think about the world in human terms. And, like all human thoughts, mathematical objects exist only in the minds of human beings.

Hart claims that, since Euclid's proof" of the theorem that there are infinitely many primes "makes no reference to living creatures, there would have been infinitely many primes even if life had never evolved. So the objects required by the truth of his theorem cannot be mental" (Hart 1996, 3). This, however, is unjustified. The Rhind Mathematical Papyrus suggests that prime numbers were introduced in pre-Greek mathematics in connection with such concrete activities as evenly dividing loaves of bread among workers. So, if life had never evolved, there would have been no prime numbers. This confirms that the reality of objects required by Euclid's theorem is mental-cultural.

The view that mathematical objects exist only in the minds of the mathematicians who hypothesize them and in the minds of the people who make use of them, must not be confused with Brouwer's view that mathematical objects are constructions of an idealized mathematician, "the creating subject" (Brouwer 1975, 511). Mathematicians may think of actual infinite sets and find some of their properties, while Brouwer's creating subject cannot.

19.7 Features of Mathematical Objects as Hypotheses

Mathematical objects, meant as hypotheses human beings make to solve mathematical problems by the analytic method, have the following features.

1) Hypotheses characterize the identity of mathematical objects, but this does not mean that mathematical objects exist literally, in the physical or in a superphysical world – just as characterizing hippogriffs as animals having the foreparts of a griffin and the body of a horse does not mean that hippogriffs exist exist literally, in the physical or in a superphysical world. Identity does not entail literal existence.

2) That mathematical objects are introduced as hypotheses by some mathematician, does not mean that such mathematician knows all their properties. Gödel claims that "the creator necessarily knows all properties of his creatures, because they can't have any others except those he has given to them" (Gödel 1986–2002, III, 311). But this is unjustified. A painter does not necessarily know all properties of his creatures, otherwise there would be no need for art critics. It is one thing to introduce mathematical objects as hypotheses and another thing to discover all their properties.

3) Some mathematical objects, introduced as hypotheses to solve mathematical problems, fail to become objects of study themselves. They do not gain stability and, after some time, disappear from mainstream mathematics. For example, Cavalieri introduced indivisibles as hypotheses to solve problems of quadrature or cubature of figures. He assumed that a surface could be conceived as made up of

infinitely many equidistant parallel lines, the indivisibles of the surface, and a solid could be conceived as made up of infinitely many equidistant parallel planes, the indivisibles of the volume. He showed that the indivisibles led to the solution of some problems of quadrature or cubature of figures. But the indivisibles failed to become objects of study themselves, because they could not be conceived as objects independent of the figures which they were supposed to compose (see, for example, Giusti 1980).

19.8 Mathematical Fictionalism

That mathematical objects are hypotheses human beings make to solve mathematical problems by the analytic method, does not mean that, as mathematical fictionalism assumes, mathematics is like fiction and mathematical objects are like characters in fiction.

According to mathematical fictionalism, all there is to mathematics is that "we have a good story about natural numbers, another good story about sets, and so forth" (Field 1989, 22). A proposition such as '2+2=4' is true much in the same sense in which the proposition "'Oliver Twist lived in London' is true: the latter is true only in the sense that it is true according to a certain well-known story, and the former is true only in the sense that it is true according to standard mathematics" (ibid., 3). Talk of mathematical entities is a convenience that permits us to develop science faster than we otherwise might, but "mathematics is not really indispensable" (ibid., 26). The conclusions we arrive at by adding a mathematical theory to a physical theory which makes no reference to mathematical entities, "are not genuinely new, they are already derivable in a more long-winded fashion from the premises" of the physical theory, "without recourse to the mathematical entities" (Field 1980, 10–11). Thus mathematics is conservative over physics. In fact "there are some fairly general strategies that can be employed to purge" physical "theories of all reference to mathematical entities" (Field 1989, 18). The paradigmatic example is "a rewritten version of Newtonian physics" in which individual variables range over concrete objects, which amounts to "formulating it without mathematical entities" (ibid., 18, footnote 11).

Mathematical fictionalism, however, has some serious limitations which make it untenable.

1) In order to make a rewritten version of Newtonian physics in which individual variables range over concrete objects, mathematical fictionalism introduces entities having properties of mathematical entities, rather than properties of concrete objects. For example, the postulates of its theory of space-time entail that there is a one-to-one correspondence between space-time points, conceived of as physical entities, concrete and not abstract, and quadruples of real numbers. Since real numbers are uncountable, this involves postulating uncountably many physical entities. According to mathematical fictionalism, this is not objectionable, nor "does it become any more objectionable when one postulates that these physical entities

obey structural assumptions analogous to the ones that platonists postulate for the real numbers" (Field 1980, 31). But, then, what mathematical fictionalism calls 'physical entities' are really mathematical entities, and hence its rewritten version of Newtonian physics does not really purge Newtonian physics of all reference to mathematical entities.

2) Mathematical fictionalism is faced with the problem that, if M is a mathematical theory that is added to a physical theory N which makes no reference to mathematical entities, the fictionalist program of purging N of all reference to mathematical entities would require to show that, if a sentence A of N is provable in $N+M$, then A is provable already in N. Now, as Shapiro argues, since there is a one-to-one correspondence between space-time points and quadruples of real numbers, "it is possible to model the natural numbers in space-time and, in effect, to do arithmetic in N. To put it differently, the natural number structure is exemplified in the universe of space-time" (Shapiro 1983, 526). But, then, "the Gödel's incompleteness theorems apply" (ibid., 527). Thus there is a sentence A of N which is provable in $N+M$ but not in N. This refutes the fictionalist "argument for the conservativeness of mathematics over physics" (ibid., 528).

3) Since mathematics M is not conservative over physics N, mathematical fictionalism is faced with the problem of making sure that the use of mathematics M in physics N does not lead to false conclusions about the physical world since M might be inconsistent. But, by Gödel's second incompleteness theorem, this is impossible.

19.9 Hypotheses vs. Fictions

The view that mathematical objects are hypotheses human beings make to solve mathematical problems by the analytic method, is not subject to the limits of mathematical fictionalism. For, between mathematics and fictions there are at least two basic differences.

1) Hypotheses are directed towards reality, because they help us to make the world understandable to ourselves. Conversely, fictions are made with the full consciousness that what is assumed, even if it may have been originally suggested by some aspect of reality, does not exist in reality, either because it is not a constituent of reality or because it conflicts with it. Thus fictions are not directed towards reality.

2) When we make hypotheses, although we are still uncertain as to their plausibility, we hope they will turn out to be plausible. Hypotheses are not merely thinkable, they must be plausible, and hence must agree with all facts of experience. They are accepted until strong and convincing counterexamples emerge, then they are replaced with other hypotheses. Conversely, fictions are not supposed to have any compatibility with experience. They are accepted despite our being aware that they have no basis in experience. We would not be disturbed by conflicts of fictions with experience, and would not infer any implication about the nature of the world from these conflicts.

As Vaihinger – the founder of fictionalism – acknowledges, "whereas every hypothesis seeks to be an adequate expression of some reality still unknown and to mirror this objective reality correctly, the fiction is advanced with the consciousness that it is an inadequate" manner "of conception, whose coincidence with reality is, from the start, excluded" (Vaihinger 2008, 268).

19.10 Mathematical Definitions

Like mathematical objects, the mathematical definitions by which they are introduced are hypotheses human beings make to solve mathematical problems by the analytic method. This may be called 'the heuristic view of mathematical definitions'.

This view is opposed to the stipulation view of mathematical definitions, according to which the latter are arbitrary stipulations; they can never be mistaken; they are merely abbreviations; and they can always be eliminated substituting the definition for the defined thing.

The stipulation view of mathematical definitions goes back to Pascal, who states that mathematical definitions are "only the impositions of names to things which have been clearly designated in perfectly known terms" (Pascal 1904–1914, IX, 242–243). Their only "usefulness and use is to clarify and abbreviate discourse, expressing by the single name that has been imposed what could otherwise be only expressed by several terms" (ibid., IX, 243). They "are very free" and "can never be subject to contradiction" (ibid., IX, 243–244). They can always be eliminated, "substituting the definition for the defined thing" (ibid., IX, 244).

Since Pascal's time, the stipulation view of mathematical definitions has become the prevailing one.

Thus Frege states that a mathematical "definition is an arbitrary stipulation by which a new sign" is "introduced to take the place of a complex expression whose sense we know" (Frege 1979, 211). Its only purpose "is to bring about an extrinsic simplification by stipulating an abbreviation" (Frege 1967, 55). A mathematical definition is "something wholly inessential and dispensable" because, "if the definiens occurs in a sentence and we replace it by the definiendum, this does not affect the thought at all" (Frege 1979, 208). Therefore, a mathematical definition "makes no addition to our knowledge," namely, "it is not possible to prove something new from" the "definition alone that would be unprovable without it" (ibid.).

Peano states that "definitions are arbitrary" (Peano 1973, 246). A "definition expresses the convention of using the defined in place of the longer defining phrase" (ibid., 235). Definitions "are useful, but not necessary, since the definiens may always be substituted for the definiendum, thus eliminating the definiendum completely form the theory" (ibid., 245).

Russell and Whitehead state that a mathematical "definition is a declaration that a certain newly-introduces symbol or combination of symbols is to mean the same as a certain other combination of symbols of which the meaning is already known"

(Withehead and Russell 1925–1927, I, 11). Mathematical definitions are "not true or false" (ibid.). They are "mere typographical conveniences. Practically, of course, if we introduced no definitions, our formulas would very soon become so lengthy as to be unmanageable; but theoretically, all definitions are superfluous" (ibid.). For, "we might always use the definiens instead, and thus wholly dispense with the definiendum" (ibid.).

Hilbert and Bernays state that a mathematical definition is "the introduction of an abbreviating symbol for a complex expression" (Hilbert and Bernays 1968–1970, I, 292). Theoretically, a mathematical definition is superfluous, because the new "symbol, wherever it occurs, can be eliminated" by "replacement of the new symbol with the defining expression" (ibid., I, 293).

The stipulation view of mathematical definitions, however, has several shortcomings. Here are some of them.

1) By assuming that mathematical definitions are arbitrary stipulations, the stipulation view is incapable of accounting for the fact that mathematicians often spend so much time in finding a definition. For example, as Grabiner points out, "the derivative was first used; it was then discovered; it was then explored and developed; and it was finally defined" (Grabiner 1983, 195). This is "a complete reversal of the usual order of textbook exposition in mathematics where," in accordance with the stipulation view, "one starts with a definition, then explores some results, and only then suggests applications" (ibid., 205). Conversely, the heuristic view can account for the fact that mathematicians often spend so much time in finding a definition, because finding a definition is finding a hypothesis which may lead to the solution of a problem. Thus, mathematical definitions are not starting points, but rather arrival points for research.

2) By assuming that mathematical definitions can never be mistaken, the stipulation view is incapable of accounting for the fact that, as a matter of fact, mathematical definitions can be mistaken. For example, a curve of constant width is one that, when rotated inside a square, always makes contact with all four sides of the square. Then a circle is a curve of constant width. Now, suppose we define a circle as follows: A circle is a curve of constant width. This definition is mistaken. For, let us consider a Reuleaux triangle, that is, a figure which is obtained from an equilateral triangle by centering a compass on each vertex and drawing the minor arc between the other two vertices.

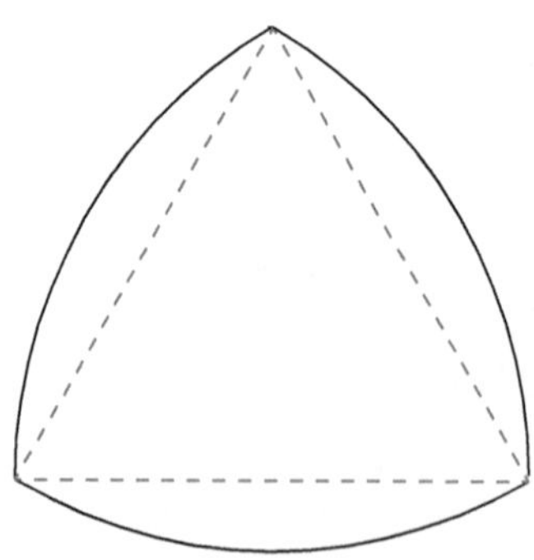

It can be easily seen that a Reuleaux triangle is a curve of constant width, and yet is not a circle. So the definition of a circle as a curve of constant width is mistaken. Conversely, the heuristic view can account for the fact that mathematical definitions can be mistaken because, according to it, a mathematical definition is a hypothesis, and hypotheses can be mistaken.

3) By assuming that mathematical definitions are merely abbreviations, the stipulation view is incapable of accounting for the fact that, of two mathematical definitions of the same concept, one may be preferable to the other. For example, by analogy with Euclid's definition of a circle, we might define a sphere as follows: a) A sphere is a solid figure contained by one surface such that all the straight lines falling upon it from one point among those lying within the figure are equal to one another. Euclid, however, defines it differently: b) "A sphere is the figure enclosed when, the diameter of a semicircle remaining fixed, the semicircle is carried around, and restored again to the same position from which it began to be moved" (Euclid, *Elements*, XI, Definition 14). Clearly, definitions a) and b) are extensionally equivalent, and yet they are not equally adequate to solve a problem. For example, to solve the problem to discover the surface of a sphere, Archimedes finds it preferable to use definition b) rather than definition a). Indeed, he inscribes a half-polygon in a semicircle, then he rotates both about the diameter of the semicircle, thus generating a sphere and a solid figure composed of two cones with vertices at the extremities of the diameter of the semicircle and of several frustrums of cones and of several truncated cones in between. As the number of sides of the half-polygon increases, the surface of the solid figure approximates the surface of the sphere, and by means of this Archimedes shows that "the surface of any sphere is equal to four times the greatest circle in it" (Archimedes 2013, I, 120). Conversely, the heuristic view can account for the fact that, of two mathematical definitions of the same concept, one may be preferable to the other, because they may offer different perspectives on the concept. Therefore, they need not be equally adequate to solve a problem.

4) By assuming that mathematical definitions can always be eliminated substituting the definition for the defined thing, the stipulation view is incapable of accounting for the fact that several mathematical definitions cannot be eliminated. For example, usually division is defined as follows: a) If $b \neq 0$, then $a/b = c$ is an abbreviation for $a = b \cdot c$. This definition of division, however, cannot be eliminated from contexts such as $1/0 \neq 2$. Suppes claims that to give a definition of division which can be eliminated "is a routine matter" (Suppes 1972, 18). One need only replace definition a) with definition: b) $a/b = c$ is an abbreviation for: if $b \neq 0$ then $a = b \cdot c$, while if $b = 0$ then $c = 0$. But thus Suppes overlooks that, if we replace definition a) with definition b), then $a/0 = 0$ for all a. This leads to the violation of some basic properties of division, such as: $a/b + c/d = ad + cb/bd$. For, by b) we have $1/0 + 1/1 = 0 + 1 \neq 0 = 1/0 = (1 \cdot 1 + 1 \cdot 0)/0 \cdot 1$. Conversely, the heuristic view can account for the fact that several mathematical definitions cannot be eliminated because, according to it, a mathematical definition is a hypothesis to solve a problem, and it is constitutive of its being a hypothesis to solve a problem that it cannot be eliminated.

19.11 Hybrids

That mathematical definitions are hypotheses human beings make to solve mathematical problems by the analytic method, has implications for the mathematical objects defined by them.

As pointed out in Chap. 12, solving a problem by the analytic method may require hypotheses which need not belong to the field of the problem, but may belong to other fields. Since a mathematical definition is a hypothesis to solve a problem by the analytic method, it follows that the mathematical definition of a concept may involve concepts which need not belong to the field of that concept, but may belong to other fields.

When this is the case, the objects defined by a mathematical definition are what Grosholz calls "hybrids" (Grosholz 2000, 88). Namely, they are objects which simultaneously exhibit features of different fields, and "their multivalence gives them their characteristic manageable inconsistency and suggestiveness" (ibid., 89). They "take on a life of their own, and become crucial to the growth of knowledge" (Grosholz 1992, 118). They may even lead to "the possible emergence of a new" field, "growing out of the hypotheses posed with respect to the hybrids" (ibid.).

An example of this is Descartes' account of curves as geometric-algebraic hybrids in his solution of Pappus' problem. Descartes assumes a "double vision in which the objects under consideration are treated both algebraically and geometrically," so "both algebraic results about finding the roots of equations, and geometric results about finding relations among lines," can be "brought to bear on the problem at hand" (Grosholz 2000, 85). This is at the origin of the new field of analytic geometry.

As Grosholz states, the multivalence of hybrids requires a mathematician to be able to work at the overlap of multiple fields, and in particular, "to tolerate the peculiar ambiguity of the hybrids which are generated there" (ibid., 89). It is on this peculiar ambiguity that the fruitfulness of hybrids depends. Being "objects which exist in the overlap" of multiple fields, hybrids may lead to a partial unification of those fields, and "some of the most significant advances in mathematical knowledge take place in the context of such partial unification" (ibid., 82). For, hybrids may "provoke discovery in unexpected ways" (ibid.).

19.12 Mathematical Diagrams

A main feature of mathematics is the use of diagrams, or figures. Such use is often associated with intuition but, as we will see, this association is unjustified.

The current philosophical literature on diagrams tries to fit them into the axiomatic method. For example, Barwise and Etchemendy state that "diagrams and other forms of visual representation can be essential and legitimate components in valid deductive reasoning" (Barwise and Etchemendy 1996, 12). That the current philosophical literature on diagrams tries to fit diagrams into the axiomatic method is also clear from Mancosu 2008, a manifesto for the attempt to fit not only diagrams, but all of mathematical practice, into an axiomatic framework. (For critical comments on this manifesto, see Cellucci 2013b, sec. 2).

But the attempt to fit diagrams into the axiomatic method contrasts with the view of several mathematicians, who think that, in the axiomatic method, demonstrations must be completely independent of diagrams. For example, Dieudonné states that "a strict adherence to axiomatic methods" involves "abstaining from introducing any diagram" (Dieudonné 1969, ix). Hence Dieudonné's famous cry, 'A bas Euclide! Mort aux triangles!' ['Down with Euclid! Death to Triangles!'], which became the slogan of Bourbaki. Cartier explains Bourbaki's opposition to diagrams by saying that "the Bourbaki were Puritans, and Puritans are strongly opposed to pictorial representations of truths of their faith" (Senechal 1998, 27).

Rather than fitting into the axiomatic method, diagrams naturally fit into the analytic method, where their use can be described as follows. To solve a problem, we draw a diagram and, from it, we acquire data about the objects involved. Then we look for some hypothesis that is a sufficient condition for solving the problem. The hypothesis is obtained from the problem, the data acquired from the diagram, and possibly other data already available, by some non-deductive rule. Then we check that the hypothesis is a sufficient condition for solving the problem, not only with respect to the particular diagram that has been drawn, but also with respect to any other diagram similar to it. Moreover, we check that the hypothesis is plausible. But the hypothesis is in turn a problem that must be solved, and will be solved in the same way. And so on, *ad infinitum*.

Already Proclus underlined that we need to check that the hypothesis is a sufficient condition to solve a problem, not only with respect to the particular diagram that has been drawn, but also with respect to any other diagram similar to it. Indeed, he says that we must use "the set out diagrams not as these particular diagrams but as they are similar to others" (Proclus 1992, 207.13–16). Much in the same way, Peirce says that the use of diagrams "consists in constructing a diagram," in "observing certain relations between parts of that diagram," in "showing that these relations will hold for all such diagrams, and in formulating this conclusion in general terms" (Peirce 1931–1958, 1.54).

19.13 Objections Against the Use of Mathematical Diagrams

Several objections have been raised against the use of diagrams in mathematics. Here are the main ones.

1) *Diagrams are based on intuition, so they cannot be used in rigorous demonstrations*. Thus, Ayer states that "the use made of diagrams in geometrical treatises shows that geometrical reasoning is not purely abstract and logical, but depends on our intuition of the properties of figures" (Ayer 1990, 78–79). But "the appeal to intuition" is "a source of danger to the geometer" and must be avoided in "a truly

rigorous geometrical proof" (ibid., 79). In particular, "Euclid himself was guilty" of making appeal to intuition, and consequently "the presence of the figure is essential to some of his proofs. This shows that his system is not, as he presents it, completely rigorous" (ibid.)

This, however, is unjustified, because diagrams are not based on intuition. As argued in the previous section, they naturally fit into the analytic method, and intuition has no role in the analytic method. Indeed, diagrams are not based on intuition but on inference.

2) *Diagrams do not yield genuine demonstrations, but only provide visual clues to stimulate mathematical thought.* Thus, Nelsen states that demonstrations essentially based on diagrams "are not really proofs," they simply provide "visual clues to the observer to stimulate mathematical thought" (Nelsen 1993, vi).

This, however, is unjustified. For example, one of the oldest demonstrations of which we have direct record, the solution of the problem of constructing a square which is double the size of a given square in Plato's *Meno*, is a demonstration essentially based on diagrams (see Chap. 23). And Plato presents it not as merely providing visual clues to stimulate mathematical thought, but as a genuine demonstration that "it is the diagonal that will produce the double-sized figure" (Plato, *Meno*, 85 b 5–6).

3) *Diagrams can be used in demonstrations, but the force of demonstrations must be completely independent of diagrams.* Thus, Hilbert states that a demonstration "can also be given by calling on a suitable figure," but this merely "makes the interpretation easier," indeed, "a theorem is only demonstrated when the demonstration is completely independent of the figure. The demonstration must call step by step on the preceding axioms" (Hilbert 2004a, 75).

This, however, is unjustified, because the force of several demonstrations essentially depends on diagrams. This applies even to the very first Hilbert's demonstration in his *Grundlagen der Geometrie*, which establishes the proposition: "For any two points *A* and *C* there always exists at least one point *D* on the line *AC* that lies between *A* and *C*" (Hilbert 1987, 5).

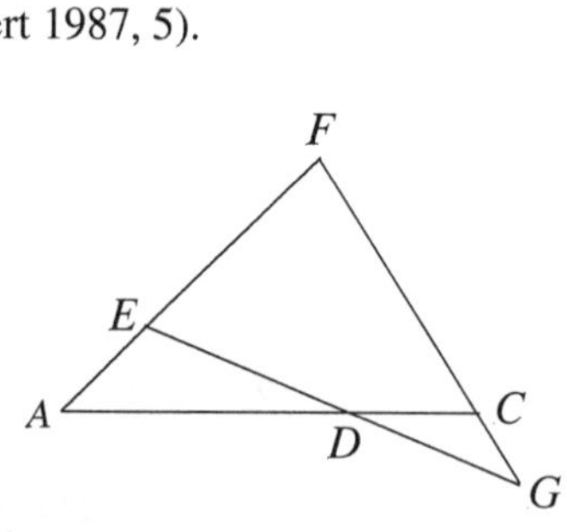

Hilbert's demonstration is as follows. There exists a point *E* not on the line *AC* (Axiom I.3: There exist at least three points not on a line). On the line *AE* there exists a point *F* such that *E* is a point of the segment *AF* (Axiom II.2: For any two points *A*, *E* there exists at least a point *F* on the line *AE* such that *E* is a point of the segment *AF*). On the line *FC* there exists a point *G* such that *C* is a point of the segment *FG* (Axiom II.2). The point *G* does not lie on the segment *FC* (Axiom II.3: Of any three points on a line, there exists no more than one that lies between the other two). Then the line *EG* must intersect the segment *AC* at a point *D* (Axiom II.4: If three points *A*, *F*, *C* do not lie on a line and if *a* is a line in the plane *AFC* which does not pass through any of the points *A*, *F*, *C* but passes through a point of the segment *AF*, then the line *a* also passes through either a point of the segment *AC* or a point of the segment *FC*).

Hilbert's demonstration is not a deduction, because Hilbert makes use of various facts which he does not demonstrate but tacitly assumes on the basis of the diagram. For example, Hilbert uses Axiom II.4 to conclude that, since the three points *A*, *F*, *C* do not lie on a line, and *EG* is a line in the plane *AFC* which does not pass through any of the points *A*, *F*, *C* but passes through a point of the segment *AF*, then *EG* passes through a point of the segment *AC*. However, Axiom II.4 does not enable him to conclude this, but only that *EG* passes either through a point of the segment *AC* or through a point of the segment *FC*. In order to conclude that *EG* passes through a point of the segment *AC* it is essential to demonstrate that *EG* cannot pass through any point of the segment *FC*. But Hilbert does not demonstrate this, he tacitly assumes it on the basis of the diagram. Thus, Hilbert's demonstration fails to call step by step on the preceding axioms.

Russell claims that "Kant, having observed that the geometers of his day" required "an appeal to the figure, invented a theory of mathematical reasoning according to which the inference is never strictly logical" (Russell 1920, 145). However, "the whole trend of modern mathematics" has been "against this Kantian theory" (ibid.). But, contrary to Russell's claim, the very first Hilbert's demonstration in his *Grundlagen der Geometrie*, allegedly the paradigm of modern geometry, is not strictly logical.

4) *Diagrams are unreliable because, when incorrectly drawn, they can lead to errors*. Thus, Hilbert states that figures "can easily be misleading" (Hilbert 2004a, 75). Therefore, "we will never rely on them" (Hilbert 2004c, 541). A popular example of this is the following 'proof' that all triangles are isosceles. Let *ABC* be any triangle. Draw the bisector of the angle *BAC*, and the perpendicular to the side *BC* at its middle point *D*. Let these two lines meet at point *O*. Draw *OF* and *OE* perpendicular to *AB* and *AC*, respectively, and join *O* to *B* and to *C*.

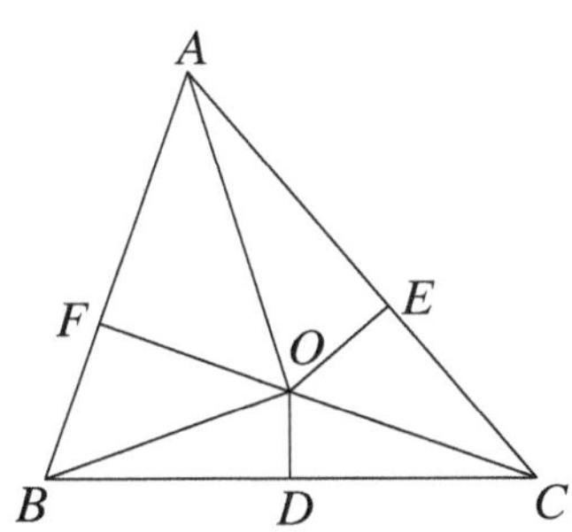

The triangles AOF and AOE are equal, because the side AO is common, angle OAF is equal to angle OAE, and the right angles are equal. Hence $AF=AE$ and $OF=OE$. The triangles OBD and OCD are equal, because the side OD is common, $DB=DC$, and the right angles are equal. Hence $OB=OC$. Then the triangles OFB and OEC are equal, because $OB=OC$, $OF=OE$, and the right angles are equal. Hence $FB=EC$. From $AF=AE$ and $FB=EC$ it follows that $AF+FB=AE+EC$, so $AB=AC$. Therefore the triangle ABC is isosceles.

From examples like this, Hilbert concludes that we can never rely on figures. This conclusion, however, is unjustified. The argument used in the above 'proof' is correct, it is the diagram that is incorrectly drawn. If we draw it correctly, then we will find that the meeting point O falls outside the triangle ABC, and that one of F and E comes inside, the other outside, the triangle ABC.

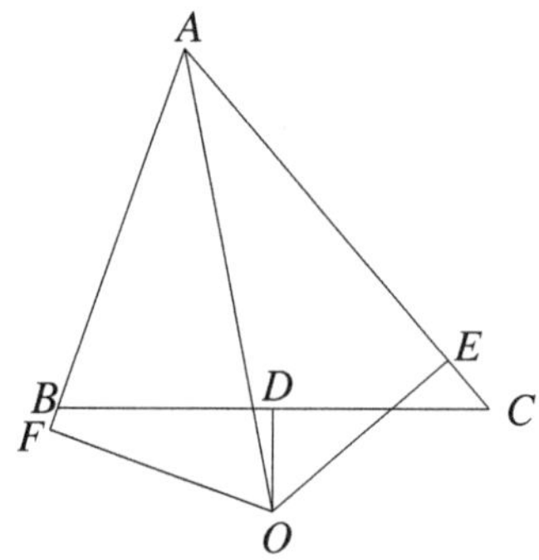

We will also find that $AF=AE$. Then $AB=AF-BF<AF+EC=AE+EC=AC$. Hence $AB<AC$, therefore the triangle ABC is not isosceles.

In this, as in many other 'proofs' of false propositions based on diagrams, errors arise from a misuse of diagrams, and can be avoided by taking care that the diagrams involved are carefully drawn and properly analysed. Much in the same way, errors such as Russell's paradox arise from a misuse of discursive reasoning, and can be avoided by taking care that the concepts involved in discursive reasoning are carefully expressed and properly analysed. Then, if one asserts that the use of diagrams must be barred because diagrams are unreliable, one must also be prepared to assert that the use of discursive reasoning must be barred because discursive reasoning is unreliable. But this is absurd.

5) *Diagrams can be unreliable even when they are correctly drawn.* Thus, Giaquinto states that "visual thinking is unreliable for situations in which limits are involved, and so it is not a means of discovery in those situations, let alone a means of proof" (Giaquinto 2008, 35). For example, consider a semicircle of radius r and divide the diameter into n equal parts. On the resulting n segments of the diameter, construct n semicircles placing them in turn on one and then on the other side of the diameter.

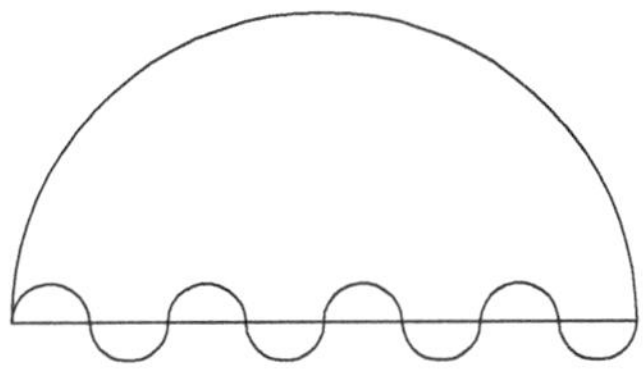

As n increases, the curved line consisting of the n semicircles get closer and closer to the diameter, and for very large n, the curved line becomes barely distinguishable from the diameter. The length L_n of the curved line also seems to approach the length of the diameter, $2r$, and in the limit seems to coincide with it, that is:

$$\lim_{n\to\infty} L_n = 2r \tag{19.1}$$

But it is not so. As n increases, the length of the curved line remains the same. For, each of the n semicircles has diameter $\frac{2r}{n}$, so radius $\frac{r}{n}$, and hence length $\left(2\pi\cdot\frac{r}{n}\right):2=\frac{\pi r}{n}$. Thus $L_n=\frac{\pi r}{n}\cdot n=\pi r$. Then:

$$\lim_{n\to\infty} L_n = \pi r. \tag{19.2}$$

From (19.1) and (19.2) it follows that $\pi r=2r$, whence $\pi=2$, which is implausible.

On the basis of examples like this, Giaquinto concludes that visualizing is never reliable when used to discover the nature of the limit of an infinite process. This, however, is unjustified. From the fact that the curved line gets closer and closer to the diameter, it does not follow that the limit of the length of the curved line is the length of the diameter. Here we are not within the condition of applicability of the concept of limit, because the difference between the length of the curved line and its assumed limit, namely the length of the diameter, is neither infinitely small nor zero, being $\pi r-2r$. Therefore, the error of concluding that the length of a semicircle is equal to the length of its diameter is not due to the diagram being unreliable, but to reading into the diagram something which is not contained therein. The visualization of the diagram does not tell us that the limit of the length of the curved line is the length of the diameter, $2r$, but only that the the curved line get closer and closer to the diameter.

6) *Diagrams do not permit to establish general propositions, since they are particular figures*. Thus, Tennant states that a diagram "is only an heuristic to prompt certain trains of inference," it "has no proper place in the proof as such" (Tennant 1986, 304). One must be "cautioned, and corrected, about an important misuse" of a "drawn triangle: the mistake of assuming as given information that is true only of the triangle that one has happened to draw, but which could well be false of other triangles that one equally well might have drawn in its stead" (ibid.).

This, however, is unjustified. As stated above, already Proclus underlined that, when using diagrams, we need to check that the hypothesis obtained from the diagram is a sufficient condition for solving the problem, not only with respect to the particular diagram that has been drawn, but also with respect to any other diagram similar to it. This guarantees that the demonstration holds, without any structural change, for any other diagram similar to the drawn one, and hence the proposition established is general.

References

Archimedes. 2013. *Opera omnia*. Cambridge: Cambridge University Press.

Ayer, Alfred Jules. 1990. *Language, truth and logic*. London: Penguin Books.

Barwise, Jon, and John Etchemendy. 1996. Visual information and valid reasoning. In *Logical reasoning with diagrams*, ed. Gerard Allwein and Jon Barwise, 3–25. Oxford: Oxford University Press.

Brannon, Elizabeth. 2005. The independence of language and mathematical reasoning. *Proceedings of the National Academy of Sciences of the United States of America* 102: 3177–3178.

Brouwer, Luitzen Egbertus. Jan. 1975. *Collected works*. Vol. 1. Amsterdam: North-Holland.

Cellucci, Carlo. 2013a. *Rethinking logic: Logic in relation to mathematics, evolution, and method*. Dordrecht: Springer.

———. 2013b. Top-down and bottom-up philosophy of mathematics. *Foundations of Science* 18: 93–106.

Descartes, René. 1996. *Oeuvres*. Paris: Vrin.

Dieudonné, Jean. 1969. *Foundations of modern analysis*. New York: Academic Press.

Field, Hartree. 1980. *Science without numbers*. Oxford: Blackwell.

———. 1989. *Realism, mathematics and modality*. Oxford: Blackwell.

Frege, Gottlob. 1967. *Begriffsschrift*, a formula language, modeled upon that of arithmetic, for pure thought. In *From Frege to Gödel: A source book in mathematical logic, 1879–1931*, ed. Jean van Heijenoort, 5–82. Cambridge: Harvard University Press.

———. 1979. *Posthumous writings*. Oxford: Blackwell.

Giaquinto, Marcus. 2008. Visualizing in mathematics. In *The philosophy of mathematical practice*, ed. Paolo Mancosu, 22–42. Oxford: Oxford University Press.

Giusti, Enrico. 1980. *Bonaventura Cavalieri and the theory of indivisibles*. Rome: Cremonese.

Gödel, Kurt. 1986–2002. *Collected works*. Oxford: Oxford University Press.

Grabiner, Judith. 1983. The changing concept of change: The derivative from Fermat to Weierstrass. *Mathematics Magazine* 56: 195–206.

Grosholz, Emily. 1992. Was Leibniz a mathematical revolutionary? In *Revolutions in mathematics*, ed. Donald Gillies, 117–133. Oxford: Oxford University Press.

———. 2000. The partial unification of domains, hybrids, and the growth of mathematical knowledge. In *The growth of mathematical knowledge*, ed. Emily Grosholz and Herbert Breger, 81–91. Dordrecht: Springer.

Hart, Wilbur Dyre. 1996. Introduction. In *The philosophy of mathematics*, ed. Wilbur Dyre Hart, 1–13. Oxford: Oxford University Press.

Hersh, Reuben. 2014. *Experiencing mathematics: What do we do, when we do mathematics?* Providence: American Mathematical Society.

Hilbert, David. 1987. *Grundlagen der Geometrie*. Stuttgart: Teubner.

———. 2004a. Die Grundlagen der Geometrie. In *David Hilbert's lectures on the foundations of geometry 1891–1902*, ed. Michael Hallett and Ulrich Majer, 72–123. Berlin: Springer.

———. 2004c. Grundlagen der Geometrie. In *David Hilbert's lectures on the foundations of geometry 1891–1902*, ed. Michael Hallett and Ulrich Majer, 540–602. Berlin: Springer.
Hilbert, David, and Paul Bernays. 1968–1970. *Grundlagen der Mathematik*. Berlin: Springer.
Hume, David. 1975. *Enquiries concerning human understanding and concerning the principles of morals*. Oxford: Oxford University Press.
Kant, Immanuel. 1900–. *Gesammelte Schriften (Akademie Aausgabe)*. Berlin: Königlich Preußischen Akademie der Wissenschaften.
———. 1992. *Lectures on logic*. Cambridge: Cambridge University Press.
———. 1998. *Critique of pure reason*. Cambridge: Cambridge University Press.
Locke, John. 1975. *An essay concerning human understanding*. Oxford: Oxford University Press.
Mancosu, Paolo. 2008. Introduction. In *The philosophy of mathematical practice*, ed. Paolo Mancosu, 1–21. Oxford: Oxford University Press.
Nelsen, Roger B. 1993. *Proofs without words: Exercises in visual thinking*. Washington: The Mathematical Association of America.
Pascal, Blaise. 1904–1914. *Oeuvres*. Paris: Hachette.
Peano, Giuseppe. 1973. Definitions in mathematics. In Giuseppe Peano, *Selected works*, 235–246. London: Allen & Unwin.
Peirce, Charles Sanders. 1931–1958. *Collected papers*. Cambridge: Harvard University Press.
Proclus. 1992. *In primum Euclidis Elementorum librum commentarii*. Hildesheim: Olms.
Russell, Bertrand. 1920. *Introduction to mathematical philosophy*. London: Allen & Unwin.
Senechal, Marjorie. 1998. The continuing silence of Bourbaki – An interview with Pierre Cartier. *The Mathematical Intelligencer* 20 (1)22–28.
Shapiro, Stewart. 1983. Conservativeness and incompleteness. *The Journal of Philosophy* 80: 521–531.
Suppes, Patrick. 1972. *Axiomatic set theory*. Mineola: Dover.
Tennant, Neil. 1986. The withering away of formal semantics? *Mind and Language* 1: 302–318.
Vaihinger, Hans. 2008. *The philosophy of 'as if'*. London: Routledge.
von Neumann, John. 1955. *Mathematical foundations of quantum mechanics*. Princeton: Princeton University Press.
Wang, Hao. 1996. *A logical journey: From Gödel to philosophy*. Cambridge: The MIT Press.
Weyl, Hermann. 1950. *The theory of groups and quantum mechanics*. Mineola: Dover.
Whitehead, Alfred North, and Bertrand Russell. 1925–1927. *Principia mathematica*. Cambridge: Cambridge University Press.

Chapter 20
Mathematics: Problem Solving or Theorem Proving?

Abstract The view that mathematics is problem solving has been challenged by the claim that, in the twentieth century, mathematics has been reduced to theorem proving. This raises the question: Is mathematics theorem proving, or problem solving? The purpose of the present chapter is to answer this question, which is a philosophical question about the nature of the method of mathematics. Indeed, since antiquity, saying that mathematics is problem solving has been an expression of the view that the method of mathematics is the analytic method, while saying that mathematics is theorem proving has been an expression of the view that the method of mathematics is the axiomatic method. The chapter argues that only the view that mathematics is problem solving is defensible, and the claim that, in the twentieth century, mathematics has been reduced to theorem proving is unjustified.

20.1 Problem Solving vs. Theorem Proving

In Chap. 18 it has been argued that mathematics is problem solving by the analytic method. Thus mathematics is problem solving. This view is opposed to the view that mathematics is theorem proving.

Saying this may seem peculiar. For, if you asked a mathematician whether mathematics is problem solving or theorem proving, the mathematician would most likely reply: Why, mathematics is both problem solving and theorem proving! Mathematicians solve problems and prove theorems, so why should it occur to anyone that mathematics is one of these things to the exclusion of the other?

From the mathematician's point of view, this answer would be perfectly understandable because, in their everyday practice, mathematicians both solve problems and prove theorems. However, the question whether mathematics is problem solving or theorem proving is not supposed to be a sociological or statistical question about the everyday practice of mathematicians, but rather a philosophical question about the nature of mathematics, and specifically about the nature of the method of mathematics. Since antiquity, saying that mathematics is problem solving has been an expression of the view that the method of mathematics is the analytic method, while saying that mathematics is theorem proving has been an expression of the view that the method of mathematics is the axiomatic method. The philosophical question whether mathematics is problem solving or theorem proving has been

© Springer International Publishing AG 2017

C. Cellucci, *Rethinking Knowledge*, European Studies in Philosophy of Science 4,
DOI 10.1007/978-3-319-53237-0_20

considered to be so important that it set the two philosophical giants of antiquity, Plato and Aristotle, against each other, and has been disputed ever since.

However, Mäenpää claims that, with his *Grundlagen der Geometrie*, Hilbert "reduced geometry to theorem proving," and "Hilbert's model has spread throughout mathematics in this century, reducing it to theorem proving. Problem solving, which was the primary concern of Greek mathematicians, has been ruled out" (Mäenpää 1997, 210). Thus, according to Mäenpää, in the past century the question whether mathematics is problem solving or theorem proving has received a conclusive answer: Mathematics is theorem proving. Actually, the story is somewhat more complicated. In his 1900 Paris lecture on mathematical problems – delivered the year after the publication of the work that, according to Mäenpää, reduced geometry to theorem proving – Hilbert tries to find a balance between the view that mathematics is problem solving and the view that mathematics is theorem proving. But the balance is an unstable one and breaks down in the Twenties, leading to the victory of the view that mathematics is theorem proving.

This challenges the claim, made in Chap. 18, that mathematics is problem solving, and raises the question: Is mathematics theorem proving, or problem solving? The purpose of this chapter is to answer this question.

20.2 Mathematicians' Views on the Method of Mathematics

Mäenpää's claim that in the past century problem solving has been ruled out is motivated by the fact that many mathematicians maintain that the method of mathematics is the axiomatic method, and hence mathematics is theorem proving.

Thus Bourbaki states that "'mathematical truth' resides" uniquely "in logical deduction starting from premises arbitrarily set by axioms" (Bourbaki 1994, 17).

Kac and Ulam state that "mathematics owes its unique position to its adherence to the axiomatic method," which "consists in starting with a few statements (axioms) whose truth is taken for granted and then deriving other statements from them by the application of rules of logic alone" (Kac and Ulam 1992, 139).

Gowers states that what mathematicians do is that they "start by writing down some axioms and deduce from them a theorem" (Gowers 2006, 183).

Other mathematicians, however, disagree. Their positions appear to be more akin to the view that the method of mathematics is the analytic method, and hence mathematics is problem solving.

Thus Halmos states that mathematics "is never deductive in its creation. The mathematician at work makes vague guesses, visualizes broad generalizations, and jumps to unwarranted conclusions" (Halmos 1968, 380).

Rota states that "we often hear that mathematics consists mainly in 'proving theorems'," but really "a mathematician's work is mostly a tangle of guesswork, analogy, wishful thinking and frustration, and proof, far from being the core of discovery, is more often than not a way of making sure that our minds are not playing tricks" (Rota 1981, xviii).

Hamming states that "mathematics is not simply laying down some arbitrary postulates and then making deductions" from them, on the contrary, "you start with some of the things you want and you try to find postulates to support them" (Hamming 1998, 645). Thus, "Euclid's postulates came from the Pythagorean theorem, not the other way" (Hamming 1980, 87).

20.3 The Top-Down and the Bottom-Up Approach to Mathematics

The view that the method of mathematics is the axiomatic method, and the view that the method of mathematics is the analytic method, correspond to two different approaches to mathematics, the top-down approach and the bottom-up approach, respectively.

According to the top-down approach, a field of mathematics is developed from above, namely from general principles, and is developed by the axiomatic method, which is a downward path from principles to conclusions derived deductively from them. An example of the top-down approach is Leibniz's approach to the infinitesimal calculus, as presented by de l'Hospital's in the very first textbook on the subject (see de l'Hospital 1696, 1–2).

According to the the bottom-up approach, a field of mathematics is developed from below, namely from problems, and is developed by the analytic method, which is an upward path from problems to hypotheses obtained non-deductively from them. An example of the bottom-up approach is Newton's is Newton's approach to the infinitesimal calculus (see Cellucci 2013a, Chap. 21).

20.4 Analytic vs. Axiomatic Method

Since antiquity, the view that the method of mathematics is the analytic method, and the view that the method of mathematics is the axiomatic method, have been considered to be opposed. The nature of the opposition is made quite clear by Plato and Aristotle.

On the one hand, Plato criticizes the view that the method of mathematics is the axiomatic method, arguing that the mathematicians who proceed by that method take for granted certain propositions, they "assume them as hypotheses, and don't feel any further need to give a justification for them, either to themselves or to anyone else, as if they were axiomatic principles; then, starting from these, they draw consequences from them, and finally reach the conclusions that they had set out to reach in their research" (Plato, *Respublica*, VI 510 c 6–d 3). Those mathematicians "are only dreaming about what is, and cannot possibly have any awaking awareness of it, as long as they leave untouched the hypotheses they use and remain incapable of accounting for them" (ibid., VII 533 b 8–c 3). For, "if your starting-point," being unjustified, "is unknown, and the conclusion and intermediate steps are also constructed out of unknown material," then the starting-point, the intermediate steps and the conclusion will be mere conventions; and "how can you imagine that such a fabric of convention can ever become science?" (ibid., VII 533 c 3–6).

In opposition to the view that the method of mathematics is the axiomatic method, Plato puts forward the view that the method of mathematics is the analytic method, arguing that the latter "does not treat hypotheses as axiomatic principles" for which no justification is given, but "truly as hypotheses" for which a justification is given, and "as stepping stones from which to take off and proceed" (ibid., VI 511 b 4–5). Only the analytic method, "doing away with the hypotheses, proceeds this way up," it "gently pulls the eye of the soul from the barbarian mud in which it is buried, and leads it upwards" (ibid., VII 533 c 8–d 3). It proceeds "passing through all attempts to disprove" the hypotheses, "and trying to disprove them not according to opinion," hence not on the basis of other hypotheses, "but according to the reality of things" (ibid., VII 534 c 1–3). This is a formidable task, but is a necessary one because, "without passing this exploration of all possibilities and in all directions, it is impossible for the mind to attain the truth" (Plato, *Parmenides*, 136 e 1–3).

On the other hand, Aristotle criticizes the view that the method of mathematics is the analytic method, arguing that the latter "does not allow one to know anything in an absolute way, but only on the basis of a hypothesis" (Aristotle, *Analytica Posteriora*, A 22, 84 a 5–6). But one cannot have scientific knowledge on the basis of a hypothesis, because a principle "must be that which is most completely known" and "must be unhypothetical. For, a principle which everyone must have who wants to know anything cannot be a hypothesis; and that which everyone must know who wants to know anything at all, he must already have before learning anything" (Aristotle, *Metaphysica*, Γ 3, 1005 b 13–17).

In opposition to the view that the method of mathematics is the analytic method, Aristotle puts forward the view that the method of mathematics – meaning by this the method of mathematics as an accomplished science – is the axiomatic method, arguing that the latter makes us know things on the basis of necessary principles. For, "scientific knowledge is universal and proceeds through what is necessary" (ibid., A 33, 88 b 31–32). Therefore, "scientific knowledge proceeds from necessary principles" (ibid., A 6, 74 b 5–6).

From Plato and Aristotle it is clear that, between the view that the method of mathematics is the analytic method and the view that the method of mathematics is the axiomatic method, there is really an opposition. On the one hand, in terms of Plato's criticism of the axiomatic method, the axiomatic method is what results from the analytic method when the process of justification of hypotheses is artificially stopped at a certain stage, and the hypotheses reached at that stage are taken as axioms and left unjustified. As a result, the propositions obtained by the axiomatic method are mere conventions. On the other hand, in terms of Aristotle's criticism of the analytic method, in such method there is nothing stable, being based on provisional hypotheses which are bound to be replaced eventually by other provisional hypotheses. Therefore, the analytic method does not allow one to know anything in an absolute way.

What Plato criticizes is a version of the axiomatic method which does not provide a justification of the principles. Aristotle's requirement that, in the axiomatic method, the principles must be true, is a response to Plato's criticism. But the version of the axiomatic method that Plato criticizes is not confined to Plato's time. For example, Curry states that, "we shall always have uses for systems whose consistency is unproved" and "neither the question of intuitive evidence nor that of a consistency proof has any bearing on the matter" (Curry 1951, 62).

20.5 Problems vs. Theorems

Since antiquity, the opposition between the view that the method of mathematics is the analytic method and the view that the method of mathematics is the axiomatic method, has been in particular an opposition between problems and theorems. In ancient Greece, there was a lengthy tradition of discussion on the nature of problems and theorems, and their relative significance for mathematics. Thus Pappus states that, "among the ancients, some say that all things are problems, others say that all things are theorems" (Pappus 1876–1878, III, 30.7–8).

Pappus tells no more about this division of opinion among the ancients, but Proclus says that it first arose within Plato's Academy. Within the latter, on the one hand, "the mathematicians of the school of Menaechmus thought it correct to call all inquiries problems" (Proclus 1992, 78.8–9). For, all inquiries involve "acts of production" (ibid., 78.21). On the other hand, other people, "such as the followers of Speusippus," thought it correct "to call all propositions 'theorems', holding the designation 'theorems' more appropriate than the designation 'problems' for the theoretical sciences" (ibid., 77.15–19). For, a problem proposes "to bring into being or to produce something not previously existing" (ibid., 77.22–78.1). But the theoretical sciences "deal with eternal things; and there is no coming to be among eternal things, so a problem has no place here" (ibid., 77.19–22).

That Menaechmus was a mathematician and Speusippus a philosopher is significant, because mathematicians generally endorsed the view that the method of mathematics is the analytic method, while philosophers generally endorsed the view that the method of mathematics is the axiomatic method.

The reason why mathematicians generally endorsed the view that the method of mathematics is the analytic method, is that they were primarily interested in promoting the growth of mathematics, and thought that it could be promoted only through the analytic method. Thus Carpus of Antioch states that "problems are prior in rank to theorems, because the subjects about which properties are sought are found through the problems" (ibid., 242.1–4). Indeed, "for problems one common procedure has been invented, that is, the method of analysis," namely the analytic method, "by proceeding according to which we can reach a solution. For, thus even the most obscure problems can be pursued" (ibid., 242.14–17). Conversely, with respect to theorems, no one to this day "has been able to give us a common method of approaching them" (ibid., 242.19–20).

On the other hand, the reason why philosophers generally endorsed the view that the method of mathematics is the axiomatic method, is that they were primarily interested in showing that mathematical knowledge is firmly grounded, and thought that this could be achieved only through the axiomatic method. Thus, Aristotle states that problems are "questions in regard to which there are conflicting reasonings, the difficulty being whether something is so-and-so or not, there being convincing arguments for both views" (Aristotle, *Topica*, I, 11, 104 b 12–14). In "regard to them we have no conclusive argument, because they are so vast, and we find it difficult to give our reasons" (ibid., I, 11, 104 b 14–16). Therefore, we cannot achieve firmly grounded knowledge through the analytic method. We can achieve it only through the axiomatic method, because knowledge is firmly grounded only "when we possess it in virtue of having" an axiomatic "demonstration" (Aristotle, *Analytica Posteriora*, A 4, 73 a 23).

Of course, Plato is a notable exception to the fact that philosophers generally endorsed the view that the method of mathematics is the axiomatic method. As we have seen, he criticizes the axiomatic method and strongly supports the analytic method, which he assumes to be the method not only of mathematics, but also of philosophy.

20.6 Opposition or Different Emphasis?

It might be objected that the distinction between problems and theorems is not really an opposition, it is not due to a difference between the aims of mathematicians and philosophers, but only to a different emphasis on the relative significance of problems and theorems for mathematics.

This is the opinion of Proclus who, with regard to the division between the mathematicians of the school of Menaechmus and the followers of Speusippus, says that "both parties are right" (Proclus 1992, 78.13–14). The mathematicians of the school of Menaechmus are right, because problems are "those propositions whose aim is to produce" (ibid., 201.6–7). Now, the discovery of theorems occurs with the movement of our thought, and "the movement of our thought in projecting its own ideas is a production. For it is in imagination that their constructions, sectionings, superpositions, comparisons, additions, and subtractions take place" (ibid., 78.22–79.1). On the other hand, the followers of Speusippus are right, because "the problems of geometry" are concerned with that which always is, not "with perceptible objects that come to be and undergo all sorts of change" (ibid., 78.15–17).

By stating that both parties are right, Proclus means to say that the distinction between problems and theorems is not really an opposition, it is simply due to a different emphasis on the relative significance of problems and theorems for mathematics. Proclus' view, however, depends on the fact that, when Proclus distinguishes between problems and theorems, he merely refers to Euclid's distinction between problems and theorems in the *Elements*. Indeed, Proclus states the distinction between problems and theorems by saying that, in the *Elements*, the "propositions that follow from the first principles are divided into problems and theorems" (ibid., 77.7–8). Now, in the *Elements* there is no opposition between problems and theorems, as apparent from the fact that, as Proclus points out, Euclid "sometimes interweaves theorems with problems and uses them in turn" (ibid., 81.18–20). This explains why Proclus states that both parties are right.

Contrary to Proclus' opinion, however, the distinction between problems and theorems is really an opposition. For the reasons spelled out by Plato and Aristotle, there is a genuine opposition between the view that the method of mathematics is the analytic method and the view that the method of mathematics is the axiomatic method. Proclus' distinction between problems and theorems merely refers to a distinction between problems and theorems within the axiomatic method. Such distinction is in rough agreement with Aristotle's distinction between productive and theoretical thinking. According to Aristotle, all thinking that is not practical is either "productive or theoretical" (Aristotle, *Metaphysics*, E 1, 1025 b 25). Productive thinking is aimed at creation, in particular, at producing the solution to a problem, "because everyone who produces aims at some goal" (Aristotle, *Nicomachean Ethics*, Z 2, 1139 b 1–2). Theoretical thinking is aimed at knowing truth, because, "of the theoretical thinking, which is neither practical nor productive, the good and the bad state are truth and falsity respectively" (ibid., Z 2, 1139 a 27–28).

20.7 Hilbert on the Method of Mathematics

Since there is really an opposition between the view that the method of mathematics is the analytic method and the view that the method of mathematics is the axiomatic method, a balance between these two views is bound to be a very unstable one. Each side of the alternative will tend to cancel the other one. A significant example of this is Hilbert. In his 1900 Paris address, he makes some statements which can be ascribed to the view that the method of mathematics is the analytic method, and other statements which can be ascribed to the view that the method of mathematics is the axiomatic method. His position can be described as follows.

1) *Mathematics advances by problem posing and problem solving*. Indeed, "mathematical research requires its problems. It is by the solution of problems that the strength of the investigator is hardened; he finds new methods and new outlooks, and gains a wider and freer horizon" (Hilbert 2000, 241).

2) *There is an unlimited supply of mathematical problems*. The "supply of problems in mathematics is inexhaustible, and as soon as one problem is solved, numerous others come forth in its place" (ibid., 248).

3) *Mathematical problems spring from an interplay between thought and experience*. The "first and oldest problems in every branch of mathematics stem from experience and are suggested by the world of external phenomena" (ibid., 243). Then, in the further development of that branch of mathematics, the human mind

"evolves new and fruitful problems from itself alone" (ibid.). Afterwards, the outer world "forces upon us new questions from actual experience, opens up new branches of mathematics; and while we seek to conquer these new fields of knowledge for the realm of pure thought, we often find the answers to old unsolved problems" (ibid.). Thus mathematical problems spring from an "ever-recurring interplay between thought and experience" (ibid.).

4) *A solution to a problem is reached by finding suitable hypotheses.* We obtain a solution to a problem "by means of a finite number of steps based upon a finite number of hypotheses which are implied [*liegen*] in the statement of the problem and which must be exactly formulated" (ibid., 244). Thus the solution is reached starting from the problem, and extracting the hypotheses for its solution that are implied in the statement of the problem.

5) *We find hypotheses by means of certain logical procedures.* Specifically, we find them "by means of logical combination, generalization, specialization, by separating and collecting ideas in fortunate ways" (ibid., 243).

6) *Among such logical procedures, generalization plays an important role.* Indeed, "if we do not succeed in solving a mathematical problem, the reason frequently consists in our failure to recognize the more general standpoint from which the problem before us appears as only a single link in a chain of related problems" (ibid., 246). This "way for finding general methods is certainly the most practical and the most secure; for he who seeks for methods without having a definite problem in mind seeks for the most part in vain" (ibid., 247).

7) *Specialization plays an even more important role.* It plays "a still more important part than generalization" because, "in most cases where we unsuccessfully seek the answer to a question," everything depends on finding "problems simpler and easier than the one in hand" and "on solving them" (ibid.).

8) *However, a problem can be considered to be solved only when the solution has been deduced from axioms by means of a finite number of steps.* Indeed, for any science, "we must set up a system of axioms which contains an exact and complete description of the relations subsisting between the elementary ideas of that science" (ibid., 250). No solution to a problem of that science "is held to be correct unless it can be derived from those axioms by means of a finite number of logical steps" (ibid.).

9) *The axioms from which a solution to a problem is deduced must be consistent, namely non-contradictory, and must be proved to be consistent by absolutely reliable means.* The axioms must be consistent, because a contradiction "acts like a spark in the gunpowder barrel and destroys everything. Therefore, every science must have an interest in dealing with a contradiction, no matter how far removed" (Hilbert 1905, 217). Moreover, the axioms must be proved to be consistent by absolutely reliable means, namely, without using "any dubious or problematical mode of inference" (Hilbert 1996d, 1139). Otherwise, one really could not be sure of the consistency of the axioms. As we have seen in Chap. 3, proving consistency by absolutely reliable means is Hilbert's consistency program.

Here 1) – 7) can be ascribed to the view that the method of mathematics is the analytic method, and 8) – 9) to the view that the method of mathematics is the axiomatic method. While 1) – 7) concern the process of discovery of a solution to a problem, 8) – 9) concern the justification of an already found solution to a problem. Indeed, as argued in Chap. 12, according to Hilbert, the purpose of the axiomatic method is not to discover a solution to a problem and hence to obtain new knowledge, but only to present and justify an already found solution to a problem, and to teach it through textbooks such as Hilbert's *Grundlagen der Geometrie*.

It has been stated above that Aristotle's requirement that, in the axiomatic method, principles, or axioms, must be true, is a response to Plato's criticism of the axiomatic method. Hilbert's requirement 9) can be seen as another response to Plato's criticism.

On the basis of the above description of Hilbert's position, it seems fair to say that, in his 1900 Paris address, Hilbert tries to find a balance between the view that the method of mathematics is the analytic method and the view that the method of mathematics is the axiomatic method.

20.8 Breaking the Balance

The balance between the view that the method of mathematics is the analytic method and the view that the method of mathematics is the axiomatic method, however, is an unstable one, and breaks down in the Twenties, when Weyl declares to approve Brouwer's attempt to replace classical mathematics with intuitionistic mathematics by saying: "Brouwer – that is the revolution!" (Weyl 1998, 99). Since then, Weyl joins Brouwer and becomes "an 'apostle of his intuitionism'" (Weyl 2009, 169). Indeed, he turns into "a staunch supporter of Brouwer," and will remain "sympathetic to intuitionism throughout his life" (Hesseling 2003, 105).

That Weyl turns into a staunch supporter of Brouwer angers Hilbert, who sees his former student, Weyl, desert him for Brouwer. He states that "what Weyl and Brouwer do" is that "they seek to ground mathematics by throwing overboard all phenomena that make them uneasy" (Hilbert 1996c, 1119). Specifically, they "calumniate the general concept of irrational number, of function, even of number-theoretic function, the Cantorian numbers of the higher number-classes, etc." (ibid.). They forbid propositions such as the "proposition that among infinitely many integers there is always a smallest," and even modes of inference such as "the logical *tertium non datur*" (ibid.). But "this means to dismember and mutilate our science" (ibid.). In fact, "Brouwer is not, as Weyl believes, the revolution, but only a repetition" of Kronecker's "attempted coup," and now that the power of the state has been armed and strengthened by Frege, Dedekind, and Cantor, this coup is doomed to fail" (ibid.). The aim "of finding a secure foundation for mathematics is also" Hilbert's aim, but Hilbert thinks that "this can be done while fully preserving" the accomplishments of mathematics through his program, which recognizes that the method adequate to this aim "is none other than the axiomatic" (ibid.).

For this reason, in Hilbert's works of the Twenties, the view that the method of mathematics is the axiomatic method rules out the view that the method of mathematics is the analytic method. As a matter of fact, the latter view no longer occurs in his writings. The process of discovery of a solution to a problem disappears from Hilbert's concern, and he totally focuses on the justification of a solution already found. Hilbert no longer wants to account for how mathematics actually proceeds, instead he wants "to endow mathematical method with the definitive reliability that the critical era of the infinitesimal calculus did not achieve" (Hilbert 1967a, 370).

20.9 The Axiomatic Ideology

Not only, in Hilbert's works of the Twenties, the view that the method of mathematics is the axiomatic method rules out the view that the method of mathematics is the analytic method, but Hilbert becomes an apostle of the axiomatic ideology, which he articulates in the following points.

1) *Every scientific discipline must be able to be put in axiomatic form*. In fact, "anything at all that can be the object of scientific thought," as soon as "it is ripe for the formation of a theory," becomes "dependent on the axiomatic method" (Hilbert 1996b, 1115).

2) *The axiomatic method is the method of research in any field*. Indeed, "the axiomatic method is and remains the indispensable tool" for "all exact research in any field whatsoever" (Hilbert 1996c, 1120).

3) *Every scientific discipline is indirectly dependent on mathematics*. For, every scientific discipline, insofar as it must be able to be put in axiomatic form, becomes dependent "indirectly on mathematics" (Hilbert 1996b, 1115).

4) *In the sign of the axiomatic method, mathematics has a leading role in all sciences*. Indeed, "in the sign of the axiomatic method, mathematics is summoned to a leading role in science" (ibid.). So, mathematics does not stand beside the other sciences, but above them.

5) *Mathematics is the supreme court that will decide all questions*. In fact, "mathematics in a certain sense develops into a tribunal of arbitration, a supreme court that will decide questions of principle – and on such a concrete basis that universal agreement must be attainable and all assertions can be verified" (Hilbert 1967a, 384).

6) *Mathematics is the foundation not only of all science, but of all our culture*. For, "our entire modern culture, in so far as it rests on the penetration and utilization of nature, has its foundation in mathematics" (Hilbert 1996f, 1163). Therefore, mathematics has a leading role in all our culture.

In mathematics, the axiomatic method is a tool alongside other ones, but this is not enough for the supporters of the axiomatic ideology. Their aim is to establish the supremacy of mathematics over all sciences, and even over all culture, and the axiomatic ideology is instrumental to this. Of course, the desire of supremacy is a legitimate one, but a philosophy of mathematics should attend to tasks more fruitful for the development of the discipline.

20.10 Hilbert on the Regressive Task

It has been argued above that, in Hilbert's works of the Twenties, the view that the method of mathematics is the axiomatic method rules out the view that the method of mathematics is the analytic method. Peckhaus disagrees with this view. He argues that, in his 1919–1920 lectures, Hilbert distinguishes between two tasks of mathematics, the "regressive task" and the "progressive task" (Hilbert 1992, 17–18).

Peckhaus describes the regressive task as "the procedure which starts with the formulation of the problem and ends with the determination of the conditions for its solution" (Peckhaus 2002, 106). So, the regressive task is the analytic method. On the other hand, Peckhaus describes the progressive task as "the way from the conditions to the actual solution to the problem" (ibid.). It requires that the solution "be deduced from definitions and principles (i.e., axioms or postulates) set at top of the system" (ibid., 104). So, the progressive task is the axiomatic method. Then Peckhaus interprets Hilbert, in the Twenties, as still trying to find a balance between the view that mathematics is problem solving and the view that mathematics is theorem proving.

This interpretation, however, seems unconvincing, because Hilbert describes "the regressive task" as the task "of providing a more firm structure and the simplest possible foundation for theories" (Hilbert 1992, 17). So, the regressive task is the task of finding a foundation for theorems already known. Now, as we have seen above, this is just what Hilbert considers to be the task of the axiomatic method. Then, the regressive task is not the task of discovering a solution to a problem, but rather the task of providing a justification for a solution already found. So, what Hilbert calls 'the regressive task' has not to do with the analytic method but rather with the axiomatic method. In fact, Hilbert states that "this regressive method finds its perfect expression in what is called today 'axiomatic method'" (ibid., 18).

Indeed, Hilbert sees the axiomatic method as the method required to give a secure foundation for mathematics. He trusts that this goal can be achieved, and hence that his program can be carried out, since he writes: "I believe that I can attain this goal completely" (Hilbert 1967b, 464). In 1930 he even goes so far as saying: "I believe" that "I have fully attained what I desired and promised: The world has thereby been rid, once and for all, of the question of the foundations of mathematics as such" (Hilbert 1996e, 1157).

Ironically, on September 7th, 1930, at a symposium in Königsberg, Hilbert's birthplace, Gödel announces his incompleteness theorems by which Hilbert's aim of giving a secure foundation for mathematics cannot be achieved. This is quickly realized by von Neumann, who states that Gödel's second incompleteness theorem "has solved negatively the foundational question: there is no rigorous justification for classical mathematics" (von Neumann 2003, 339). For his part, Gödel declares that, despite all the attempts to find a secure foundation for mathematics "undertaken by Hilbert and his disciples," the "hope of succeeding along these lines has vanished entirely in view of some recently discovered facts" (Gödel 1986–2002, III, 52). Afterwards, Gödel even calls Hilbert's program a "curious hermaphroditic thing" (ibid., III, 279). Thus Gödel comes "to ultimately speaking of" Hilbert's program "with something like disdain" (Davis 2005, 198).

20.11 Axiomatic Method and Gödel's Incompleteness Theorems

Gödel's's incompleteness theorems are highly relevant to the question of the method of mathematics. They refute the view that the method of mathematics is the axiomatic method. This can be seen as follows.

1) The view that the method of mathematics is the axiomatic method requires mathematics to consist in the deduction of propositions from given axioms. But this requirement cannot be satisfied by Gödel's first incompleteness theorem, by which, for any consistent, sufficiently strong, formal system there is a sentence of the system which is true but cannot be deduced from the axioms of the system. Thus mathematics cannot consist in the deduction of propositions from given axioms.

2) The view that the method of mathematics is the axiomatic method requires that the mathematical knowledge resulting from the deduction of propositions from given axioms be proved to be firmly grounded. But this requirement cannot be satisfied by Gödel's second incompleteness theorem, by which, for any consistent, sufficiently strong, formal system, it is impossible to prove, by any reliable means, that the axioms of the system are consistent. Therefore, the mathematical knowledge resulting from the deduction of propositions from given axioms cannot be proved to be firmly grounded.

20.12 Gödel's Incompleteness Theorems and Recalcitrant Mathematicians

Despite the fact that Gödel's incompleteness theorems refute the view that the method of mathematics is the axiomatic method, this view remains deeply rooted in mathematicians.

Already in 1941, Post expressed his "continuing amazement that, ten years after Gödel's remarkable achievement," the "current views on the nature of mathematics are thereby affected only to the point of seeing the need of many formal systems, instead of a universal one" (Post 1965, 345). On the contrary, "has it seemed to us to be inevitable" that Gödel's achievement "will result in a reversal of the entire axiomatic trend of the late nineteenth and early twentieth centuries," and axiomatic "thinking will then remain as but one phase of mathematical thinking" (ibid.).

Today, we must still express our continuing amazement that, many decades after Gödel's remarkable achievement, the current views on the nature of mathematics are thereby affected only to the point of seeing the need of many formal systems, instead of a universal one.

Thus Curry states that although, by Gödel's first incompleteness theorem, "it is hopeless to find a single formal system which will include all of mathematics as ordinarily understood," and hence "the essence of mathematics" cannot lie "in any particular kind of formal system," nevertheless the essence of mathematics lies in a formal systems, so "mathematics is the science of formal systems" (Curry 1951, 56).

Mac Lane states that although, by Gödel's first incompleteness theorem, "we cannot realistically constrain mathematics to be a single formal system," nevertheless we can "view mathematics as an elaborate tightly connected network of formal systems" (Mac Lane 1986, 417). Indeed, "mathematics consists of formal rules, formal systems, and formal definitions of concepts" (ibid., 442).

In particular, instead of acknowledging that Gödel's incompleteness theorems refute the view that the method of mathematics is the axiomatic method, several mathematicians maintain that they are irrelevant to mathematics.

Thus Davies states: "I got the dread seeing yet another discussion of Gödel's theorems and their importance, when I knew that they had almost no relevance to the work of most mathematicians" (Davies 2008, 88).

Macintyre states: "In the last thirty-five years, number theory has made sensational progress, and the Gödel phenomenon has surely seemed irrelevant" (Macintyre 2011, 6). Generally, "as far as incompleteness is concerned, its remote presence has little effect on current mathematics" (ibid., 14).

But this ignores that, as Post states, if properly understood, Gödel's results would involve a reversal of the entire axiomatic trend of the late nineteenth and early twentieth centuries. Therefore, it would have a substantial impact not only on how the nature of mathematics is conceived, but also on how mathematics is practiced.

20.13 The Impossibility of Achieving Hilbert's Aim

It has been already stated above that, by Gödel' incompleteness theorems, Hilbert's aim of giving a secure foundation for mathematics cannot be achieved. In particular, by Gödel's second incompleteness theorem, for any consistent, sufficiently strong, formal system, it is impossible to prove, by any reliable means, that the axioms of the system are consistent. Specifically, what is impossible to prove is a sentence canonically expressing the consistency of the system, where a sentence is said to canonically express the consistency of the system if it satisfies certain conditions. The requirement that the sentence should satisfy such conditions is essential for the validity of Gödel's second incompleteness theorem. (For details see, for example, Cellucci 2007, 180–184).

On this account, Detlefsen maintains that the requirement that a sentence expressing the consistency of the system should satisfy the conditions in question is not "something to which the Hilbertian is committed by the nature of his enterprise"

(Detlefsen 1992, 202). Removing this unnecessary requirement "may afford a way for the Hilbertian to carry out his program" (ibid.).

This argument, however, overlooks that, even admitting that the requirement that a sentence expressing the consistency of the system should satisfy the conditions in question is not something to which the Hilbertian is committed, there are other reasons why Hilbert's aim of giving a secure foundation for mathematics cannot be achieved. By Gödel's third incompleteness theorem, for any consistent, sufficiently strong, formal system, the outer consistency of the system cannot be proved by absolutely reliable means, where a formal system is said to be outer consistent if every sentence of the Goldbach type that is provable in the system is true. (For Gödel's third incompleteness theorem, see Gödel 1986–2002, II, 305–306; or Cellucci 2007, 185–186). Now, as Gödel states, in order to realize Hilbert's program, "it is necessary to prove this 'outer' consistency (which for the usual systems is trivially equivalent with consistency)" (Gödel 1986–2002, II, 305). With respect to Gödel's second incompleteness theorem, Gödel's third incompleteness theorem has the advantage that the sentence expressing the outer consistency of the system is not required to satisfy any special conditions. Indeed, in a sense, Gödel's third incompleteness theorem is a version of Gödel's first incompleteness theorem (see Cellucci 2007, 186–187). For this reason, Gödel describes his third incompleteness theorem as "the best and most general version of the unprovability of consistency in the same system" (Gödel 1986–2002, II, 305).

Hilbert's aim of giving a secure foundation for mathematics cannot be achieved also by a corollary of Gödel's first incompleteness theorem, already mentioned in Chap. 8. By such corollary, for any consistent formalization of mathematics, there will be a consistent extension of such formalization in which a false sentence is demonstrable. Thus consistency is no guarantee against falsity. As we have seen in Chap. 12, Hilbert claims that 'consistent' is identical to 'true'. This claim is refuted by the above corollary of Gödel's first incompleteness theorem.

Actually, Hilbert need not have waited for Gödel to realize that consistency is no guarantee against falsity. Already Kant had warned that it is, "to be sure, a necessary logical condition" that in "a concept no contradiction must be contained;" but this is "far from sufficient for the objective reality of the concept, i.e., for the possibility of such an object as is thought through the concept" (Kant 1998, A220/B268). Indeed, it is not enough to assume, as "condition of all our judgments whatsoever," that "they do not contradict themselves," because "for all that a judgment may be free of any internal contradiction, it can still be either false or groundless" (ibid., A150/B190).

20.14 Analytic Method and Gödel's Incompleteness Theorems

While Gödel's incompleteness theorems refute the view that the method of mathematics is the axiomatic method, they not only do not refute, but even provide evidence for the view that the method of mathematics is the analytic method. This can be seen as follows.

1) The view that the method of mathematics is the analytic method is unaffected and even confirmed by Gödel's first incompleteness theorem, because this view does not confine mathematics within the closed space of an axiomatic system. It permits mathematics to develop in an open space, where problem solving can make use of interactions among various knowledge systems. Indeed, in the analytic method, the solution to a problem is obtained from the problem, and possibly other data already available, by means of hypotheses not necessarily belonging to the same field as the problem. Since Gödel's first incompleteness theorem implies that solving a problem of a given field may require hypotheses from other fields, Gödel's result even provides evidence for the view that the method of mathematics is the analytic method.

2) The view that the method of mathematics is the analytic method is unaffected and even confirmed by Gödel's second incompleteness theorem, because this view does not assume that the solution to a problem is definitive, true and certain. Indeed, in the analytic method, the hypotheses for the solution to a problem are always provisional, plausible and uncertain, therefore no solution to a problem can be absolutely certain. Since Gödel's second incompleteness theorem implies that no solution to a problem can be absolutely certain, Gödel's result even provides evidence for the view that the method of mathematics is the analytic method.

20.15 Other Shortcomings of the Axiomatic Method

Gödel's incompleteness theorems are not the only reason why the view that the method of mathematics is the axiomatic method is untenable.

1) The view that the method of mathematics is the axiomatic method does not account for the way mathematical problems are solved. For example, as already pointed out in Chap. 12, according to the current prevailing view, Fermat's Problem was solved by Wiles and Taylor. However, what Wiles and Taylor really solved was not Fermat's Problem, but rather the problem posed by the Taniyama-Shimura hypothesis. As Friend states, "one of the problems with the axiomatic view is that it does not reflect the *modus operandi* of the working mathematician in search of proofs" (Friend 2014, 213–214).

2) The view that the method of mathematics is the axiomatic method does not account for the fact that solving a problem of a given field may require hypotheses from other fields. For example, to solve Fermat's problem, that is a problem about

the natural numbers, Ribet used the Taniyama-Shimura hypothesis, that is a hypothesis about modular forms in hyperbolic space. On the contrary, according to the axiomatic method, a solution to a problem of a given field must be deduced from the axioms of that field.

3) The view that the method of mathematics is the axiomatic method does not account for the fact that a demonstration may produce something new. For, according to the axiomatic method, a solution to a problem of a given field must be deduced from the axioms of that field, and, since deductive rules are non-ampliative, deduction produces nothing essentially new with respect to the axioms from which it starts. Thus a theorem is already implicitly contained in the axioms. Since deduction can teach us nothing essentially new with respect to the axioms, Rota states that the "identification of mathematics with the axiomatic method has led to a widespread prejudice among scientists that mathematics is nothing but a pedantic grammar, suitable only for belaboring the obvious" (Rota 1997, 142).

4) The view that the method of mathematics is the axiomatic method does not account for the fact that, once a mathematician has found a solution to a mathematical problem, other mathematicians look for other solutions. For example, several hundred demonstrations have been produced for the Pythagorean theorem and the quadratic residue theorem. A Fields Medal has been awarded to Selberg for producing a new demonstration of a theorem, the prime-number theorem, for which a demonstration was already known. (For other examples of alternative demonstrations of the same theorem, see Dawson 2015). Now, if the purpose of demonstration were to provide a justification for the solution of a mathematical problem, once a solution has been found there would be no point in looking for other solutions, let alone for hundreds of them.

20.16 Other Advantages of the Analytic Method

The view that the method of mathematics is the analytic method is not subject to the further limitations of the view that the method of mathematics is the axiomatic method considered above.

1) The view that the method of mathematics is the analytic method accounts for how mathematical problems are actually solved. For example, to solve Fermat's problem, Ribet used the Taniyama-Shimura hypothesis. To solve the problem posed by the Taniyama-Shimura hypothesis, Wiles and Taylor used other hypotheses. And so on.

2) The view that the method of mathematics is the analytic method accounts for the fact that solving a problem of a given field may require hypotheses from other fields. For, in the analytic method, the hypotheses to solve a problem need not belong to the field of the problem, but may belong to other fields.

3) The view that the method of mathematics is the analytic method accounts for the fact that a demonstration yields something new. Being obtained from the problem, and possibly other data, by means of some non-deductive rule, the hypotheses

for the solution to a problem contain something essentially new with respect to them, because non-deductive rules are ampliative.

4) The view that the method of mathematics is the analytic method accounts for the fact that, once a mathematician has found a solution to a mathematical problem, other mathematicians look for other solutions. Wittgenstein asks: "Every proof, even of a proposition which has already been proved, is a contribution to mathematics," but "why is it a contribution if its only point was to prove the proposition?" (Wittgenstein 1978, III, § 60). The view that the method of mathematics is the analytic method provides an answer to this question. A mathematical problem has many faces, each of which may suggest different hypotheses, so it may lead to different solutions to the problem. Each such solution establishes new relations between the problem and problems of other fields, thus showing the problem in a new light.

20.17 A Problematic View

All the above evidence points to the conclusion that the view that the method of mathematics is the axiomatic method is untenable, and only the view that the method of mathematics is the analytic method is defensible.

Many mathematicians, however, would object to this conclusion because, according to them, while non-deductive reasoning is essential in finding hypotheses, it cannot be a substitute for rigorous demonstration. In their opinion, finding hypotheses by non-deductive reasoning is mathematics in the making, but mathematics in finished form, as it is presented in textbooks, journals, and lectures, is rigorous demonstration.

Thus Mac Lane states that "conjecture has long been accepted and honored in mathematics," and indeed, "if a mathematician has really studied the subject and made advances therein, then he is entitled to formulate an insight as a conjecture" (Mac Lane 1994, 191–192). But "the next step must be proof," because something "is not mathematics until it is finally proved" (ibid., 192). The "heaven of mathematics requires much more" than conjecture, it requires proof, because, "in theological terms, we are not saved by faith alone, but by faith and works" (ibid., 191). Works are proofs, so "mathematics rests on proof – and proof is eternal" (ibid., 193).

This view, however, is untenable because the ultimate basis of rigorous demonstration are axioms which cannot be justified in any absolute sense and are only plausible, so they have the same status as the hypotheses on which the analytic method is based. Then the distinction between mathematics based on rigorous demonstration, and mathematics based on the analytic method, becomes blurred. (For more on this, see Chap. 21).

20.18 Answering the Dilemma

On these grounds, it seems fair to conclude that the view that the method of mathematics is the axiomatic method is untenable, and only the view that the method of mathematics is the analytic method is tenable. So, the view that mathematics is theorem proving is unsustainable, and only the view that mathematics is problem solving is defensible.

This gives an answer to the question posed at the beginning of this chapter, whether mathematics is theorem proving or problem solving, confirming the view put forward in Chap. 18. Mathematics is problem solving, and specifically problem solving by the analytic method. Since, as already pointed out, there is nothing subjective and psychological about the analytic method, the same applies to the claim that mathematics is problem solving by the analytic method.

To state that mathematics is problem solving by the analytic method is to put problems and solutions at the heart of mathematics, and to characterize mathematical knowledge as the result of problem solving by the analytic method.

As Halmos states, although "mathematics could surely not exist without" axioms, theorems and proofs, which "are all essential," nevertheless "none of them is at the heart of the subject," because "the mathematician's main reason for existence is to solve problems," and therefore, "what mathematics really consists of is problems and solutions" (Halmos 1980, 519).

Therefore, problems and solutions, not axioms, are at the heart of mathematics, and are the engines of mathematical knowledge.

References

Bourbaki, Nicholas. 1994. *Elements of the history of mathematics*. New York: Springer.

Cellucci, Carlo. 2007. *La filosofia della matematica del Novecento*. Rome: Laterza.

———. 2013a. *Rethinking logic: Logic in relation to mathematics, evolution, and method*. Dordrecht: Springer.

Curry, Haskell Brooks. 1951. *Outlines of a formalist philosophy of mathematics*. Amsterdam: North-Holland.

Davies, Edward Brian. 2008. Interview. In *Philosophy of mathematics: 5 questions*, ed. Vincent Hendricks and Hannes Leitgeb, 87–99. New York: Automatic Press/VIP.

Davis, Martin. 2005. What did Gödel believe and when did he believe it? *The Bulletin of Symbolic Logic* 11 : 194–206.

Dawson, John W. 2015. *Why prove it again? Alternative proofs in mathematical practice*. Basel: Birkhäuser.

de l'Hospital, Guillaume. 1696. *Analyse des infiniment petits*. Paris: Imprimerie Royale.

Detlefsen, Michael. 1992. On an alleged refutation of Hilbert's program using Gödel's first incompleteness theorem. In *Proof, logic and formalization*, ed. Michael Detlefsen, 199–235. London: Routledge.

Friend, Michèle. 2014. *Pluralism in mathematics: A new position in philosophy of mathematics*. Dordrecht: Springer.

Gödel, Kurt. 1986–2002. *Collected works*. Oxford: Oxford University Press.

Gowers, Timothy. 2006. Does mathematics need a philosophy? In *18 unconventional essays on the nature of mathematics*, ed. Reuben Hersh, 182–200. New York: Springer.
Halmos, Paul Richard. 1968. Mathematics as a creative art. *American Scientist* 56 : 375–389.
———. 1980. The heart of mathematics. *The American Mathematical Monthly* 87 : 519–524.
Hamming, Richard. 1980. The unreasonable effectiveness of mathematics. *The American Mathematical Monthly* 87 : 81–90.
———. 1998. Mathematics on a distant planet. *The American Mathematical Monthly* 105 : 640–650.
Hesseling, Dennis E. 2003. *Gnomes in the fog: The reception of Brouwer's intuitionism in 1920s*. Basel: Springer.
Hilbert, David. 1905. *Logische Prinzipien des mathematisches Denkens*. Unpublished manuscript. Göttingen: Bibliothek, Mathematischen Institut, Universität Göttingen.
———. 1967a. On the infinite. In *From Frege to Gödel: A source book in mathematical logic, 1879–1931*, ed. Jean van Heijenoort, 369–392. Cambridge: Harvard University Press.
———. 1967b. The foundations of mathematics. In *From Frege to Gödel: A source book in mathematical logic, 1879–1931*, ed. Jean van Heijenoort, 464–479. Cambridge: Harvard University Press.
———. 1992. *Natur und mathematischen Erkennen: Vorlesungen, gehalten 1919–1920 in Göttingen*. Basel: Birkhäuer.
———. 1996b. Axiomatic thought. In *From Kant to Hilbert: A source book in the foundations of mathematics*, II, ed. William Ewald, 1107–1115. Oxford: Oxford University Press.
———. 1996c. The new grounding of mathematics: First report. In *From Brouwer to Hilbert*, II, ed. William Ewald, 1117–1134. Oxford: Oxford University Press.
———. 1996d. The logical foundations of mathematics. In *From Kant to Hilbert: A source book in the foundations of mathematics*, II, ed. William Ewald, 1134–1148. Oxford: Oxford University Press.
———. 1996e. The grounding of elementary number theory. In *From Kant to Hilbert: A source book in the foundations of mathematics*, II, ed. William Ewald, 1149–1157. Oxford: Oxford University Press.
———. 1996f. Logic and the knowledge of nature. In *From Kant to Hilbert: A source book in the foundations of mathematics*, II, ed. William Ewald, 1157–1165. Oxford: Oxford University Press.
———. 2000. Mathematical problems. Appendix to Jeremy Gray, *The Hilbert challenge*, 240–282. Oxford: Oxford University Press.
Kac, Mark, and Stanislaw M. Ulam. 1992. *Mathematics and logic*. Mineola: Dover.
Kant, Immanuel. 1998. *Critique of pure reason*. Cambridge: Cambridge University Press.
Macintyre, Angus. 2011. The impact of Gödel's incompleteness theorems on mathematics. In *Kurt Gödel and the foundations of mathematics: Horizons of truth*, ed. Matthias Baaz et al., 3–25. Cambridge: Cambridge University Press.
Mac Lane, Saunders. 1986. *Mathematics: Form and function*. New York: Springer.
———. 1994. Response. In Michael Atiyah et al., Responses to 'Theoretical mathematics: Toward a cultural synthesis of mathematics and theoretical physics', by A. Jaffe and F. Quinn, 190–193. *Bulletin of the Americal Mathematical Society* 30 : 178–207.
Mäenpää, Petri. 1997. From backward reduction to configurational analysis. In *Analysis and synthesis in mathematics: History and philosophy*, ed. Michael Otte and Marco Panza, 201–226. Dordrecht: Springer.
Pappus. 1876–1878. *Collectio*. Berlin: Weidmann.
Peckhaus, Volker. 2002. Regressive analysis. *Logical Analysis and History of Philosophy* 5 : 97–110.
Post, Emil Leon. 1965. Absolutely unsolvable problems and relatively undecidable propositions: Account of an anticipation. In *The undecidable*, ed. Martin Davis, 340–433. New York: Raven Press.
Proclus. 1992. *In primum Euclidis Elementorum librum commentarii*. Hildesheim: Olms.

Rota, Gian-Carlo. 1981. Introduction. In Philip J. Davis and Reuben Hersh, *The mathematical experience*, xvii–xix. Boston: Birkhäuser.

———. 1997. *Indiscrete thoughts*. Boston: Birkhäuser.

von Neumann, John. 2003. Letter to Gödel, 29 November 1930. In Kurt Gödel, *Collected works*, vol. 5, 339–340. Oxford: Oxford University Press.

Weyl, Hermann. 1998. On the new foundational crisis of mathematics. In *From Brouwer to Hilbert: The debate on the foundations of mathematics in the 1920s*, ed. Paolo Mancosu, 86–118. Oxford: Oxford University Press.

———. 2009. *Mind and nature: Selected writings on philosophy, mathematics, and physics*. Princeton: Princeton University Press.

Wittgenstein, Ludwig. 1978. *Remarks on the foundations of mathematics*. Oxford: Blackwell.

Chapter 21
Concepts of Demonstration

Abstract The view that mathematics is theorem proving, hence the method of mathematics is the axiomatic method, and the view that mathematics is problem solving, hence the method of mathematics is the analytic method, lead to two different concepts of demonstration: axiomatic demonstration and analytic demonstration. This chapter highlights the limitations of axiomatic demonstration. It maintains that analytic demonstration is the basic concept of demonstration and that, in terms of this concept, it is possible to deal with the question of the depth of demonstrations and theorems. On this basis, the chapter characterizes a view of mathematics, called the heuristic view, in terms of the following features: mathematics is problem solving by the analytic method; mathematical objects are hypotheses human beings make to solve mathematical problems by the analytic method; and analytic demonstration is the basic concept of demonstration.

21.1 Axiomatic and Analytic Demonstration

The view that the method of mathematics is the axiomatic method, and the view that the method of mathematics is the analytic method, lead to two different concepts of demonstration: axiomatic demonstration and analytic demonstration.

An axiomatic demonstration consists in a deduction of a proposition from given axioms which are true, either in the strong sense that there is a system of things, specified in advance, for which the axioms are true, or in the weak sense that they are consistent. The purpose of axiomatic demonstration is to justify and teach an already acquired proposition.

An analytic demonstration consists in a non-deductive derivation of a hypothesis from a problem and possibly other data, where the hypothesis is a sufficient condition for the solution of the problem, namely, such that a solution to the problem can be deduced from it, and is plausible; then it consists in a non-deductive derivation of a new hypothesis from the previous hypothesis, considered in turn as a problem, and possibly from other data, where the new hypothesis is a sufficient condition for the solution of the problem posed by the previous hypothesis and is plausible; and so on, *ad infinitum*. The purpose of analytic demonstration is to discover hypotheses that are sufficient conditions for the solution of a problem, and are plausible.

© Springer International Publishing AG 2017

C. Cellucci, *Rethinking Knowledge*, European Studies in Philosophy of Science 4,
DOI 10.1007/978-3-319-53237-0_21

Axiomatic demonstration and analytic demonstration essentially differ as to their role in knowledge. Axiomatic demonstration, being based on the axiomatic method, can only justify and teach an already acquired proposition, so it is only a means to justify and teach already acquired knowledge. Analytic demonstration, being based on the analytic method, can discover hypotheses that are sufficient conditions for the solution of a problem and are plausible, so it is a means to acquire knowledge.

This depends on the fact that, since deductive rules are non-ampliative, axiomatic demonstration cannot produce new knowledge, it simply makes explicit what is implicit in the axioms. As Hintikka states, axioms "are supposed to tell you everything there is to be told," and "the rest of your work will consist in merely teasing out the logical consequences of the axioms" (Hintikka 1996, 1). Conversely, since non-deductive rules are ampliative, analytic demonstration can produce new knowledge.

A remark about the use of the term 'demonstration'. Throughout this book, 'demonstration' is preferably used in place of 'proof'. This is motivated by the fact that 'proof' is commonly used to indicate deductive arguments.

For example, Gowers states that mathematicians "are rarely satisfied with the phrase 'it seems that'. Instead, they demand a proof" (Gowers 2002, 36). A proof is an argument which "starts with axioms that are universally accepted and proceed to the desired conclusion by means of only the most elementary logical rules (such as 'if *A* is true and *A* implies *B* then *B* is true')" (ibid., 39).

Conversely, 'demonstration' does not have such strict connotation, because it comes from the Latin *demonstrare*, Greek *apodeiknumi*, which means 'to show'. Then 'demonstration' seems more suitable, not only for demonstrations based on diagrams, but as a general term for both axiomatic and analytic demonstration.

21.2 *A Priori* and *A Posteriori* Demonstration

The concepts of axiomatic demonstration and analytic demonstration correspond to the medieval concepts of *a priori* demonstration and *a posteriori* demonstration, respectively. An *a priori* demonstration is one that descends from the cause to the effect, and an *a posteriori* demonstration is one that ascends from the effect to the cause.

Thus Aquinas says that "demonstration can be made in two ways: one is through the cause, and is called *a priori*, and this is to argue from what is prior absolutely. The other is through the effect, and is called a demonstration *a posteriori*; this is to argue from what is prior relatively only to us" (Aquinas, *Summa Theologica*, Part I, Question 2, Article 2). By the expressions 'what is prior absolutely' and 'what is prior relatively only to us' Aquinas designates what Aristotle designates by the expressions 'what is clearer and more knowable by nature' and 'what is more knowable and clearer to us', respectively (see Chap. 12).

A priori demonstration is also called demonstration *propter quid*, and *a posteriori* demonstration is also called demonstration *quia*.

Thus Albert of Saxony says that demonstration is twofold, namely, that "which proceeds from causes to effect, and is called *a priori* demonstration and demonstration *propter quid*," and that "which proceeds from effects to causes," and "is called *a posteriori* demonstration and demonstration *quia*" (Albert of Saxony, *Questiones Subtilissime in Libros Aristotelis de Caelo et Mundo – Questiones Subtilissime super Libros Posteriorum*, Book I, Question IX, f. 8 r).

21.3 Axiomatic and Analytic Theory

In addition to leading to two different concepts of demonstration, the view that the method of mathematics is the axiomatic method and the view that the method of mathematics is the analytic method lead to two different concepts of mathematical theory: axiomatic theory and analytic theory.

An axiomatic theory is a closed set of axioms and propositions deduced from them. A closed set, because the axioms are given once and for all, and the propositions deduced from them are completely determined by the axioms. Thus, an axiomatic theory is a closed system, in the sense stated in Chap. 13. The axioms must be true, either in the strong sense that there is a system of things, specified in advance, for which the axioms are true, or in the weak sense that they are consistent. The development of an axiomatic theory consists in deducing propositions from the axioms. A deduction of a proposition from the axioms is an argument that makes the proposition absolutely certain.

An analytic theory is an open set of problems and hypotheses that permit to solve them. An open set, because the hypotheses are not given once and for all but new hypotheses can always be introduced, or the existing ones can be modified. Thus an analytic theory is an open system, in the sense stated in Chap. 13. The development of an analytic theory consists in introducing new hypotheses, or modifying the existing ones, and deducing solutions to problems from them. Hypotheses are introduced by some non-deductive rule and must be plausible, namely, the arguments for them must be stronger than the arguments against them, on the basis of the existing knowledge. The solution to a problem is not absolutely certain but only plausible.

21.4 A Limitation of Axiomatic Demonstration

According to the current prevailing view, axiomatic demonstration is the basic concept of demonstration. It is superior to analytic demonstration since it establishes irrefutably the truth of the theorem under consideration, once and for all.

Thus Pólya states that there are two concepts of proof, namely, "properly so-called logical" proof, and "heuristic proof" (Pólya 1941, 450). Logical proof "establishes the truth" of "the theorem under consideration," conversely, heuristic proof "cannot demonstrate the truth" of a theorem, it can only augment "our confidence in a theorem which is still only a conjecture" (ibid.). Pólya's 'logical proof' and 'heuristic proof' correspond to 'axiomatic demonstration' and 'analytic demonstration', respectively. According to Pólya, logical proof "is definitive, it establishes irrefutably the truth of the theorem – once for all" (ibid.). Conversely, heuristic proof "is provisional; the one I find today increases my confidence which may be shaken tomorrow by another," and "definitely shattered the following day by the rigorous refutation of the theorem under consideration" (ibid.). Logical proof "appears generally by itself on the pages of mathematical treatises," conversely, heuristic" proof, "which in general guided the invention of the logical" proof, "is omitted" (ibid.).

The current prevailing view, however, is unjustified because axiomatic demonstration has important limitations. As argued in Chap. 20, by Gödel's incompleteness theorems and for other reasons, the view that the method of mathematics is the axiomatic method is untenable.

In particular, the claim that axiomatic demonstration establishes irrefutably the truth of the theorem under consideration, once and for all, is untenable because it is impossible to know that the axioms are true, even in the weak sense that they are consistent. In order to know that the axioms are consistent, we should be able to give a demonstration that they are consistent. But, by Gödel's second incompleteness theorem, the sentence 'The axioms are consistent' will not be demonstrable from these axioms, but only from a proper extension of them. Again by Gödel's second incompleteness theorem, the sentence 'The axioms of the proper extension are consistent' will not be demonstrable from these axioms, but only from a proper extension of them. And so on, *ad infinitum*. Therefore, it is impossible to know that the axioms are consistent. We can only know that they are plausible. But then axioms have the same status as the hypotheses of analytic demonstration. So, the concept of axiomatic demonstration collapses into that of analytic demonstration, and our reliance on axiomatic demonstration ultimately rests on plausibility. Pólya himself ends up admitting that "a good part of our reliance on" axiomatic demonstration "may come from plausible reasoning" (Pólya 1954, II, 168).

Not only it is impossible to know that the axioms are consistent, but it is also impossible to justify the deductive inferences by which propositions are derived from axioms, in an absolute sense. One can justify deductive rules only as much, or as little, as non-deductive rules (for details, see Cellucci 2013a, Chap. 18).

21.5 An Alleged Way Out for Axiomatic Demonstration

In the previous section it has been argued that the claim that axiomatic demonstration establishes irrefutably the truth of the theorem under consideration, once and for all, is untenable because it is impossible to know that the axioms are true, even in the weak sense that they are consistent. It might be objected that there is another criterion of the truth of mathematical axioms, namely, that axioms are justified if true consequences follow from them. Then it is the consequences that give the reasons why we believe the axioms.

Thus Zermelo claims that "principles must be judged from the point of view of science", namely, from the point of view of their consequences, "and not science from the point of view of principles fixed once and for all" (Zermelo 1967, 189). For example, the principle of choice is justified because there is "a number of elementary and fundamental theorems and problems" that "could not be dealt with at all without the principle of choice" (ibid., 188). So long as it "cannot be definitely refuted, no one has the right to prevent the representatives of productive science from continuing to use this 'hypothesis'" (ibid., 189).

As we have already seen, Russell and Gödel support this criterion of the truth of mathematical axioms (see Chaps. 3 and 19, respectively). But the criterion fails, because from a false axiom one can deduce true consequences. In order to assert that an axiom is true, one ought to be able to show that all of its logical consequences are true, but this is generally unfeasible.

This is already made quite clear by Kant, who states that "inferring the truth of a cognition from the truth of its consequences, would be allowed only if all of the possible consequences are true," but "this procedure is unusable, because to have insight into all possible consequences of any proposition that is assumed exceeds our powers" (Kant 1998, A790/B818).

Evidence for Kant's statement, is given by the fact that the set of all consequences of the axioms of second-order Peano arithmetic PA^2 is not algorithmically enumerable. Indeed, since the axioms of PA^2 are categorical, a second-order sentence is a consequence of the axioms of PA^2 if and only if it is true of the natural numbers. Then, if the set of all consequences of the axioms of PA^2 were algorithmically enumerable, so would be the set of all second-order sentences true of the natural numbers. But, by Tarski's theorem for second-order sentences, the set of all second-order sentences true of the natural numbers is not definable in the set of all natural numbers by any second-order formula, and hence *a fortiori* is not algorithmically enumerable. Therefore, the set of all consequences of the axioms of PA^2 is not algorithmically enumerable. This means that there exists no algorithm, *a fortiori* no feasible procedure, for enumerating all consequences of the axioms of PA^2. So, as Kant states, inferring the truth of a cognition from the truth of its consequences is unfeasible because it exceeds our powers.

Moreover, inferring the truth of a cognition from the truth of its consequences is incompatible with the concept of axiomatic demonstration, which involves that it is the axioms that give the reason for believing the consequences, and not the other way round.

21.6 A Further Limitation of Axiomatic Demonstration

A further limitation of axiomatic demonstration is that, with an axiomatic demonstration, we never know what we are talking about. Indeed, with axiomatic demonstration, one is led to Russell's conclusion that "mathematics may be defined as the subject in which we never know what we are talking about, nor whether what we are saying is true" (Russell 1994, 76).

That, with an axiomatic demonstration, we never know what we are talking about, is implicit in Hilbert's statement that, in the case of the axioms of geometry, if, instead of points, lines, planes, we think of a system of love, law, chimney-sweep which satisfies all axioms, then Pythagoras' theorem also applies to these things (see Chap. 12). Indeed, if a system of love, law, chimney-sweep satisfies the axioms of geometry, why should not we consider such entities as geometrical entities?

Hilbert's argues that the circumstance that the axioms of geometry can be satisfied by non-geometrical objects "can never be a defect in a theory, and it is in any case unavoidable" (Hilbert 1980a, 41). Moreover, "it takes a very large amount of ill will to want to apply" the axioms of geometry "to other appearances than the ones for which they were meant;" indeed, "the application of a theory to the world of appearances always requires a certain measure of good will and tactfulness" (ibid.).

But Hilbert's argument is unsatisfactory, because appealing to good will and tactfulness is not part of the conception of axiomatic demonstration, and in any case it does not contribute to clarify what we are talking about in mathematics.

21.7 The Demand for Purity of Methods

The sense of axiomatic demonstration is expressed by the demand for purity of methods – the requirement that demonstrations should be pure, namely, a demonstration should not use concept, axioms, or results belonging to fields different from the field to which the conclusion of the demonstration belongs.

The demand for purity of methods goes back to Aristotle, who states that "a proposition of one science cannot be demonstrated by another science" (Aristotle, *Analytica Posteriora*, A 7, 75 b 14). For example, it is not possible to demonstrate "a geometrical proposition by arithmetic" (ibid., A 7, 75 a 38–39). For, each science has a kind as its subject matter, kinds are separated, and principles "must be in the same kind as the things demonstrated" (ibid., A 28, 87 b 2–3). So, "arithmetical demonstrations always keep to the kind which is the subject of the demonstration, and similarly with all other sciences. Hence the kind must be the same" (ibid., A 7, 75 b 6–9).

The demand for purity of methods has been made several times since Aristotle.

Thus Newton states that, while "the antients did so industriously distinguish" geometry and arithmetic "from one another, that they never introduced arithmetical terms into geometry," the "moderns, by confounding both, have lost the simplicity in which all the elegancy of geometry consists" (Newton 1720, 230). Thus, "multiplications, divisions, and such sort of computations, are newly received into geometry" (ibid., 229). But "equations are expressions of arithmetical computation, and properly have no place in geometry" (ibid.). Therefore, "these two sciences ought not to be confounded" (ibid., 230).

Frege states that it was with great reluctance that numbers such as "the complex numbers were finally introduced. The overcoming of this reluctance was facilitated by geometrical interpretations; but with these, something foreign was introduced into arithmetic" (Frege 1984, 116). Indeed, "it appeared contrary to all reason that purely arithmetical theorems should rest on geometrical axioms;" therefore "the task of deriving what was arithmetical by purely arithmetical means" could "not be put off" (ibid., 117).

Hilbert states that it is necessary "everywhere to guarantee the purity of method," since "in many cases our understanding is not satisfied when, in a demonstration of a proposition of arithmetic, we draw on geometry, or in a demonstration of a geometrical truth we draw on function theory" (Hilbert 2004b, 236). This is "the demand for 'purity' of methods" (Hilbert 1987, 125).

21.8 Impossibility of Satisfying the Demand for Purity of Methods

Despite its large support, by Gödel's first incompleteness theorem the demand for purity of methods cannot be satisfied. As Gödel himself points out, "a special case of" the first incompleteness theorem "is that there are arithmetic propositions which can be proved only by analytical methods, and, further, that there are arithmetic propositions which cannot be proved even by analysis but only by methods involving extremely large infinite cardinals and similar things" (Gödel 1986–2002, III, 48).

As a matter of fact, in the historical development of mathematics, the demand for purity of methods has been largely disregarded. A recent example is provided by the solution of Fermat's problem. As already mentioned in Chap. 12, to solve Fermat's problem, that is a problem about the natural numbers, Ribet used the Conjecture of Taniyama-Shimura, that is a hypothesis about modular forms in hyperbolic space. Also, to solve the problem posed by the Conjecture of Taniyama-Shimura hypothesis, Wiles and Taylor used hypotheses from various fields of mathematics, ranging from differential geometry to complex analysis.

The reason why, in the historical development of mathematics, the demand for purity of methods has been largely disregarded, is that restricting demonstrations to pure demonstration may yield very inefficient demonstration systems. An extreme example of this is given by Boolos, who considers the following axioms:

(1) $D(1)$

(2) $\forall x(D(x) \rightarrow D(s(x)))$

(3) $\forall n(f(n,1)=s(1))$

(4) $\forall x(f(1,s(x))=s(s(f(1,x))))$

(5) $\forall n \forall x(f(s(n),s(x))=f(n,f(s(n),x)))$

In the intended interpretation, the variables x and n range over the positive integers, 1 denotes the number one, s denotes the successor function, and f denotes an

Ackermann-style function, namely a function whose value grows rapidly, even for small inputs.

Now, consider the sentence $D(f(s(s(s(s(1)))), s(s(s(s(1))))))$. There is a very short 'impure' axiomatic demonstration of this sentence from the axioms (1)–(5) plus the axioms of set theory, so short that a demonstration "whose every symbol can easily be written down" (Boolos 1998, 376). On the contrary, "it is well beyond the bounds of physical possibility that any actual or conceivable creature or device could ever write down all the symbols" of a 'pure' axiomatic demonstration of the sentence, namely an axiomatic demonstration from the axioms (1)–(5) alone, since "there are far too many symbols in any such" axiomatic demonstration "for this to be possible" (ibid., 376–377). No such axiomatic demonstration "could possibly be written down in full detail, in this universe" (ibid., 377). Indeed, writing it down in full detail would require at least

$$2^{\cdot^{\cdot^{\cdot^{2}}}}$$

symbols, where the stack of 2 s contains 65,536 items. This is a gigantic number, much larger than 2^{79}, the number of electrons, protons, and neutrons in the universe.

Nevertheless, the demand for purity of methods has been made even after the discovery of Gödel's first incompleteness theorem. Thus, Bourbaki states that to set up the theory of a given structure "amounts to the deduction of the logical consequences of the axioms of the structure, excluding every other hypothesis on the elements under consideration" (Bourbaki 1950, 226). For example, to set up theory of the group structures, amounts to the formulation of "the axioms of the group structures" and to "the development of their consequences" (ibid., 225).

But this conflicts with the fact that only the first stages in the development of the theory of the group structures amount to the formulation of the axioms of the group structures and to the development of their consequences. Modern developments of the theory of the group structures require significant uses of 'impure' methods of geometry, topology, probability, measure theory, etc. This shows that the development of a mathematical theory cannot amount to the deduction of the logical consequences of the axioms of the theory, but requires significant uses of methods of other theories.

21.9 The Point of Analytic Demonstration

An alternative to the current prevailing view that axiomatic demonstration is the basic concept of demonstration, is that analytic demonstration is the basic concept of demonstration. This alternative is motivated by the fact that, as argued in Chap. 20, the view that the method of mathematics is the analytic method is unaffected, and even confirmed, by Gödel's incompleteness theorems, and is not subject to the further limitations of the axiomatic method considered there.

In particular, the point of analytic demonstration can be seen in terms of Gödel's first incompleteness theorem. By the latter, the hypotheses to solve a problem of a given field may have to be sought in another field. This is compatible with the concept of analytic demonstration since, in an analytic demonstration, the hypotheses need not belong to the field of the problem but may belong to other fields. Conversely, it is incompatible with the concept of axiomatic demonstration since, in an axiomatic demonstration, the axioms are axioms of the field of the proposition being proved.

The point of analytic demonstration can also be seen in terms of Gödel's second incompleteness theorem. By the latter, the mathematical knowledge resulting from the deduction of propositions from given axioms cannot be proved to be firmly grounded. This is compatible with the concept of analytic demonstration. Conversely, it is incompatible with the concept of axiomatic demonstration, which is supposed to be "an argument that puts a statement beyond all possible doubt" (Gowers 2002, 36).

The view that analytic demonstration is the basic concept of demonstration is also supported by Friend, who states that "all proofs can be thought of as analytic, including rigorous proofs" (Friend 2014, 208). And "all proofs are better viewed as analytic" (ibid., 213). For, analytic proofs "lead us to much more fundamental types of exploration than we would have engaged in had we viewed proofs as axiomatic" (ibid., 205).

A corollary of the view that analytic demonstration is the basic concept of demonstration is that mathematical theories are analytic theories. As Goodman states, in most respects "mathematical theories resemble other scientific theories. They are constructed to solve particular problems," and "are not originally deductive structure based on axioms but, rather, informal bodies of reasoning based on conjectures and bold extrapolation" (Goodman 1991, 124).

21.10 Analytic Demonstration and Intuition

In analytic demonstration, intuition has no role. It has no role in the discovery of hypotheses, because the latter are obtained from the problem, and possibly other data already available, by some non-deductive rule, so not by intuition but by inference. It has no role in the justification of hypotheses, because their plausibility is established by means of the plausibility test procedure described in Chap. 12, so not by intuition but by inference. Since in analytic demonstration intuition has no role, we may conclude that mathematics is not based on intuition.

The mathematicians who claim that intuition has a central role in mathematics, by 'intuition' simply mean the feeling of 'almost knowing' some hypothesis, without having consciously gone through a step-by-step reasoning process to get there. This feeling can be explained in terms of the fact that they arrived at the hypothesis through some unconscious non-deductive inference, which led them to trust that the hypothesis might be plausible. Indeed, why do only mathematicians have those intuitions? Because they have sufficient background knowledge, and have been thinking and rethinking intensively about the problem. This provides them with the data, on the basis of which they make an expert guess through some unconscious non-deductive inference.

21.11 Analytic Demonstration and Depth

In terms of analytic demonstration, we can deal with the question of depth. Mathematicians often say that certain demonstrations or theorems are deep, but there is no widely shared view of what a deep demonstration or theorem consists in.

Some people believe that a demonstration is deep if it is very long, and a theorem is deep if its demonstration is very long. Thus Shanks states that "a deep theorem" is "one whose proof requires a great deal of work – it may be long, or complicated" (Shanks 1993, 64). But a counterexample to this view is provided, for example, by the Appel-Haken demonstration of the four colour theorem, which is very long but most mathematicians would agree that it is not deep, being very long only because it requires the consideration of a "large number of cases" (Appel and Haken 1989, 487). Moreover, by Gödel's speed-up theorem, there are theorems whose demonstrations are very long but can be drastically shortened using more powerful axioms. Since their demonstrations are very long, such theorems ought to be classified as deep. But, since their demonstrations can be drastically shortened using more powerful axioms, the theorems in question ought to be classified as not deep at all.

Other people believe that a demonstration is deep if it is fruitful, in the sense that it leads to yet further demonstrations, particularly to important demonstrations. Thus Arana states that "a consequentialist view of depth measures the depth of a theorem by some quality of its consequences, or of the consequences of its proofs. Typically, the quality in question is fruitfulness, the degree to which a theorem (or a proof of a theorem) leads to yet further theorems and proofs, particularly to important theorems and proofs" (Arana 2015, 171). But a counterexample is provided by the fact that demonstrations formerly judged important recede in importance as mathematics develops new machinery.

Instead, a demonstration or a theorem can be said to be deep if it ties together disparate areas, or opens new areas, of mathematics. Analytic demonstrations can be deep, because they may reveal unexpected relations between different areas of mathematics, which may suggest new perspectives and new problems, and even open new areas of mathematics. Conversely, axiomatic demonstrations cannot be deep, because they remain entirely within the boundaries of a closed system, so they cannot tie together disparate areas, or open new areas of mathematics. Indeed, since axiomatic demonstrations are deductions from the axioms and, as argued in Chap. 12, deductive rules are non-ampliative, axiomatic demonstrations merely make explicit what is already contained in the axioms.

Byers states that "the depth of a mathematical situation" – demonstration or theorem – "is a measure of the creativity that accompanies its birth" (Byers 2007, 374). In fact, analytic demonstrations, which can be deep, can be creative because they may tie together disparate areas of mathematics, or open new areas of mathematics. Byers also states that a mathematical situation is non-creative if it involves "juggling a number of predetermined elements according to predetermined rules" (ibid., 388). In fact, axiomatic demonstrations, which are not deep, are non-creative because their conclusions are obtained from predetermined axioms, according to predetermined deductive rules.

21.12 Analytic Demonstration and Published Demonstrations

Since, in analytic demonstration, hypotheses are obtained by rules of discovery consisting of non-deductive rules, in order to understand the nature of mathematics it is essential to analyse such rules. (An analysis of several of them is given in Cellucci 2013a, Chaps. 20 and 21).

Against this, it could be objected that, instead of analysing rules of discovery, in order to understand the nature of mathematics it would be more fruitful to work up case studies from the history of mathematics.

Thus Grosholz states that, instead of analysing "rules of discovery," the efforts of philosophers of mathematics would be "better spent on studying the history of mathematics and the actual development of important recent problem-solutions" (Grosholz 2015, 139).

This, however, is problematic. Studying the history of mathematics is possible only on the basis of published demonstrations, namely demonstrations as presented, justified and taught in journals and textbooks. But the way mathematical results are presented, justified, and taught in journals and textbooks has little or nothing to do with the way they were discovered. Therefore, studying the history of mathematics does not provide an adequate basis for understanding the real mathematical processes.

That the way mathematical results are presented, justified, and taught in journals and textbooks has little or nothing to do with the way they were discovered, is declared by many great mathematicians.

Thus Newton states that his own propositions "were invented by analysis" (Newton 1967–1981, VIII, 647). Namely, they were invented by analytic demonstration. But, since "the ancients for making things certain admitted nothing into geometry before it was demonstrated synthetically," namely by axiomatic demonstration, Newton "demonstrated the propositions synthetically," and "this makes it now difficult for unskilful men to see the analysis by which those propositions were found out" (Newton 1714–1716, 206). Newton "could have written analytically what" he "had found out analytically," but he "was writing for philosophers steeped in the elements of geometry," namely in Euclid's *Elements*, "and putting down geometrically demonstrated bases for physical science" (Newton 1967–1981, VIII, 451). This "is the reason why" Newton demonstrated his propositions synthetically, "after the manner of the ancients" (ibid., 648).

Descartes states that "in their writings the ancient geometers usually employed" axiomatic demonstration alone, "not because they were utterly ignorant" of analytic demonstration, which was actually the way they discovered their results, but because "they had such a high regard for it that they kept it to themselves like a sacred mystery" (Descartes 1996, VII, 156). They did so "with a kind of pernicious cunning" since, "as notoriously many inventors are known to have done where their own discoveries were concerned, they have perhaps feared" that their method, "just because it was so easy and simple, would be depreciated if it were divulged" (ibid., X, 376).

Much in the same vein, Grothendieck states that the creative work in mathematics "is not reflected virtually to any extent in the texts or talks that are intended to present such work," whether "textbooks and other didactic texts, or articles and original memoirs, or oral courses and seminar presentations, etc." (Grothendieck 1985, 84). This is because, "from the very beginning of mathematics," there has been "a sort of 'conspiracy of silence' about these 'ineffable works' that prelude to the hatching of any new idea, great or small, which comes to renew our knowledge of a part of this world, in perpetual creation, in which we live" (ibid.).

Since the way mathematical results are presented, justified, and taught in journals and textbooks has little or nothing to do with the way they are discovered, studying the history of mathematics can teach us about the sequence of mathematical results and theories, but not about the real mathematical processes.

21.13 The Purpose of Axiomatic Demonstration

One of the reasons for the claim that axiomatic demonstration is the basic concept of demonstration is that many people take textbooks such as Euclid's *Elements* as a paradigm for mathematics.

But this is unjustified. Admittedly, as Hersh points out, "a naive non-mathematician" who "looks into Euclid" and "observes that axioms come first," understandably "concludes that in mathematics, axioms come first. First your assumptions, then your conclusions, no? But anyone who has done mathematics knows what comes first – a problem" (Hersh 1997, 6). In mathematics, "problems, and solutions come first. Later come axiom sets," therefore "the view that mathematics is in essence derivations from axioms is backward. In fact, it's wrong" (ibid.).

Textbooks such as Euclid's *Elements* are not a paradigm for mathematics. Kuhn states that paradigms are fixed by certain "textbooks, elementary and advanced" which serve "for a time implicitly to define the legitimate problems and methods of a research field for succeeding generations of practitioners" (Kuhn 1996, 10). But this claim is misleading, because textbooks do not fix paradigms for research even in normal science, namely in "puzzle-solving within the tradition that the textbooks define" (ibid., 166). For, they say nothing as to how new results are discovered, even in normal science, so they do not serve to define the legitimate methods of a research field.

Contrary to the view that textbooks such as Euclid's *Elements* are a paradigm for mathematics, as we have seen in Chap. 12, already Aristotle stated that the purpose of the axiomatic method, and hence of axiomatic demonstration, is not to obtain new knowledge, but only to justify and teach an already acquired proposition. In fact, Aristotle does not use axiomatic demonstration in his own scientific research work, where he proceeds by the analytic-synthetic method.

That the purpose of axiomatic demonstration is not to obtain new knowledge but only to justify and teach an already acquired proposition, is also apparent from Euclid's *Elements*. The latter, which are commonly considered the prototype of axiomatic demonstration, are a compilation and organization of already acquired propositions for didactic purposes. As Proclus informs us, Euclid "put together the *Elements* collecting many of the theorems of Eudoxus, perfecting many others by Theaetetus, and bringing to incontrovertible demonstration things which had only been somewhat loosely established by his predecessors" (Proclus 1992, 68.7–10). On the other hand, in the *Elements* Euclid "did not bring in everything he could have collected," but only "theorems and problems that are worked out for the instruction of beginners" (ibid., 69.6–9). He omitted matters that "are unsuitable for a selection of elements because they lead to great and unlimited complexity" (ibid., 74.21–22).

In fact, Euclid did not use axiomatic demonstration in his own research work, in which instead he proceeded by the analytic-synthetic method. As Pappus informs us, Euclid's *Data*, *Porisms*, and *Surface Loci* were part of "the so-called *Treasure of Analysis*," a "special body of doctrine provided for the use of those who, after finishing the ordinary elements, are desirous of acquiring the power of solving problems" (Pappus 1876–1878, VII, 634.3–7). This special body of doctrine is "the work of three men, Euclid, the author of the *Elements*, Apollonius of Perga, and Aristaeus the Elder, and proceeds by the method of analysis and synthesis" (ibid., VII, 634.8–11). This explains why it is unjustified to take Euclid's *Elements* as a paradigm for mathematics.

That the purpose of axiomatic demonstration is to justify and teach an already acquired proposition is motivated by the fact that, in textbook writing, axiomatic demonstration has the advantage of compactness. For example, Pappus asserts that he writes his comparison of the five Platonic bodies "not by the so-called analytic investigation," namely by the analytic method, "by which some of the ancients produced the proofs," but "by the synthetic method," namely by the axiomatic method, "in order to achieve the clearest and most concise form" (ibid., V, 410.27–412.3).

On the other hand, however, axiomatic demonstration often appears somewhat artificial and difficult to understand, because it hides the real mathematical processes, and hence misrepresents the nature of mathematics. Indeed, "textbook and monograph presentations of mathematics" are "so difficult to follow" because the presentation "is often 'backward'. The discovery process is eliminated from the description and is not documented" (Davis and Hersh 1981, 281–282).

Strictly speaking, it is even unjustified to consider Euclid's *Elements* the prototype of axiomatic demonstration. As already pointed out in Chap. 17, even Euclid's demonstration of the first proposition of the *Elements* is not a deduction.

21.14 An Objection to the Heuristic Purpose of Analytic Demonstration

It has been argued above that, while the purpose of axiomatic demonstration is to justify and teach an already acquired proposition, the purpose of analytic demonstration is to discover hypotheses that are sufficient conditions for the solution of a problem and are plausible. For, unlike synthesis, analysis is a method of discovery.

Against this, Netz objects that, if one says that analysis is a method of discovery, "in exactly the same way, one could even say that synthesis is a method of discovery," because, "just as an analysis ends up with the added construction of the solution, so a synthesis ends up with the solution" (Netz 2000, 143). Actually, analysis is not "a method of discovery in any privileged sense" (ibid.). Like synthesis, analysis is "a tool for the presentation of results" rather than "a tool for their discovery" (ibid., 156). With respect to synthesis, analysis creates "the illusion that the solution is necessary and emerges naturally out of the problem," but this is only an illusion because, "given a synthesis," we are always "capable of producing, *post factum*, an analysis which makes the synthesis appear necessary" (ibid.).

These objections, however, are unwarranted. It is unjustified to assert that, if one says that analysis is a method of discovery, in exactly the same way one could say that synthesis is a method of discovery, because both of them end up with the solution. Synthesis starts from axioms which must be discovered, but cannot be discovered by synthesis since this would involve an infinite regress, they can be discovered only by analysis.

Moreover, there is no algorithm for discovering hypotheses, and hence for obtaining the solution by analysis, while, as already mentioned in Chap. 11, there is an algorithm for enumerating all deductions from given axioms, and hence for obtaining the solution by synthesis.

Furthermore, it is unjustified to assert that, like synthesis, analysis is a tool for the presentation of results. While analysis is effective as a tool for discovery, analysis is space-consuming as a tool for the presentation of results, and "the economics of the book business demands maximum information in minimum space" (Davis and Hersh 1981, 282). Therefore, analytic demonstration is almost completely absent in textbook writing.

In addition, it is unjustified to assert that, given a synthesis, we are always capable of producing, *post factum*, an analysis which makes the synthesis appear necessary. As Newton points out, when propositions are demonstrated synthetically, this makes it difficult to see the analysis by which those propositions were found out. In fact, given a synthesis, we will be generally incapable of producing an analysis which makes the synthesis appear necessary.

21.15 Axiomatic Demonstration and Formal Demonstration

In the past century, a special kind of axiomatic demonstration has been considered, namely formal demonstration – deduction of a formula from given axioms by formal deduction rules. This raises the question of the relation between axiomatic demonstration and formal demonstration.

According to a widespread view, every axiomatic demonstration can be represented by a formal demonstration. For example, Macintyre states that "one could go on to translate" all "classical informal proofs into formal proofs of some accepted formal system", where such translations "do map informal proofs to formal proofs" (Macintyre 2005, 2420).

To a certain extent, this view is implicit in Frege. To a certain extent only, since Frege states that "every inference is non-formal in that the premises as well as the conclusions have their thought-contents which occur in this particular manner of connection only in that inference" (Frege 1984, 318).

Anyway, the view that every axiomatic demonstration can be represented by a formal demonstration is explicitly formulated by Hilbert, and hence may be called 'Hilbert's Thesis'.

In fact, Hilbert states that formal demonstrations are capable "to express the entire thought-content of the science of mathematics in a uniform manner" (Hilbert 1967b, 475). They "are the images of the thoughts that make up the usual procedure of traditional mathematics" (Hilbert 1996e, 1153). Formal demonstrations do "nothing other than to imitate the intimate activity of our understanding, and to make a protocol of the rules whereby our thinking actually proceeds" (ibid., 1156). They are "carried out according to certain definite rules, in which the technique of our thinking is expressed. These rules form a closed system that can be discovered and definitively stated," and are "the rules according to which our thinking actually proceeds" (Hilbert 1967b, 475).

Hilbert's Thesis, however, is invalid. Indeed, even the very first Hilbert's demonstration in his *Grundlagen der Geometrie* cannot be represented by a formal demonstration since, as we have seen in Chap. 19, it is not a deduction. It would be possible to give a formal demonstration of the same proposition, but this would involve replacing Hilbert's use of a diagram by additional axioms (see Meikle and Feuriot 2003). The resulting formal demonstration would be much more complex and essentially different from Hilbert's demonstration, so it could not be considered to be a representation of it.

Hilbert's Thesis is usually motivated by the claim that the correctness of formal demonstrations can be mechanically verified by computer. Thus, Worrall and Zahar state that, in the case of formal demonstrations, "whether a string of sentences is a proof or not can be checked in a finite number of steps," and "there is no serious sense in which such proofs are fallible" (Worrall and Zahar 1976, 57). Therefore, axiomatic proof can attain complete reliability through formalization.

This claim, however, is unjustified. The process through which an axiomatic demonstration is translated into a formal demonstration is not a mechanical one, so one cannot rule out the possibility of an incorrect translation leading to a pseudo validation of an alleged theorem. Moreover, as already pointed out in Chap. 16, mechanically verifying the correctness of formal demonstrations is generally impossible. And, even if it were possible, there would remain the problem of demonstrating the consistency of the axioms on which formal demonstrations are based, and of demonstrating it by absolutely reliable means. But, by Gödel's second incompleteness theorem, this is impossible.

21.16 Analytic Demonstration and Subformula Property

A special kind of formal demonstration is Gentzen's proof with the "subformula property" (Gentzen 1969, 88). By a proof with the subformula property Gentzen means a proof such that "no concepts enter into the proof other than those contained in its final result" (ibid., 69). That is, no concepts enter into the proof other than those contained in the conclusion.

Sometimes analytic demonstration, namely demonstration based on the analytic method, is confused with Gentzen's proof with the subformula property.

Thus Hintikka claims that, from the comparison between "the old heuristic method known as analysis (geometrical analysis) famous in antiquity" and certain "new techniques in symbolic logic," it is apparent that "the method of analysis is almost a special case of these" techniques, because "the logic of the method"of analysis "satisfies the so-called subformula property" (Hintikka and Remes 1976, 253). Since, by what is "known as the cut elimination problem," every proof in Gentzen's proof system can be transformed into a proof with the subformula property, "Greek geometers' analytical practice thus relied tacitly on the possibility of cut elimination" (Hintikka 2012, 54–55).

Poggiolesi claims that an analytic proof is a proof obtained by "the analytic method," a method with "a long and venerable history" which "extends back to ancient Greece" (Poggiolesi 2012, 445). Analytic proof has the analyticity property, namely, "every element which occurs in the proof will also occur in the conclusion" (ibid.). We can "use 'analyticity property' and 'subformula property' as synonymous" (ibid., 447, footnote 2). For, a proof with the subformula property is a proof such that "every formula which occurs in it is a subformula of the formulas which occur in the conclusion" (ibid., 445). Thus the notion of proof with the "subformula property" can "be thought of as a formalisation of the" notion of "analytic proof" (ibid., 444). On this basis, since, by the cut elimination theorem, every proof in Gentzen's proof system can be transformed into a proof with the subformula property, Gentzen can be said to "prefer and support the analytic method" (ibid., 446).

But analytic demonstration has nothing to do with Gentzen's proof with the subformula property because, in the analytic method, the hypotheses to solve a problem need not belong to the field of the problem, but may belong to other fields (see Chap. 12). So, concepts may enter into an analytic demonstration other than those contained in the conclusion.

Poggiolesi also claims that "a methodological requirement strictly connected with" analytic demonstration consists "in the so-called 'purity of methods'," namely the demand that "mathematicians should not use in their proofs concepts belonging to a theoretical domain different from that employed initially" (Poggiolesi 2011, 13).

But the demand for purity of methods has nothing to do with analytic demonstration, because an analytic demonstration may use concepts belonging to theoretical domains different from that employed initially, namely, different from the theoretical domain to which the conclusion of the demonstration belongs.

Poggiolesi even claims that the property "that a proof system needs to satisfy to be considered as good" is "the analyticity property" (Poggiolesi 2012, 443–444). Namely, the proof system must be such that every element which occurs in a proof will also occur in the conclusion. This implies that the proof system must satisfy the demand for purity of methods. For, if every element which occurs in a proof will also occur in the conclusion, then the proof will not contain concepts belonging to theoretical domains different from the theoretical domain to which the conclusion of the proof belongs.

But a proof system which satisfies the demand for purity of methods cannot be considered as good. For, as already pointed out above, restricting proofs to pure proofs may yield very inefficient proof systems. What is more important, the demand for purity of methods cannot be satisfied by Gödel's first incompleteness theorem.

21.17 Deductive Demonstration

Because of Gödel's first incompleteness theorem, some attempts have been made to distinguish between axiomatic demonstration and deductive demonstration – meaning by the latter, deduction from obvious truths.

Thus Prawitz states that all "mathematical knowledge is acquired by deductive proofs from obvious truths" (Prawitz 2014, 90). For, all mathematical assertions ultimately rest "on deductive proofs whose initial premisses are assertions of obviously true sentences" (ibid., 88). Of course, "because of Gödel's incompleteness theorem we cannot expect in general to be able to solve mathematical problems by deducing answers from some given set of axioms. But this is not an argument against the general view that in mathematics one solves problems by deductive proofs from" initial "premisses held to be" obviously "true" (ibid., 75). For, "this view is not tied to the idea that one can specify once and for all a set of axioms from which all deductive proofs are to start" (ibid., 90). For example, Wiles and Taylor did not give "an axiomatic proof" of Fermat's last theorem, they "inferred the theorem deductively from initial premisses that were agreed by mathematicians to express known truths" (ibid., 88). Generally, deductive demonstration is mathematicians' demonstration, namely, demonstration as it is understood by mathematicians. Indeed, "since the Greeks, most mathematicians understand themselves as not making mathematical assertions unless they believe it has been established deductively, which means that there is a deductive proof of the assertion whose initial premisses are established deductively or express obvious truths" (ibid.).

But the concept of deductive demonstration does not adequately account for mathematicians' demonstration. For example, when Cantor demonstrated that to every transfinite cardinal there exist still greater cardinals, he did not deduce this result from truths already known, let alone from obvious truths, because it could not be demonstrated within the bounds of traditional mathematics. Demonstrating it required formulating new concepts and new hypotheses about them. So, not all mathematical knowledge is acquired by deductive proofs from obvious truths.

Moreover, as Prawitz himself points out, even if ultimately all mathematical knowledge were acquired by deductive proofs from obvious truths, there would remain "the challenge to explain in what way the ultimate starting points for mathematical proofs are obvious truths" (ibid., 90). Prawitz says that explaining this is "an open question" (ibid.). However, this is an open question only if deductive rules are ampliative. Indeed if, as argued in Chap. 12, deductive rules are non-ampliative, then the view that ultimately all mathematical knowledge is acquired by deductive proofs from obvious truths is doubtful. The view implies that all mathematical knowledge ultimately consists of obvious truths, which is hardly the case. It also implies that there can be no revolutions in mathematics, which is also hardly the case. As a matter of fact, there are revolutions in mathematics, not in the sense that an old order is turned down, but in the sense that new discoveries are made which would not have been possible within the old order. Cantor's creation of set theory is an example of revolution in mathematics in this sense (For more on deductive demonstration, see Cellucci 2017).

21.18 Analytic Demonstration and Evolution

While axiomatic demonstration is simply a means to justify and teach an already acquired proposition, analytic demonstration goes deeply into the nature of organisms. It reflects the way in which organisms mainly solve their problems, starting from the most basic one, survival. For example, New Caledonian crows make a wide variety of tools, by means of which they develop techniques that help them to solve their survival problem (see Hunt and Gray 2006). Much in the same way, human beings make analytic demonstrations, by means of which they develop techniques that help them to solve their survival problem. Even if there are obvious differences between analytic demonstrations and tools made by non-human animals, viewing analytic demonstrations as having a biological role helps to make sense of the phenomenon of demonstration.

This phenomenon is hardly understandable if ‘demonstration’ is intended as ‘axiomatic demonstration’, namely, as a means to justify and teach an already acquired proposition. It is understandable only if ‘demonstration’ is intended as ‘analytic demonstration’, namely, as a means to discover hypotheses that are sufficient conditions for the solutions of a problem that meets some need of human beings, and are plausible.

Rota states that, “of all escapes from reality, mathematics is the most successful ever”, all other escapes, “sex, drugs, hobbies, whatever”, being “ephemeral by comparison”, and he speaks of “the mathematician’s feeling of triumph as he forces the world to obey the laws his imagination has freely created” (Rota 1997, 70). But mathematics is no escape from reality, on the contrary, it is an answer to needs, even basic needs, of human beings, that are deeply rooted in reality. Mathematicians don’t force the world to obey the laws their imagination has freely created, since such laws are just the way mathematicians make the world understandable to themselves, and the working of the world does not depend on mathematicians. Moreover, their creations are not completely free, because they are a product of the mathematician’s biological structure and hence are bound to it.

21.19 Mathematical Styles

The concepts of demonstration which have been discussed in this chapter yield different mathematical styles, namely, different standards for how to reason in mathematics.

Lakatos distinguishes between two mathematical styles: the deductivist style and the heuristic style. The deductivist style “starts with a painstakingly stated list of axioms” and “definitions,” then “the list of axioms and definitions is followed by” theorems, where the “theorem is followed by the proof” (Lakatos 1976, 142). There are “two arguments for deductivist style” (ibid., 144). The first argument is that “the logic of discovery is deduction,” the second argument is that mathematics proceeds deductively, but “mathematical discovery is a completely non-rational affair” (ibid., 143).

These two arguments, however, are faulty. Indeed, they are, respectively, Musgrave’s argument and Popper’s argument, already discussed and found wanting in Chap. 11.

To the deductivist style, Lakatos opposes the heuristic style. In the deductivist style “the axioms and definitions frequently look artificial and mystifyingly complicated;” as to theorems, “it seems impossible that anyone should ever have guessed them” (ibid., 142). The “deductivist style hides the struggle, hides the adventure. The whole story vanishes, the successive tentative formulations of the theorem in the course of the proof-procedure are doomed to oblivion” (ibid.). Conversely, the heuristic style “highlights these factors. It emphasizes the problem-situation” (ibid., 144).

Lakatos, however, provides no adequate formulation for the heuristic style. This is due to his conviction that there can be no logic of discovery, so heuristics can only have subjective and psychological rules (see Chap. 11). He even asserts that "attempting to turn heuristic into a system of rules which claim to take account of the art of discovery" is "pathological" (Lakatos 1961, 75).

Conversely, on the basis of what has been said in this chapter, an adequate formulation of the heuristic style is given by analytic demonstration. On the other hand, the deductivist style can be articulated into four styles: axiomatic demonstration, formal demonstration, Gentzen's, Gentzen's proof with the subformula property, and deductive demonstration.

21.20 The Heuristic View vs. the Foundationalist View of Mathematics

In Chap. 18 it has been argued that mathematics is problem solving by the analytic method, which has a heuristic role. In Chap. 19 it has been argued that mathematical objects are hypotheses human beings make to solve mathematical problems by the analytic method, so mathematical objects have a heuristic role. In this chapter it has been argued that analytic demonstration, which has a heuristic role, is the basic concept of demonstration.

These features can be used to characterize the heuristic view of mathematics. According to it, mathematics is problem solving by the analytic method; mathematical objects are hypotheses human beings make to solve mathematical problems by that method; analytic demonstration is the basic concept of demonstration; and mathematical theories are analytic theories.

The heuristic view is opposed to the foundationalist view of mathematics. According to it, mathematics is theorem proving by the axiomatic method; mathematical objects are entities on their own; axiomatic demonstration is the basic concept of demonstration; and mathematical theories are axiomatic theories.

The heuristic view and the foundationalist view of mathematics express two different approaches to the nature of mathematics, the bottom-up approach and the top-down approach, respectively. The bottom-up approach tries to give an account of mathematics in terms of the real processes through which problems are solved. The top-down approach tries to give an account of mathematics in terms of some general philosophical assumptions about the nature of mathematics and mathematical objects. (For more on the distinction between the top-down approach and the bottom-up approach to the nature of mathematics, see Cellucci 2013b; for an extension of this distinction to approaches to the natural sciences, see Galavotti 2014).

References

Appel, Kenneth, and Wolfgang Haken. 1989. *Every planar map is four colorable*. Providence: American Mathematical Society.

Arana, Andrew. 2015. On the depth of Szemerédi's theorem. *Philosophia Mathematica* 23: 163–176.

Boolos, George. 1998. *Logic, logic, and logic*. Cambridge: Harvard University Press.

Bourbaki, Nicholas. 1950. The architecture of mathematics. *The American Mathematical Monthly* 57: 221–232.

Byers, William. 2007. *How mathematicians think*. Princeton: Princeton University Press.

Cellucci, Carlo. 2013a. *Rethinking logic: Logic in relation to mathematics, evolution, and method*. Dordrecht: Springer.

———. 2013b. Top-down and bottom-up philosophy of mathematics. *Foundations of Science* 18: 93–106.

———. 2017. Varieties of maverick philosophy of mathematics. In *Humanizing mathematics and its philosophy*, ed. Bharath Sriraman, forthcoming. Basel: Birkhäuser.

Davis, Philip J., and Reuben Hersh. 1981. *The mathematical experience*. Boston: Birkhäuser.

Descartes, René. 1996. *Oeuvres*. Paris: Vrin.

Frege, Gottlob. 1984. *Collected papers on mathematics, logic, and philosophy*. Oxford: Blackwell.

Friend, Michèle. 2014. *Pluralism in mathematics: A new position in philosophy of mathematics*. Dordrecht: Springer.

Galavotti, Maria Carla. 2014. For a bottom-up approach to the philosophy of science. In *From a heuristic point of view*, ed. Cesare Cozzo, and Emiliano Ippoliti, 195–211. Newcastle upon Tyne: Cambridge Scholars Publishing.

Gentzen, Gerhard. 1969. Investigations into logical deduction. In *The collected papers of Gerhard Gentzen*, ed. Manfred E. Szabo, 68–131. Amsterdam: North-Holland.

Gödel, Kurt. 1986–2002. *Collected works*. Oxford: Oxford University Press.

Goodman, Nicolas D. 1991. Modernizing the philosophy of mathematics. *Synthese* 88: 119–126.

Gowers, Timothy. 2002. *Mathematics: A very short introduction*. Oxford: Oxford University Press.

Grosholz, Emily. 2015. Review of Carlo Cellucci, *Rethinking logic: Logic in relation to mathematics, evolution and method*. *Philosophia Mathematica* 23: 136–140.

Grothendieck, Alexander. 1985. *Récoltes et semailles. Réflexions et témoignage sur un passé de mathématicien*. Montpellier: Université des Sciences et Techniques du Languedoc.

Hersh, Reuben. 1997. *What is mathematic, really?* Oxford: Oxford University Press.

Hilbert, David. 1967b. The foundations of mathematics. In *From Frege to Gödel: A source book in mathematical logic, 1879–1931*, ed. Jean van Heijenoort, 464–479. Cambridge: Harvard University Press.

———. 1980a. Letter to Frege, 29 December 1899. In Gottlob Frege, *Philosophical and mathematical correspondence*, ed. Gottlob Frege, 38–41. Oxford: Blackwell.

———. 1987. *Grundlagen der Geometrie*. Stuttgart: Teubner.

———. 1996e. The grounding of elementary number theory. In *From Kant to Hilbert: A source book in the foundations of mathematics*, II, ed. William Ewald, 1149–1157. Oxford: Oxford University Press.

———. 2004b. Grundlagen der Euklidischen Geometrie. In *David Hilbert's lectures on the foundations of geometry 1891–1902*, eds. Michael Hallett and Ulrich Majer, 221–286. Berlin: Springer.

Hintikka, Jaakko, and Unto Remes. 1976. Ancient geometrical analysis and modern logic. In *Essays in memory of Imre Lakatos*, eds. Robert Cohen, Paul Feyerabend, and Max Wartofsky, 253–276. Dordrecht: Springer.

Hintikka, Jaakko. 1996. *The principles of mathematics revisited*. Cambridge: Cambridge University Press.

———. 2012. Method of analysis: A paradigm of mathematical reasoning? *History and Philosophy of Logic* 33: 49–67.

Hunt, Gavin R., and Russell D. Gray. 2006. Tool manufacture by New Caledonian crows: Chipping away at human uniqueness. *Acta Zoologica Sinica* 52(Supplement): 622–625.
Kant, Immanuel. 1998. *Critique of pure reason*. Cambridge: Cambridge University Press.
Kuhn, Thomas Samuel. 1996. *The structure of scientific revolutions*. Chicago: The University of Chicago Press.
Lakatos, Imre. 1961. *Essays in the logic of mathematical discovery*. Cambridge: University of Cambridge.
———. 1976. *Proofs and refutations: The logic of mathematical discovery*. Cambridge: Cambridge University Press.
Macintyre, Angus. 2005. The mathematical significance of proof theory. *Philosophical Transactions of the Royal Society* 363: 2419–2435.
Meikle, Laura I., and Jacques D. Fleuriot. 2003. Formalizing Hilbert's *Grundlagen* in Isabelle/Isar. In *Theorem proving in higher order logics*, eds. D. Basin and B. Wolff, 319–334. Berlin: Springer.
Netz, Reviel. 2000. Why did Greek mathematicians publish their analyses? In *Ancient & medieval traditions in the exact sciences*, eds. Patrick Suppes, Julis M. Moravcsik, and Henry Mendell, 139–157. Stanford: CSLI Publications.
Newton, Isaac. 1714–1716. An account of the book entituled 'Commercium Epistolicum Collinii et aliorum, de analysi promota'. *Philosophical Transactions of the Royal Society of London* 29: 173–224.
———. 1720. *Universal arithmetick: Or, a treatise of arithmetical composition and resolution. To which is added, Dr. Halley's method of finding the roots of equations arithmetically*. London: J. Senex, W. Taylor, T. Warner, and J. Osborn.
———. 1967–1981. *The mathematical papers*. Cambridge: Cambridge University Press.
Pappus. 1876–1878. *Collectio*. Berlin: Weidmann.
Poggiolesi, Francesca. 2011. *Gentzen calculi for modal propositional logic*. Dordrecht: Springer.
———. 2012. On the importance of being analytic: The paradigmatic case of the logic of proofs. *Logique et Analyse* 55: 443–461.
Pólya, George. 1941. Heuristic reasoning and the theory of probability. *The American Mathematical Monthly* 48: 450–465.
———. 1954. *Mathematics and plausible reasoning*. Princeton: Princeton University Press.
Prawitz, Dag. 2014. The status of mathematical knowledge. In *From a heuristic point of view*, ed. Cesare Cozzo, and Emiliano Ippoliti, 73–90. Newcastle upon Tyne: Cambridge Scholars Publishing.
Proclus. 1992. *In primum Euclidis Elementorum librum commentarii*. Hildesheim: Olms.
Rota, Gian-Carlo. 1997. *Indiscrete thoughts*. Boston: Birkhäuser.
Russell, Bertrand. 1994. *Mysticism and logic*. London: Routledge.
Shanks, Daniel. 1993. *Solved and unsolved problems in number theory*. New York: Chelsea.
Worrall, John, and Elie Zahar. 1976. Editor's note. In Imre Lakatos, *Proofs and refutations: The logic of mathematical discovery*, 56–57. Cambridge: Cambridge University Press.
Zermelo, Ernst. 1967. A new proof of the possibility of a well-ordering. In *From Frege to Gödel: A source book in mathematical logic, 1879–1931*, ed. Jean van Heijenoort, 183–198. Cambridge: Harvard University Press.

Chapter 22
Mathematical Explanations

Abstract The distinction between axiomatic demonstration and analytic demonstration provides a basis for dealing with the question of mathematical explanations. One may speak of mathematical explanations in two different senses: mathematical explanations of mathematical facts and mathematical explanations of empirical facts. This chapter mainly deals with mathematical explanations of mathematical facts. It argues that there is an objective distinction between explanatory and non-explanatory demonstrations, and distinguishes between two different approaches to explanatory demonstrations: the static approach, based on axiomatic demonstration, and the dynamic approach, based on analytic demonstration. The chapter maintains that explanatory demonstrations in the static approach are intended to convince the audience that a proposition should be accepted, while explanatory demonstrations in the dynamic approach are intended to reveal how the demonstration was discovered. The chapter also deals with mathematical explanations of empirical facts, maintaining that it is unjustified to claim that there are such explanations.

22.1 Mathematical Explanations of Mathematical Facts

The distinction between axiomatic demonstration and analytic demonstration, made in Chap. 21, provides a basis for dealing with the question of mathematical explanations.

Mathematical explanations come in two varieties: mathematical explanations of mathematical facts and mathematical explanations of empirical facts. This chapter mainly deals with mathematical explanations of mathematical facts, but also discusses the question whether there are mathematical explanations of empirical facts.

The main issue about mathematical explanations of mathematical facts is whether there is an objective distinction between explanatory and non-explanatory demonstrations. It is often denied that such a distinction exists.

Thus Wittgenstein states that, in mathematics, "what we want is to describe, not to explain" (Wittgenstein 1978, III, § 78). Therefore, "we must do away with all explanation, and description alone must take its place" (Wittgenstein 1958, I, § 109).

Resnik and Kushner state that, from the fact that several proofs "leave many of our why-questions unanswered," we "derive the mistaken idea that there is an

© Springer International Publishing AG 2017

C. Cellucci, *Rethinking Knowledge*, European Studies in Philosophy of Science 4,
DOI 10.1007/978-3-319-53237-0_22

objective distinction between explanatory and nonexplanatory proofs" (Resnik and Kushner 1987, 154).

This, however, contrasts with the view of several mathematicians, who think that there is an objective distinction between explanatory and non-explanatory demonstrations.

Thus Atiyah states: "I remember one theorem that I proved and yet I really couldn't see why it was true," so "I kept worrying about it, and five or six years later I understood why it had to be true. Then I got an entirely different proof" and, "using quite different techniques, it was quite clear why it had to be true" (Atiyah 1984, 17).

Gowers and Nielsen state that "for mathematicians, proofs are more than guarantees of truth: they are valued for their explanatory power, and a new proof of a theorem can provide crucial insights" (Gowers and Nielsen 2009, 879).

One of the main aims of this chapter is to justify the view that there is an objective distinction between explanatory and non-explanatory demonstrations.

22.2 The Deductive View of Explanation

An approach to mathematical explanations of mathematical facts is provided by the deductive view of explanation. In the past century, the latter has been the most widespread view of scientific explanations of empirical facts, but immediately extends to mathematical facts. Although often credited to Hempel and Oppenheim (1948), the deductive view of explanation was actually introduced by Popper in 1934, under the name of 'causal explanation'.

According to Popper, "to give a causal explanation of an event means to deduce a statement which describes it, using as premisses of the deduction one or more universal laws, together with certain singular statements, the initial conditions" (Popper 1959, 59). The universal laws are "universal statements" having "the character of natural laws," while the initial conditions are "singular statements," which "apply only to the specific event in question" (ibid., 60). It is "from universal statements in conjunction with initial conditions that we deduce" the statement which describes the event being explained, which is also a "singular statement" (ibid., 60). The "initial conditions describe what is usually called the 'cause' of the event in question" (ibid.). The universal law and the initial conditions are supposed to be true.

For example, suppose that a thread has a tensile strength of 1 lb. and is loaded with a weight of 2 lbs., and that the thread breaks. The problem is to explain why the thread breaks. An explanation might go as follows. If a thread is loaded with a weight exceeding its tensile strength, then it breaks (universal law); the thread has a tensile strength of 1 lb. and is loaded with a weight of 2 lbs. (initial conditions); therefore, the thread breaks. The fact that the thread has a tensile strength of 1 lb. and is loaded with a weight of 2 lbs. is "the 'cause' of its breaking" (ibid.). This is an explanation according to the deductive view, because the *explanandum* – the thread breaks – is deduced from the universal law and the initial conditions, which are supposed to be true.

The deductive view of explanation also extends to mathematical facts. Suppose that the problem is to explain why $1 + 3 + 5 + 7 = 16 = 4^2$. An explanation might go as follows. For all n, $1 + 3 + 5 + \ldots + (2n - 1) = n^2$ (universal law); $n = 4$ (initial condition); therefore, $1 + 3 + 5 + 7 = 16 = 4^2$.

The deductive view of explanation, however, is inadequate. This can be shown by the following simple example.

Suppose that a gas in a closed container of constant volume is heated strongly, and that the pressure of the gas rises. The question is to explain why the pressure of the gas rises. An answer might go as follows. If the volume of a gas is kept constant, then the temperature and pressure of the gas are directly proportional to each other (universal law); the temperature of the gas rises (initial condition); therefore, the pressure of the gas rises. This is an explanation according to the deductive view, because the *explanandum* – the pressure of the gas rises – is deduced from the universal law and the initial condition, which are supposed to be true.

So far so good. However, suppose now that a gas in a closed container of constant volume is heated strongly, and that the temperature of the gas rises. The question is to explain why the temperature of the gas rises. An answer might go as follows. If the volume of a gas is kept constant, then the temperature and pressure of the gas are directly proportional to each other (universal law); the pressure of the gas rises (initial condition); therefore, the temperature of the gas rises. This looks like a perfectly good explanation according to the deductive view, because the *explanandum* – the temperature of the gas rises – is deduced from the universal law and the initial condition, which are supposed to be true. But it seems very odd to consider this as an explanation of why the temperature of the gas rises. For, the real explanation is that the gas is heated strongly.

From this example it is clear why the deductive view of explanation is inadequate. It implies that there is a symmetric relation between the *explanandum* and the initial conditions, while in fact such relation is asymmetric. Thus, the gas pressure rises because the gas temperature rises as a result of the gas being heated strongly, and not the other way around.

22.3 Aristotle on Explanations

According to the deductive view of explanation, an explanation is a certain kind of demonstration, specifically, an axiomatic demonstration with certain universal laws and initial conditions as axioms. This is in line with the basic assumption of the proponents of the deductive view of explanation, that theories, when rigorously stated, are axiomatic systems in which "the axioms are chosen in such a way that all the other statements belonging to the theoretical system can be derived from the axioms by purely logical or mathematical transformations" (Popper 1959, 71).

The deductive view of explanation partakes in an ancient tradition going back to Aristotle. But, unlike the modern proponents of the deductive view of explanation, Aristotle was well aware that the relation between the *explanandum* and the initial conditions was asymmetric.

Indeed, Aristotle states that "to have knowledge" of a thing is "to have a demonstration of it" (Aristotle, *Analytica Posteriora*, B 3, 90 b 9–10). But there are two kinds of knowledge, 'knowing that' and 'knowing why', because "knowing that [*to oti*] is different from knowing why [*to dioti*]" (ibid., A 13, 78 a 22). For, there are two kinds of demonstration, 'demonstration that' and 'demonstration why', since there are "differences between a demonstration that [*tou oti*] and a demonstration why [*tou dioti*]" (ibid., A 13, 78 b 33–34). In a 'demonstration that' "the cause is not stated" (ibid., A 13, 78 b 14–15). In a 'demonstration why' the cause is stated, since "a demonstration through the cause is a demonstration why" (ibid., B 16, 98 b 19–20). The cause is the explanation, so a 'demonstration that' is non-explanatory, while a 'demonstration why' is explanatory.

Aristotle illustrates the distinction between 'demonstration that' and 'demonstration why' by some examples. One of them is the following:

(A) Something does not twinkle if and only if it is near; the planets do not twinkle; therefore, the planets are near.

(B) Something does not twinkle if and only if it is near; the planets are near; therefore, the planets do not twinkle.

Here, (A) is a 'demonstration that', for "it is not because the planets do not twinkle that they are near" (ibid., A 13, 78 a 37–38). Conversely, (B) is a 'demonstration why', for "it is because" the planets "are near that they do not twinkle" (ibid., A 13, 78 a 38). Thus demonstration (A) is non-explanatory, while demonstration (B) is explanatory.

Aristotle's example shows that, as stated above, he was well aware that the relation between the *explanandum* and the initial conditions was asymmetric. It is then surprising that, in the past century, Popper and others proposed the deductive view of explanation, overlooking that Aristotle had already pointed out a basic limitation of that view. Aristotle's example also shows that there is an objective distinction between explanatory and non-explanatory demonstrations, because only explanatory demonstrations show the cause.

There is, however, a problem with Aristotle's distinction. According to Aristotle, a demonstration is a deduction that proceeds from premisses which are not only true, primitive, and immediate, but also "causes of the conclusion" (ibid., A 2, 71 b 22). Now, in a 'demonstration that' the cause is not stated. Thus, according to Aristotle, properly speaking a 'demonstration that' is not a demonstration at all. For the same reason, 'knowing that' is not knowing at all. In fact, Aristotle states that "we do not have knowledge of a thing till we have grasped the 'why' of it (which is to grasp its primary cause)" (Aristotle, *Physica*, B 3, 194 b 18–20).

The difficulty can be removed by conjecturing that, for Aristotle, there are two different senses of knowledge, a weaker sense, expressed by 'knowing that', and a stronger sense, expressed by 'knowing why'. Correspondingly, there are two different senses of demonstration, a weaker sense, expressed by 'demonstration that', and a stronger sense, expressed by 'demonstration why'. In this perspective, the distinction between explanatory and non-explanatory demonstrations parallels a distinction between knowing something through its explanation, or 'knowing why', and knowing something not through its explanation, or 'knowing that'.

22.4 Descartes on Explanations

Like Aristotle, Descartes holds that there exists an objective distinction between explanatory and non-explanatory demonstrations. But Descartes draws the distinction somewhat differently from Aristotle.

According to Descartes, although axiomatic demonstration "demonstrates the conclusion clearly," it is not satisfying "nor appeases the minds of those who are eager to learn, since it does not show how the thing in question was discovered" (Descartes 1996, VII, 156). So axiomatic demonstration appears "discovered more through chance than through method," and hence by using it "we get out of the habit of using our reason" (ibid., X, 375). Therefore, Descartes does not use axiomatic demonstration in his *Geometry*. Only analytic demonstration "shows the true way by which a thing was discovered methodically" (ibid., VII, 155). Through analytic demonstration, "it is the causes which are demonstrated by the effects," and "the causes from which I deduce" the effects "serve not so much to demonstrate them as to explain them" (ibid., VI, 76). Therefore, "there is a great difference between" merely "demonstrating and explaining" (ibid., II, 198).

Thus, according to Descartes, explanatory demonstrations are those which show how the thing was discovered, and are the analytic demonstrations. On the other hand, non-explanatory demonstrations are those which do not show how the thing was discovered, and are the axiomatic demonstrations.

22.5 Static and Dynamic Approach to Explanatory Demonstrations

Aristotle's distinction between 'demonstration that' and 'demonstration why', and Descartes' distinction between 'axiomatic demonstration' and 'analytic demonstration', suggest two different approaches to explanatory demonstrations. Aristotle's distinction suggests a static approach, while Descartes' distinction suggests a dynamic approach.

According to the static approach, a demonstration of P is explanatory if it is an axiomatic demonstration that gives an answer to the question: Why is it the case that P? This means that, according to the static approach, a demonstration of P is explanatory if it is an axiomatic demonstration that shows the ground of the validity of P.

According to the dynamic approach, a demonstration of *P* is explanatory if it is an analytic demonstration that gives an answer to the question: How can one arrive at *P*? This means that, according to the dynamic approach, a demonstration of *P* is explanatory if it is an analytic demonstration that reveals the way to the discovery of *P*, and specifically reveals to the researcher how to find a solution to the problem *P*, and to the audience how the solution to the problem *P* was found.

The above formulation of the static approach, however, needs some refinement. As van Fraassen points out, "being an explanation is essentially relative" because "what is requested, by means of the interrogative 'Why is it the case that *P*?', differs from context to context" (van Fraassen 1980, 156). In fact, "it is a use of science to satisfy certain of our desires," and "the exact content of the desire, and the evaluation of how well it is satisfied, varies from context to context" (ibid.).

For example, let *P* be the Pythagorean theorem and suppose we ask: Why is it the case that *P*? By this question we might desire to know, for example: (a) Why is it the case that *P* only for right-angled triangles, and not for acute-angled or obtuse-angled triangles? Or: (b) Why is it the case that *P* only in Euclidean space, and not in certain non-Euclidean spaces? A demonstration of *P* might satisfy our desire (a) but not (b), and viceversa.

Since being an explanation is essentially relative, the above statement of the static approach must be modified as follows. According to the static approach, a demonstration of *P* is explanatory with respect to a certain context if it is an axiomatic demonstration that gives an answer to the question: Why is the case that *P* with respect to that context?

22.6 Explanatory Demonstrations and Published Demonstrations

In Chap. 21 it has been argued that published demonstrations of mathematical propositions do not express the way those propositions were actually discovered.

This does not occur because, as Descartes and Grothendieck maintain, from the very beginning of mathematics there has been a sort of conspiracy of silence (see Chap. 21). Rather, it occurs because either mathematicians are not fully aware of the processes by which the discovery came about, or feel uneasy to reveal that such processes were not rigorously deductive. Or, simply, it occurs because, as Hersh states, "if you have made a publishable discovery," although "discovering the interesting result was probably the outcome of a heuristic investigation," you "will probably omit that story from your article, if only to save yourself extra trouble. If you choose to include it, you risk a rejection note from the editor: 'We don't have space for all the good papers we are receiving, even without irrelevant heuristics'" (Hersh 2014, 82).

This shows a limitation of the present literature on mathematical explanations of mathematical facts, which uses published demonstrations as examples of explanatory demonstrations. (For a survey of this literature, see Mancosu 2008). These demonstrations do not reveal the way to discovery, so they are useless as illustrations of explanatory demonstrations in the dynamic approach. This depends on the fact that the present literature on mathematical explanations of mathematical facts does not distinguish between a static and a dynamic approach to explanatory demonstrations.

22.7 The Rhetorical Role of Axiomatic Demonstration

In Chap. 21 it has been stated that the purpose of axiomatic demonstration is to justify and teach an already acquired proposition. This serves to convince the audience – readers of research papers or textbooks, conference audiences or students in the classroom – that the proposition should be accepted. Therefore, several people have asserted that axiomatic demonstration has a rhetorical role.

Thus Hardy states: “Proofs are what Littlewood and I call ‘gas’, rhetorical flourishes designed to affect psychology, pictures on the board in the lecture, devices to stimulate the imagination of pupils” (Hardy 1929, 18).

Kitcher states that in mathematics “a rhetorical function is served by the presentation of the proof,” since “a proof presentation that is effective for one audience” can “be useless for others,” and “‘gas’ is necessary even in professional mathematics” (Kitcher 1991, 7).

Krantz states that “a proof is a rhetorical device for convincing someone else that a mathematical statement is true” (Krantz 2011, 5).

Admittedly, negative connotations are usually associated with rhetoric – “that glib and oily art to speak and purpose not” (Shakespeare, *King Lear*, Act I, Scene I, 224–225). Therefore, one might dislike saying that axiomatic demonstration has a rhetorical role, and prefer to say that it has a didactic role. But the substance is the same, both expressions referring to the capacity of the demonstration to convince the audience that a proposition should be accepted.

That axiomatic demonstration has a rhetorical role has an implication for the static approach. It means that, according to it, a demonstration of *P* that is explanatory with respect to a certain context, is an axiomatic demonstration which is capable of convincing the audience that *P* should be accepted in that context, by giving an answer to the question: Why is the case that *P* with respect to that context?

That axiomatic demonstration has a rhetorical role clarifies why, as stated above, published demonstrations are useless as illustrations of explanatory demonstrations in the dynamic approach. Such demonstrations are not intended to reveal how the demonstration was discovered, but only to convince the audience that a proposition should be accepted.

Actually, in a sense, conviction plays a role also in the dynamic approach. The conviction that a result is plausible often motivates the researcher to search for hypotheses capable of bringing about a solution. This is an aspect of the general question of the role of emotion in knowledge, already mentioned in Chap. 2.

22.8 Functions of Explanatory Demonstrations

That explanatory demonstrations in the static approach are intended to convince the audience that a proposition should be accepted, shows that such demonstrations have a social function. They can be addressed to a variety of audiences, from students to fellow researchers.

Thus Hersh states: "A proof is a conclusive argument that a proposed result follows from accepted theory. 'Follows' means the argument convinces qualified, skeptical mathematicians. Here I am giving an overtly social definition of 'proof'," a definition that "is plainly true to life" (Hersh 1997, 6).

The social function of explanatory demonstrations in the static approach is different from the creative function of explanatory demonstrations in the dynamic approach. What is essential to the latter is the capacity of the demonstration to serve as a means of extending scientific knowledge, by suggesting a hypothesis that is the key to the discovery of a solution to a problem.

In view of the different functions of explanatory demonstrations in the static and the dynamic approach, a distinction between these two approaches is quite natural. As Hersh states, "the role of proof in class isn't the same as in research" (ibid., 59).

Then the alternative is not, as Lord Rayleigh says, between demonstrations which "command assent" and demonstrations which "woo and charm the intellect," evoking "delight and an overpowering desire to say 'Amen, Amen'" (Huntley 1970, 6). The alternative is rather between demonstrations that have a social function and demonstrations that have a creative function.

22.9 Relevance to Mathematical Practice

The existence of an objective distinction between explanatory and non-explanatory demonstrations is important only if it is relevant to mathematical practice. Now, the distinction made above is relevant to mathematical practice, but the reason for its relevance is different in the static and the dynamic approach.

In the static approach, the distinction is relevant to mathematical practice, because explanatory demonstrations convince the audience that a proposition should be accepted. So they are essential to the acceptance of mathematics.

In the dynamic approach, the distinction is relevant to mathematical practice, because explanatory demonstrations extend mathematical knowledge by suggesting a hypothesis which is the key to the discovery of a solution to a problem. So they are essential to the growth of mathematics.

Actually, in the dynamic approach, explanatory demonstrations are relevant to mathematical practice also in another respect. The growth of mathematics is often viewed as being cumulative. Mathematical discoveries are then considered to be mere additions or increments to the growing stockpile of mathematical results.

Thus Kitcher states that "mathematics is cumulative in a way that natural science is not" (Kitcher 1983, 161).

Devlin states that "mathematical knowledge is cumulative" (Devlin 1990, 33). This depends on the fact that "mathematics consists in making deductions from axioms" (ibid., 34).

But the view that the growth of mathematics is cumulative does not account for the fact that, in the development of mathematics, it may happen that, to solve a problem, it is necessary to introduce new hypotheses that cannot be deduced from established mathematics – the body of mathematical theories and results that are currently accepted by the mathematical community, in mathematics journals or textbooks. This leads to new developments which cannot be deduced from established mathematics. Therefore, the growth of mathematics cannot be viewed as being cumulative.

The introduction of new hypotheses that cannot be deduced from established mathematics highlights an important feature of analytic demonstration. The new hypotheses may react upon some parts of established mathematics providing new perspectives on them, which may lead to reconstruct such parts on a new basis. In the dynamic approach, explanatory demonstrations are relevant to scientific practice also in this respect.

22.10 Global and Local View of Mathematical Explanations

The static approach and the dynamic approach to explanatory demonstrations involve a global and a local view of mathematical explanations of mathematical facts, respectively. This is because the axiomatic method has a global character, while the analytic method has a local character.

Indeed, in the axiomatic method, on which explanatory demonstrations are based in the static approach, axioms serve to demonstrate, and hence to explain, all facts of a given theory. The explanations of all these facts are based on the same axioms, and hence are global. For this reason, Sierpinska states that "the quest for explanation in mathematics cannot be a quest for proof, but it may be an attempt to find a rationale of a choice of axioms, definitions, methods of construction of a theory" (Sierpinska 1994, 76). Indeed, in the static approach, the mathematical explanation of a mathematical fact stands not so much in the demonstration itself as rather in the theory as a whole, hence in the rationale of the choice of axioms, definition, methods of construction.

On the other hand, in the analytic method, on which explanatory demonstrations are based in the dynamic approach, the hypotheses for the solution to a problem are not general principles, good for all problems, but are aimed at a specific problem. In this sense, they are local rather than global. Not being global, they need not belong to the same field as the problem, but may belong to any field. For this reason, while the axiomatic method is incompatible with Gödel's first incompleteness theorem, the analytic method is compatible with it. Moreover, in the analytic method, the hypotheses for the solution to a problem, being aimed at a specific problem, can take care of the peculiarities of the problem, and hence can spawn an explanation tailor-made for the problem. So, in the dynamic approach, the mathematical explanation of a mathematical fact can provide for the peculiarity of the problem.

22.11 Mathematical Explanations and Mathematical Understanding

A concept that is often considered to be strictly related to explanation is understanding. Von Neumann claims that "in mathematics you don't understand things. You just get used to them" (Zukav 1980, 208 footnote). However, since antiquity, it has been widely held that in mathematics there is understanding, and also that understanding is strictly related to explanation, in particular, that without explanation there is no understanding.

Thus Plato states: "As to the person who has no explanation, wouldn't you say that, to the extent that he is unable to give an account of it, to himself or to anyone else, he has no understanding of it?" (Plato, *Respublica*, VII, 534 b 4–6).

Moreover, since antiquity, it has been widely held that the aim of mathematical explanations of mathematical facts is to provide mathematical understanding.

But what does mathematical understanding consist in? A difficulty in answering this question relates to the fact that there are many kinds of mathematical items. Since, however, here we are concerned with mathematical understanding in relation to mathematical explanations, we are interested in dealing with two such items, demonstrations and theorems.

22.12 The Nature of Mathematical Understanding

With respect to demonstrations and theorems, mathematical understanding is the recognition of the fitness of the parts to each other and to the whole. To understand a demonstration or a theorem, is to recognize how the various parts fit each other and the whole.

In particular, to understand a demonstration is to recognize how the various parts of the demonstration fit each other and the whole. This means recognizing what the whole idea of the demonstration is, what the contribution of each part of the demonstration to the whole idea is, and why such contribution is essential.

On the other hand, to understand a theorem is to recognize how the various concepts involved in the theorem fit each other and the whole. This means recognizing what the content of the theorem is, what the contribution of each concept involved in the theorem to such content is, and what relation the theorem establishes between such concepts.

That this is the proper concept of mathematical understanding is implicit in the statements of several mathematicians.

Thus Poincaré states: "What is it, to understand? Has this word the same meaning for all the world ? To understand the demonstration of a theorem, is" this merely "to examine successively each of the syllogisms composing it and to ascertain its correctness, its conformity to the rules of the game?" (Poincaré 2013, 430–431). And, "to understand a definition, is this merely to recognize that one already knows the meaning of all the terms employed and to ascertain that it implies no contradiction?" (ibid., 431). The answer to all these questions is negative. For example, in order to understand a demonstration, we need "to know not merely whether all the syllogisms of a demonstration are correct, but why they link together in this order rather than another," and, if we do not know this, we do not think we "understand" (ibid.).

Weyl states that we are not content to be driven to accept "a mathematical truth by a long chain of formal inferences and calculations leading us blindfolded from link to link. We would like to be shown not only the goal but also the way and general outline we are to travel to that goal, to understand the underlying ideas of the proof and their connections" (Weyl 1985, 14).

Thus, both Poincaré and Weyl acknowledge that to understand a mathematical item is to recognize how the various parts fit each other and the whole.

22.13 Explanatory Demonstrations and Mathematical Understanding

It has been stated above that to understand a demonstration is to recognize how the various steps of the demonstration fit each other and the whole. Once again, we must distinguish between demonstrations which are explanatory in the static approach and demonstrations which are explanatory in the dynamic approach.

In the static approach, that the audience is able to follow each step of a demonstration does not mean that the audience fully understands the demonstration. The audience may not see the idea of the demonstration, and hence may not get the deeper context, therefore the audience does not have understanding of the demonstration. Conversely, when a demonstration is explanatory in the static approach, the audience is able not only to follow each step of the demonstration, but also to see how the various steps of the demonstration fit each other and the whole, and hence the audience has understanding of the demonstration.

On the other hand, in the dynamic approach, that the researcher or the audience are able to see the idea of a demonstration, does not mean that they are also able to see the full train of inferences that leads to the discovery of the demonstration. Thus, they do not have understanding of the way to discovery. Conversely, when a demonstration is explanatory in the dynamic approach, the researcher and the audience are able to see not only the idea of the demonstration, but also the full train of inferences which leads to the discovery of the demonstration, and hence they have understanding of the way to discovery.

That, when a demonstration is explanatory in the dynamic approach, the audience has understanding of the way to discovery, may give members of the audience the illusion that they could have discovered the demonstration themselves. It is only an illusion, but in a sense the members of the audience rediscover the demonstration as they repeat it. Thus Poincaré states: "It seems to me then, in repeating a reasoning learned, that I could have invented it. This is often only an illusion; but even then, even if I am not so gifted as to create it by myself, I myself re-invent it in so far as I repeat it" (Poincaré 2013, 385).

22.14 Explanatory Demonstrations and Memorability

Mathematicians often say that certain demonstrations are memorable, and that a memorable demonstration is greatly preferable to an unmemorable one. By a memorable demonstration they mean a demonstration that, once studied, can be reconstructed easily, not because one memorizes it character for character, but because one knows the key ideas of the demonstration and can simply develop them. Mathematicians also say that there is a close connection between the memorability of a demonstration, its explanatory powers, and understanding.

Thus Gowers states that certain proofs do not "need quite a lot of direct memorization" (Gowers 2007, 43). For, they are based on some key ideas and, "instead of remembering the details of the proof," one needs only "remember a few important ideas and develop the technical skill to convert them quickly into a formal proof" (ibid., 40). Therefore, "all other things being equal, a memorable proof is greatly preferable to an unmemorable one" (ibid., 39). Clearly, "memorability does seem to be intimately related to other desirable properties of proofs," such as "explanatory power," or such as understanding, because "it is obviously easier to remember a proof if one understands the argument" (ibid., 40).

Demonstrations which are explanatory in the static approach or in the dynamic approach are both memorable, though for different reasons. Demonstrations which are explanatory in the static approach are memorable because they permit the audience to see how the various steps of the demonstration fit each other and the whole, and hence to have understanding of the demonstration. On the other hand, demonstrations which are explanatory in the dynamic approach are memorable because they permit the researcher and the audience to see not only the idea of the demonstration, but also the full train of inferences which leads to the discovery of the demonstration, and hence to have understanding of the way to discovery.

22.15 Mathematical Explanations of Empirical Facts

After considering mathematical explanations of mathematical facts, let us consider mathematical explanations of empirical facts.

The main issue about mathematical explanations of empirical facts is whether there are explanations of this kind, namely, explanations of empirical facts in which mathematical facts play an essential role. Several people claim that there are such explanations.

Thus Baker states that "there are genuine mathematical explanations of physical phenomena" (Baker 2005, 236).

Pincock states that "mathematical explanations of physical phenomena exist" (Pincock 2012, 206).

Colyvan states: "I and at least some of my mathematical realist colleagues are happy to embrace mathematical explanations of physical phenomena" (Colyvan 2013, 95).

Some examples are commonly cited as evidence that there are mathematical explanations of empirical facts. The most cited ones are the honeycomb problem, the magicicada problem, and the Königsberg bridges problem.

22.16 The Honeycomb Problem

The honeycomb problem concerns the explanation of the structure of the bee's honeycomb. The problem is: Why do bees make honeycombs with cells in the shape of hexagons?

This problem was already addressed in the ancient world. Thus, Pappus proposes a mathematical explanation of the structure of the bee's honeycomb by saying that bees make honeycombs with cells "all equal, similar and contiguous to one another, and hexagonal in form," by virtue "of a certain geometrical forethought" (Pappus 1876–1878, V, 304.24–26). Indeed, "bees know just this fact which is useful to them, that the hexagon is greater than the square and the triangle and will hold more honey for the same expenditure of material in constructing each" (ibid., V, 306.29–32).

Since antiquity, this explanation of the structure of the bee's honeycomb has been reasserted several times. For example, Lyon and Colyvan maintain that the explanation of the structure of the bee's honeycomb consists of a biological part and a mathematical part. The biological part is that "hive-bees minimise the amount of wax they use to build their combs, since there is an evolutionary advantage in doing so" (Lyon and Colyvan 2008, 228). The mathematical part is the honeycomb theorem, which states that "a hexagonal grid represents the best way to divide a surface into regions of equal area with the least total perimeter" (ibid., 228–229). From the honeycomb theorem it follows that the hive-bee will divide "the honeycomb up into hexagons rather than some other shape" (ibid., 229). This is a mathematical explanation of the structure of the bee's honeycomb, because the honeycomb theorem plays an essential role in it.

This explanation, however, is problematic. What bees actually make are cells that are circular in cross section, being molded by the shape of a bee's body. Then the wax, softned by the heat of the bees' bodies, gets pulled into hexagonal cells by surface tension at the junctions where three walls meet. Thus, contrary to Pappus' claim, bees do not make the hexagonal cells of bees' honeycombs by virtue of a certain geometrical forethought. The cells "have a circular shape at 'birth'," being molded by the shape of a bee's body, "but quickly transform into the familiar rounded hexagonal shape," by "the flow of molten visco-elastic wax near the triple junction between the neighbouring circular cells," the heat for melting the wax being "provided by the 'hot' worker bees" (Karihaloo et al. 2013).

It might be objected that, even if the hexagonal cells are a result of physical forces, the fact remains that, by the honeycomb theorem, the hexagonal cells minimize the amount of wax needed to build honeycombs, and hence there is an evolutionary advantage in doing so. Thus, even though the hexagonal cells are a result of physical forces, the explanation remains a mathematical one, since the honeycomb theorem plays an essential role in it.

This objection, however, is not valid because the honeycomb theorem only applies to a two-dimensional structure, while honeycombs are three-dimensional structures, where each cell is a special type of hexagonal prism bound by six trapeziums, the hexagonal base, and a top consisting of three rhombi.

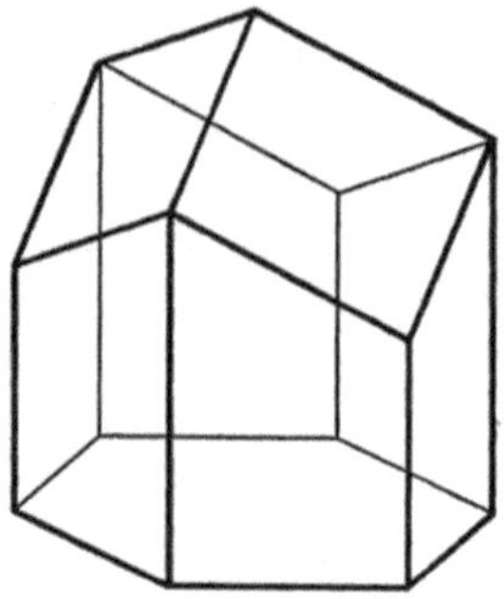

Now, these cells do not minimize the amount of wax needed to build honeycombs. In order to minimize the amount of wax needed to build honeycombs, cells should have a top that, instead of consisting of three rhombi, "consists of two hexagons and two rhombi" (Tóth 1964, 473). So, the argument that the hexagonal cells minimize the amount of wax needed to build honeycombs, and hence there is an evolutionary advantage in doing so, is not valid.

22.17 The Magicicada Problem

The magicicada problem concerns the explanation of the life cycle of a cicada species, the *Magicicada septendecim*, which spends 17 years underground before emerging briefly. The problem is: why is its life cycle a prime number?

Baker maintains that the explanation of the fact that the life cycle of the *Magicicada septendecim* is a prime number consists of a biological part and a mathematical part. The biological part is that "having a life-cycle period that minimizes intersection with other (nearby/lower) periods," in particular periods of predators, "is evolutionarily advantageous," and "cicadas in ecosystem-type E are limited by biological constraints to periods from 14 to 18 years" (Baker 2009, 614). The mathematical part is that "prime periods minimize intersection (compared to non-prime periods)" (ibid.). From this it follows that "cicadas in ecosystem-type E are likely to evolve 17-year periods" (ibid.). This is a mathematical explanation of the life cycle of the *Magicicada septendecim*, because "it is the link between primeness and minimizing intersection with other period lengths that does the explanatory work" (ibid., 616).

This explanation, however, is problematic. The prime number 17 appears in the description of the life cycle only because we measure the latter in years. But, as Daly and Langford point out, "instead of measuring it as 17 years, we could measure it as 68 seasons, for instance, or as 204 months" (Daly and Langford 2009, 652). Now, neither 68 nor 204 is a prime number. Since the year is an arbitrary time-unit, the fact that the prime number 17 appears when we measure the life cycle in years has no explanatory importance. Then, it seems unjustified to say that it is the link between primeness and minimizing intersection with other period lengths that does the explanatory work.

Baker and Colyvan maintain that measurement in years is privileged with respect to measurement in seasons or months, because "biologists do not in fact use any of these other units in describing or discussing cicada life-cycles, which suggests that years are the most salient unit in this context" (Baker and Colyvan 2011, 329). But this argument is flawed. Biologists do not use years because they are the most salient unit in this context. They use years only because the finer measurement in seasons or months is not necessary, the rougher measurement in years is enough, and is more convenient since it involves smaller numbers.

Baker and Colyvan also maintain that "once per year (in non-equatorial regions) there is a sustained period of cold weather, and this has an impact on breeding and survival of many terrestrial species," so "measurements using years as units can sensibly feature in the descriptions and explanations of biological phenomena" (ibid., 330 and footnote). But this argument is also flawed. As Daly and Langford point out, seasonal changes have an impact on breeding and survival of many terrestrial species, because they "trigger such biological phenomena as germination, migration and hibernation" (Daly and Langford 2009, 652). And, as Zimecki points out, the monthly lunar cycle also has an impact on breeding and survival of many terrestrial species, for example, it "may affect hormonal changes early in phylogenesis (insects)" (Zimecki 2006, 1). Therefore, measurements using seasons or months as units can sensibly feature in the descriptions and explanations of biological phenomena.

22.18 The Königsberg Bridges Problem

The Königsberg bridges problem concerns the explanation of the impossibility of a walk across the seven bridges, which existed in the city of Königsberg at Kant's time over the Pregel river, crossing each bridge only once. The problem is: why was it impossible to walk across all the seven bridges, crossing each bridge only once?

Pincock maintains that, in order to give an explanation of this fact, "we can use graph theory to describe the bridges of Königsberg without knowing" the "details of the bridges' physical construction" (Pincock 2007, 263). Specifically, we can describe the bridges of Königsberg by the graph with points *A*, *B*, *C*, *D* shown in the figure below.

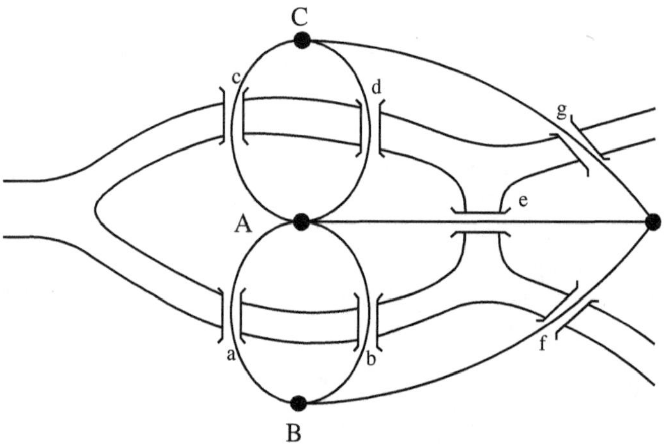

If, by degree of a point, we mean the number of lines touching that point, then the point *A* has degree 5 and each of the points *B*, *C*, *D* has degree 3, so our graph has four points with odd degrees. Now, Euler demonstrated the following theorem: A graph can be traversed with each line traversed only once if and only if it has zero or exactly two points with odd degrees. Since our graph has four points with odd degrees, by Euler's theorem it was impossible to walk across all the seven bridges crossing each bridge only once. Then Euler's theorem explains why it was "impossible to make such a crossing" (ibid., 259). Pincock maintains that the explanation "is unaffected by the physical constitution of the bridges" for, changing "the bridges to gold," would have kept the impossibility "intact" (Pincock 2012, 54). So "the graph representation not only captures the feature of interest but also has as part of its content that various aspects of the system are also irrelevant to this feature" (ibid.). This is a mathematical explanation of why it was impossible to walk across all the seven bridges crossing each bridge only once, because Euler's theorem plays an essential role in it.

This explanation, however, is problematic because it depends on the assumption that all the physical features not captured by the graph representation are irrelevant to the question of the impossibility. But then the explanation is not an explanation of a physical fact, namely a fact about a physical system, it is only an explanation of a mathematical fact, specifically, a fact about a graph. As Lange points out, "the explanation would not have been distinctively mathematical if it had been that no one ever turned left rather than right after crossing a given bridge, or the bridges were made of a corrosive material, or someone was poised to shoot anyone who tried to cross a given bridge" (Lange 2013, 489). If one of these empirical facts had been the case, such empirical fact would have been the explanation of the impossibility, and Euler's theorem would have had no part in it. Then the above explanation cannot be said to be a mathematical explanation of an empirical fact, since it essentially depends on the circumstance that certain empirical facts do not occur.

22.19 Mathematical Explanations and Pythagoreanism

It could be argued that, like the honeycomb problem, the magicicada problem, and the Königsberg bridges problem, all other examples that are commonly cited as evidence that there are mathematical explanations of empirical facts are problematic. On this basis, it seems fair to conclude that it is unjustified to claim that there are mathematical explanations of empirical facts.

Actually, the claim that there are mathematical explanations of empirical facts is a form of pythagoreanism, because "the so-called Pythagoreans, in their interest in mathematics," believed that "the principles of mathematics were the principles of all things" (Aristotle, *Metaphysica*, A 5, 985 b 23–26). In particular, they believed that "the elements of numbers were the elements of all things" (ibid., A 5, 986 a 1–2). Now, to assume that there are mathematical explanations of empirical facts is to assume that the principles of mathematics are the principles, if not of all things, at least of the empirical facts of which there are supposed to be mathematical explanations.

But to assume that the principles of mathematics are the principle of certain empirical facts is to confuse mathematics with the world itself. This confusion is at the basis of mathematical structural realism, which, as we have seen in Chap. 7, maintains that the external physical reality is a mathematical structure. But to maintain this is to commit the hammer mistake: "If the only tool you have is a hammer," then you "treat everything as if it were a nail" (Maslow 1966, 15). Indeed, if the only tool you have is mathematics, you treat everything as if it were a mathematical structure.

Galilei states that natural philosophy is written in the book of the universe, and is written in mathematical language (see Chap. 8). However, when human beings open that book, they find it filled with their own handwriting. Indeed, mathematics is not the world, it is only a means by which human beings make the world understandable to themselves, so it is just a human way of seeing and thinking about the world.

A famous Magritte's painting, *La trahison des images* [The treason of images], is an image of a pipe with the text 'Ceci n'est pas une pipe [This is not a pipe]' under it.

The text may seem wrong, but is actually correct, because an image of a pipe is not a pipe. When Magritte once was asked about this, he replied that of course the image was not a pipe, "just try to fill it with tobacco" (Spitz 1994, 47). Much in the same way, we may say that of course mathematics is not the world, just try to live in it. (On the relation of mathematics to the world, see Chap. 24).

22.20 Mathematical Explanations and Mathematical Platonism

Some of the people who claim that there are mathematical explanations of empirical facts, also claim that such explanations lend support to mathematical platonism.

Thus Baker states that, if mathematics plays "an explanatory role in science," this opens the way "to argue for the existence of mathematical entities" (Baker 2009, 611). At least, it opens the way to argue for "the existence of mathematical entities that feature in scientific explanations" (ibid., 612).

This is not surprising because, as stated above, the claim that there are mathematical explanations of empirical facts is a form of mathematical pythagoreanism, and, "like the Pythagoreans," Plato "held that the numbers are the causes of substance for other things" (Aristotle, *Metaphysica*, A 5, 987 b 23–25).

In order to show that, if mathematics plays an explanatory role in science, this opens the way to argue for the existence of mathematical entities, Baker uses the following 'indispensability argument': "We ought rationally to believe in the existence of any entity that plays an indispensable explanatory role in our best scientific theories;" but "mathematical objects play an indispensable explanatory role in science;" therefore, "we ought rationally to believe in the existence of mathematical objects" (Baker 2009, 613).

Baker's indispensability argument, however, is based on the assumption that we ought rationally to believe in the existence of any entity that plays an indispensable explanatory role in our best scientific theories. This assumption is unjustified, because certain theories which were once our best scientific theories have proven to be untenable, and the entities that could play an indispensable explanatory role in them have proven not to exist. Moreover, there is no evidence that our current best scientific theories, and the entities that could play an indispensable explanatory role in them, will fare any better in the future. Since this assumption is unjustified, it is unwarranted to conclude that we ought rationally to believe in the existence of mathematical objects.

Baker's indispensability argument is based also on another assumption which is likewise unjustified, namely, the assumption that mathematical objects play an indispensable explanatory role in science. This assumption is unjustified because it presupposes that there are mathematical explanations of empirical facts, but, by what has been argued above, this is unwarranted.

References

Atiyah, Michael. 1984. Interview by Roberto Minio. *The Mathematical Intelligencer* 6 (1): 9–19.

Baker, Alan. 2005. Are there genuine mathematical explanations of physical phenomena? *Mind* 114: 223–238.

———. 2009. Mathematical explanation in science. *The British Journal for the Philosophy of Science* 60: 611–633.

Baker, Alan, and Mark Colyvan. 2011. Indexing and mathematical explanation. *Philosophia Mathematica* 19: 323–334.

Colyvan, Mark. 2013. *An introduction to the philosophy of mathematics*. Cambridge: Cambridge University Press.

Daly, Christopher, and Simon Langford. 2009. Mathematical explanation and indispensability arguments. *The Philosophical Quarterly* 59: 641–658.

Descartes, René. 1996. *Oeuvres*. Paris: Vrin.

Devlin, Keith. 1990. *Mathematics: The new golden age*. London: Penguin Books.

Gowers, Timothy. 2007. Mathematics, memory and mental arithmetic. In *Mathematical knowledge*, ed. by Mary Leng, Alexander Paseau, and Michael Potter, 33–58. Oxford: Oxford University Press.

Gowers, Timothy, and Michael Nielsen. 2009. Massively collaborative mathematics. *Nature* 461(15 October): 879–881.

Hardy, Godfrey Harold. 1929. Mathematical proof. *Mind* 38: 1–25.

Hempel, Carl Gustav, and Paul Oppenheim. 1948. Studies in the logic of explanation. *Philosophy of Science* 15: 135–175.

Hersh, Reuben. 1997. *What is mathematic, really?* Oxford: Oxford University Press.

———. 2014. *Experiencing mathematics: What do we do, when we do mathematics?* Providence: American Mathematical Society.

Huntley, H.E. 1970. *The divine proportion: A study of mathematical beauty*. Mineola: Dover.
Karihaloo, Bhushan, K. Zhang, and J. Wang. 2013. Honeybee combs: How the circular cells transform into rounded hexagons. *Journal of the Royal Society Interface* 10 (86). doi:10.1098/rsif.2013.0299.
Kitcher, Philip. 1983. *The nature of mathematical knowledge*. Oxford: Oxford University.
———. 1991. Persuasion. In *Persuading science: The art of scientific rhetoric*, eds. Marcello Pera and William Shea, 3–27. Sagamore Beach: Science History Publications.
Krantz, Steven G. 2011. *The proof is in the pudding: The changing nature of mathematical proof.* New York: Springer.
Lange, Marc. 2013. What makes a scientific explanation distinctively mathematical? *The British Journal for the Philosophy of Science* 64: 485–511.
Lyon, Aidan, and Mark Colyvan. 2008. The explanatory power of phase spaces. *Philosophia Mathematica* 16: 227–243.
Mancosu, Paolo. 2008. Introduction. In *The philosophy of mathematical practice*, ed. Paolo Mancosu, 1–21. Oxford: Oxford University Press.
Maslow, Abraham H. 1966. *The psychology of science: A reconnaissance*. New York: Harper & Row.
Pappus. 1876–1878. *Collectio*. Berlin: Weidmann.
Pincock, Christopher. 2007. A role for mathematics in the physical sciences. *Noûs* 41: 253–275.
———. 2012. *Mathematics and scientific representation*. Oxford: Oxford University Press.
Poincaré, Henri. 2013. *The foundations of science: Science and hypothesis – The value of science – Science and method*. Cambridge: Cambridge University Press.
Popper, Karl Raimund. 1959. *The logic of scientific discovery*. London: Hutchinson.
Resnik, Michael David, and David Kushner. 1987. Explanation, independence and realism in mathematics. *British Journal for the Philosophy of Science* 38: 141–158.
Sierpinska, Anna. 1994. *Understanding in mathematics*. London: The Falmer Press.
Spitz, Ellen Handler. 1994. *Museums of the mind: Magritte's labyrinth and other essays in the arts*. New Haven: Yale University Press.
Tóth, László Fejes. 1964. What the bees know and what they do not know. *Bulletin of the American Mathematical Society* 70: 468–481.
van Fraassen, Bas. 1980. *The scientific image*. Oxford: Oxford University Press.
Weyl, Hermann. 1985. Axiomatic versus constructive procedures in mathematics. *The Mathematical Intelligencer* 7 (4): 10–17, 38.
Wittgenstein, Ludwig. 1958. *Philosophical investigations*. Oxford: Blackwell.
———. 1978. *Remarks on the foundations of mathematics*. Oxford: Blackwell.
Zimecki, Michał. 2006. The lunar cycle: Effects on human and animal behavior and physiology. *Postępy Higieny i Medycyny Doświadczalnej* 60: 1–7.
Zukav, Gary. 1980. *The dancing Wu Li Masters: An overview of the new physics*. New York: Bantam Books.

Chapter 23
Mathematical Beauty

Abstract Several mathematicians have maintained that mathematical beauty plays an important role in mathematical research. This raises the problem: What is mathematical beauty? This chapter supports the view that a piece of mathematics, demonstration or theorem, is beautiful when it provides understanding, meaning by this the recognition of the fitness of the parts to each other and to the whole. Specifically, a demonstration is beautiful when it is clear what the whole idea of the demonstration is, what the contribution of each part of the demonstration to the whole idea is, and why such contribution is essential. A theorem is beautiful when it is clear what the content of the theorem is, what the contribution of each concept to the content is, and what relations the theorem establishes between such concepts. Mathematical beauty thus meant can have a role in mathematical discovery, because it can guide the mathematician in selecting which hypothesis to consider and which to disregard.

23.1 Aesthetic Judgments and the Neuroscience of Aesthetics

In Chap. 2 it has been stated that emotions play an important role in knowledge, including mathematical and scientific knowledge. In particular, aesthetic emotions play a prominent role in mathematical knowledge. This view has been put forward by many mathematicians and scientists.

Thus, Poincaré states that mathematics has "an esthetic aim," indeed "its adepts find therein delights analogous to those given by painting and music;" and "the joy they thus feel" has "the esthetic character, even though the senses take no part therein" (Poincaré 2013, 280).

Weyl states: "My work has always tried to unite the true with the beautiful and when I had to choose one or the other, I usually chose the beautiful" (Dyson 1956, 457).

Weinberg states that "mathematicians are led by their sense of mathematical beauty to develop formal structures" (Weinberg 1993, 125).

Hardy even states that "there is no permanent place in the world for ugly mathematics" (Hardy 1992, 14). When Snow "taxed him and said that if one would present a solution which finally proved Goldbach's theorem, and if it was ugly, would

© Springer International Publishing AG 2017
C. Cellucci, *Rethinking Knowledge*, European Studies in Philosophy of Science 4,
DOI 10.1007/978-3-319-53237-0_23

he accept it? Hardy replied, 'This is impossible. It couldn't possibly be a proof of Goldbach's theorem if it were ugly'" (Snow 1973, 812).

Hardy's position was too extreme, there are permanent parts of mathematics that could hardly be said to be beautiful. For example, Hersh and John-Steiner point out that the formula which solves the general quadratic equation "is one of the most memorized formulas in math. Not beautiful!" (Hersh and John-Steiner 2011, 61). More moderately, several mathematicians assume that at least some mathematics is beautiful. In particular, they make aesthetic judgments of demonstrations and theorems, qualifying some of them as 'elegant' or 'clumsy', 'beautiful' or 'ugly'.

Even this more moderate point of view, however, has been disputed. Doubt has been raised as to whether the apparent aesthetic judgments that mathematicians often make are really aesthetic at all. For example, Todd maintains that there are "strong reasons for casting doubt on the aesthetic nature of at least many of the claims made in mathematics and science" (Todd 2008, 63).

Recent findings in the neuroscience of aesthetics are relevant to the question. They have been established using functional magnetic resonance imaging to image the activity in the brains of a number of mathematicians, when they viewed mathematical formulae which they had individually rated as beautiful, indifferent or ugly. The findings indicate that "the experience of mathematical beauty correlates with activity in the same brain area" – the medial orbito-frontal cortex – which is "active during the experience of visual, musical, and moral beauty" (Zeki et al. 2014, 8). Moreover, "the activity there is parametrically related to the declared intensity of the experience of beauty, whatever its source" (ibid.). This does not mean that the medial orbito-frontal cortex alone is responsible for, or underlies, the experience of mathematical beauty. Other brain areas are active during that experience, "distinct from the areas engaged when viewing paintings" or "listening to musical excerpts" (ibid., 10). Nevertheless, the findings suggest that "the activity in a common area of the emotional brain that correlates with the experience of beauty derived from different sources merely mirrors neurobiologically the same powerful and emotional experience of beauty that mathematicians and artists alike have spoken of" (ibid., 8).

However, while addressing the question what neural mechanisms allow us to experience beauty, the findings in the neuroscience of aesthetics do not say what mathematical beauty is, in particular, why it is that a demonstration or a theorem are beautiful. The aim of this chapter is to try to give an answer to these questions.

23.2 Two Different Traditions About Mathematical Beauty

With respect to the question what mathematical beauty is, there are two different traditions.

(A) Mathematical beauty consists of properties of certain mathematical entities which are intrinsic to such entities, and hence are independent of the subject and period of history.

(B) Mathematical beauty is a projection of the subject. If certain mathematical entities exhibit properties that are valued by the aesthetic criteria of the subject, then the subject will project beauty on those mathematical entities and will describe them as beautiful. Thus mathematical beauty is dependent upon the subject and period of history.

The first tradition goes back to antiquity, the second one has been widespread in the modern and contemporary period.

23.3 Mathematical Beauty as an Intrinsic Property

An eminent representative of the first tradition about mathematical beauty is Plato. He states that "straight lines and circles, and the plane and solid figures which are formed out of them by means of compasses, rulers and squares," are "not, as other things are, beautiful in a relative way, but they are by their very nature forever beautiful by themselves" (Plato, *Philebus*, 51 c 3–d 1). They are "not beautiful at one time and not at another, or beautiful by one standard and ugly by another, or beautiful in one place and ugly in another," being "beautiful to some people but ugly to others" (Plato, *Symposium*, 211 a 3–5). On the contrary, they are of a kind of beauty which "always is and does not come into to be or perish, nor does it grow or wane" (ibid., 211 a 1–2). We may acquire knowledge of this kind of beauty only by intellectual intuition, by which alone we may arrive at "that particular knowledge which is knowledge solely of the beautiful itself" (ibid., 211 c 7–8).

The first tradition about mathematical beauty has had a large following among mathematicians and scientists. For example, Dirac states that, while "beauty does depend on one's culture and upbringing for certain kinds of beauty, pictures, literature, poetry, and so on," mathematical beauty "is of a completely different kind and transcends these personal factors. It is the same in all countries and at all periods of time" (Dyson 1992, 305).

However, it is is somewhat problematic to say that mathematical beauty is the same in all countries and at all periods of time, because often what is considered mathematically beautiful in a certain period or culture is not so considered in another period or culture. In particular, philosophers and mathematicians have given several different characterizations of mathematical beauty.

Thus Plato states that "measure and proportion manifest themselves in all areas as beauty and virtue" (Plato, *Philebus*, 64 e 6–7). For, "nothing beautiful lacks proportion" (Plato, *Timaeus*, 87 c 5).

Aristotle states that "the supreme forms of beauty are order, symmetry, and definiteness, which the mathematical sciences demonstrate in a special degree" (Aristotle, *Metaphysica*, M 3, 1078 a 36–b 2).

Poincaré states that "the mathematical entities to which we attribute" the "character of beauty and elegance" are "those whose elements are harmoniously disposed so that the mind without effort can embrace their totality," and "this harmony is at once a satisfaction of our aesthetic needs and an aid to the mind, sustaining and guiding" (Poincaré 2013, 391).

Hardy states that in beautiful theorems "there is a very high degree of unexpectedness, combined with inevitability and economy" (Hardy 1992, 113).

Hersh and John-Steiner state that "we can point to three important elements of beauty in mathematical content: simplicity, concrete specificity, and unexpected or surprising integration or connection of disparate elements" (Hersh and John-Steiner 2011, 60).

Proportion, order, symmetry, definiteness, harmony, unexpectedness, inevitability, economy, simplicity, specificity, integration or connection are different properties. This explains why Hofstadter asserts that "there exists no set of rules which delineate what it is that makes a piece beautiful, nor could there ever exist such a set of rules" (Hofstadter 1999, 555).

23.4 Mathematical Beauty as a Projection of the Subject

An eminent representative of the second tradition about mathematical beauty is Kant. One might think that Kant leaves no room for beauty in mathematics. For, he claims that, while "it is customary to call" certain "properties of geometrical" figures "beauty" and "to speak of this or that beautiful property of, e.g., a circle," still, this "is not an aesthetic judging," so "a judging without a concept," but only "an intellectual judging in accordance with concepts," therefore, one cannot really consider such properties "a beauty of the mathematical figure" (Kant 2000, 238).

By this, however, Kant does not mean to deny that certain mathematical entities may exhibit properties that are valued by the aesthetic criteria of the subject, and hence that the subject may project beauty on those mathematical entities and describe them as beautiful. He only means to deny that mathematical beauty consists of properties of certain mathematical entities which are intrinsic to such entities, and hence are independent of the subject.

Indeed, Kant states that "beauty is not a quality of the object considered for itself" (ibid., 221). For, "beauty is nothing by itself, without relation to the feeling of the subject" (ibid., 103). If representations are "related in a judgment solely to the subject (its feeling), then they are to that extent always aesthetic" (ibid., 90). In particular, beauty of a mathematical entity "is not a property of the object outside of me, but merely a kind of representation in me" (ibid., 237). Thus mathematical beauty is dependent upon the subject.

On the other hand, however, Kant also states: "That is beautiful which pleases universally without a concept" (ibid., 104). When someone says that a thing is beautiful, "he judges not merely for himself, but for everyone, and speaks of beauty as if it were a property of things" (ibid., 98). This raises the problem: What is the ground of this universality of judgments of beauty? The ground cannot be the subject's feeling of pleasure, because "such a pleasure would be none other than mere agreeableness in sensation, and hence by its very nature could have only private validity" (ibid., 102). The ground must be another state of mind which, unlike the subject's feeling of pleasure, is universal.

Now, according to Kant, the only states of mind which are universal are those that have to do with cognition, since cognition "is the only kind of representation that is valid for everyone" (ibid., 103). On the other hand, the powers exercised in cognition are imagination and understanding, which both are necessary, "imagination for the composition of the manifold of intuition and understanding for the unity of the concept that unifies the representations" (ibid., 102). Therefore, the universality of judgments of beauty must have to do with imagination and understanding.

In particular, since beautiful is that which pleases universally without a concept, in judgments of beauty imagination and understanding must be involved "merely subjectively, namely without a concept of the object" (ibid.). Then the state of mind which is the ground of the universality of judgments of beauty "can be nothing other than the state of mind that is encountered in the relation" of imagination and understanding "to each other, insofar as they relate a given representation to cognition in general" (ibid.). This relation is subjective, since it does not concern the relation of imagination and understanding to the object, but at the same time is universal, because it is presupposed for all cognition, and hence "must be valid for everyone" (ibid., 103).

Once again, since in judgments of beauty "no determinate concept" is involved, in them imagination and understanding are "in free play" (ibid., 102). Hence, the state of mind which is the ground of the universality "in a judgment of taste, since it is supposed to occur without presupposing a determinate concept, can be nothing other than the state of mind in the free play of the imagination and the understanding," so far as they harmonize "with each other as is requisite for a cognition in general" (ibid., 103). Therefore, the ground of the universality of judgments of beauty is "the harmony of the faculties of cognition," imagination and understanding, and on this harmony "is this universal subjective validity of satisfaction, which we combine with the representation of the object that we call beautiful, grounded" (ibid.).

On this basis, we may call a mathematical entity beautiful if such entity effects a state of harmony of the imagination and the understanding "by which we feel our entire cognitive faculty (understanding and imagination) strengthened" (Kant 1996, 268). In particular, we may "call a demonstration" of certain properties of mathematical entities beautiful if by this we mean that through it "the understanding, as the faculty of concepts, and the imagination, as the faculty of exhibiting them, feel themselves strengthened *a priori* (which, together with the precision which is introduced by reason, is called its elegance)" (Kant 2000, 238–239). Here "the satisfaction, although its ground lies in concepts, is subjective" (ibid., 239).

On the other hand, since in judgments of mathematical beauty the satisfaction is subjective, there is the risk that subjectivity reduces to mere privateness. In order to avoid this risk, judgments of mathematical beauty must be built upon cultural tradition. In fact, "among all the faculties and talents, taste is precisely the one which, because its judgment is not determinable by means of concepts," is "most in need of the examples of what in the progress of culture has longest enjoyed approval if it is not quickly to fall back into barbarism and sink back into the crudity of its first attempts" (ibid., 164). Indeed, there is no use of our entire cognitive faculty (understanding and imagination), "however free it might be," which, "if every subject always had to begin entirely from the raw predisposition of his own nature, would not fall into mistaken attempts if others had not preceded him with their own" (ibid.).

For this reason, "the ancient mathematicians" have been "regarded until now as nearly indispensable models of the greatest thoroughness and elegance of the synthetic method" (ibid., 163). Namely, of the axiomatic method. This does not mean that they are "*a posteriori* sources of taste," nor that they "contradict the autonomy of taste in every subject" (ibid.). It only means that they are an antidote to the risk that, in judgments of mathematical beauty, subjectivity reduces to mere privateness. Only because of such antidote, when we call some mathematical entity beautiful, "the pleasure that we feel is expected of everyone else in the judgment of taste as necessary, just as if it were to be regarded as a property of the object that is determined in it in accordance with concepts" (ibid., 103).

23.5 Rota's Phenomenology of Mathematical Beauty

The second tradition about mathematical beauty has several representatives. Of course, they need not share all of Kant's views about mathematical beauty, but only the view that beauty is a projection of the subject rather than a property which is intrinsic to certain mathematical entities. Some recent representatives of such tradition are Breitenbach 2013; McAllister 2005; Rota 1997; Sinclair 2011. I will discuss Rota's position because this will be useful as reference in what follows.

Rota states that "the beauty of a piece of mathematics is dependent upon schools and periods. A theorem that is in one context thought to be beautiful may in a different context appear trivial" (Rota 1997, 126). Thus mathematical beauty is dependent upon the context. Appreciating it requires "familiarity with a huge amount of background material" (ibid., 129). For example, "a proof is viewed as beautiful only after one is made aware of previous clumsier proofs" (ibid., 130). Familiarity with a huge amount of background material "is arrived at the cost of time, effort, exercise and Sitzfleisch" (ibid., 128). Therefore, the appreciation of mathematical beauty cannot be instantaneous. We must avoid the "light bulb mistake" which consists in believing that mathematical beauty is "appreciated with the instantaneousness of a light bulb being lit," and hence that "the appreciation of mathematical beauty" is "an instantaneous flash" (ibid., 130). Beauty plays a positive role in the development of mathematics, because "the lack of beauty in a piece of mathematics" is "a motivation for further research" (ibid., 128). An even "cursory look at any mathematics research journal will confirm" that "much mathematical research consists precisely of polishing and refining statements and proofs of known results" (ibid., 129). This arises because, even when a mathematician has been able "to follow the proof" of a statement and hence "to verify its truth in the logical sense of the term," he is still missing something, namely, he "is missing the sense of the statement that has been verified to be true" (ibid., 131). Indeed, the mere "logical truth of a statement does not enlighten us as to the sense of the statement. Enlightenment not truth is what the mathematician seeks" (ibid.).

Nevertheless, "mathematicians seldom explicitly acknowledge the phenomenon of enlightenment" (ibid., 132). This depends on the fact that neither mathematical beauty nor mathematical truth "admits degrees" (ibid., 131). Conversely, "enlightenment admits degrees: some statements are more enlightening than others," and "mathematicians dislike concepts admitting degrees" (ibid., 132). Therefore, mathematicians talk of 'mathematical beauty', but this is only "a trick mathematicians have devised to avoid facing up to the messy phenomenon of enlightenment" (ibid.). Mathematicians "say that a proof is beautiful when it gives away the secret of the theorem, when it leads us to perceive the inevitability of the statement being proved" (ibid.). They say that a theorem is beautiful when they see "how the theorem 'fits' in its place, how it sheds light around itself, like a *Lichtung*, a clearing in the woods" (ibid.). But "they say that a theorem is beautiful when they mean to say that the theorem is enlightening" (ibid.).

23.6 Some Limitations of Rota's Views

Despite its suggestiveness, Rota's approach to mathematical beauty has some limitations.

1) Rota claims that beauty plays a positive role in the development of mathematics, because the lack of beauty in a piece of mathematics is a strong motivation for further mathematical research. But sometimes the quest for beauty may be an obstacle to the development of science. Examples of this are Galileo's rejection of Kepler's elliptic orbits and Dirac's rejection of renormalization (see Chap. 11).

2) Rota claims that, when mathematicians say that a piece of mathematics is beautiful, they really mean to say that it is enlightening. But Rota stops short of explaining what he means by 'enlightening'.

3) Rota claims that, unlike enlightenment, beauty does not admit degrees, and mathematicians dislike any concepts admitting degrees. But beauty, including mathematical beauty, does admit degrees, as it is clear from the fact that we commonly say that something is more beautiful than something else. Further evidence for the fact that mathematical beauty admits degrees is provided by a poll of readers of *The Mathematical Intelligencer* who ranked 24 theorems, on a scale from 0 to 10, for beauty (see Wells 1990).

23.7 Mathematical Beauty and Perception

Before trying to suggest how to avoid the limitations of Rota's approach to mathematical beauty, a preliminary objection must be dealt with. The objection is that perceptual experience is central to beauty, while the recognition of something that might count as mathematical beauty has nothing to do with perception.

Thus Zangwill states that, "as the etymological origins of the word 'aesthetic' suggest, aesthetic properties are those that we appreciate in perception. Lovers of beauty are 'lovers of sight and sounds'" (Zangwill 1998, 81). For, "aesthetic properties are properties which something has only if it has sensory properties" (ibid., 66). Conversely, "proofs, theories" do "not necessarily have any sensory embodiment" (ibid., 78). Therefore, when we say that a proof is elegant or beautiful, this "is not genuine aesthetic appreciation," but "aesthetic terms are metaphorically applied in these cases" (ibid., 79).

Similarly, van Gerwen states that "beauty centrally involves a perceptual experience" (van Gerwen 2011, 250). For, "beauty is proven to exist in perception" (ibid., 257). Conversely, "the recognition of something that might count as mathematical beauty" has "nothing to do with perception" (ibid., 259). Therefore, "to speak of mathematical beauty is to speak in a loose manner" (ibid., 264).

This objection, however, is unjustified. As Kant states, "that is beautiful which pleases in the mere judging (neither in sensation nor through a concept)" (Kant 2000, 185). Thus a perceptual experience need not be central to beauty.

The objection would imply not only that there is no mathematical beauty, but also that there is no literary beauty, because the recognition of something that might count as literary beauty has little to do with perception. Of course, to read a novel, I must be able to see the text in front of me, but surely my aesthetic experience in reading a novel does not consist in deciphering the words on the page in front of me. Nor do I literally see the facts narrated in the novel, I only imagine them, and imagination is the capacity to represent something even when it is not itself present before my eyes.

Zangwill claims that, "if a literary work has aesthetic properties, they derive from the particular choice of words, because of the way they sound," and "if a literary work has values which are not linked" to "the sonic properties of words, then they are not aesthetic values" (Zangwill 1998, 75). But, since generally there is no unsurmountable problem in translating a literary work in other languages so that its values are essentially preserved, a literary work has values which are not merely linked to the sonic properties of words. Thus, if Zangwill were right, no literary work would have aesthetic values. But it would be difficult to deny that literary works have aesthetic values, so the assumption that a perceptual experience is central to beauty does not seem to be justified.

Of course, sculptural beauty, painterly beauty, musical beauty, literary beauty, mathematical beauty, etc., are all different kinds of beauty. But this means that there are different kinds of beauty even in the realm of beauty derived from more perceptually based sources. As Gowers states, "beauty in mathematics is not the same as beauty in music, but neither is musical beauty the same as the beauty of a painting, or a poem, or a human face" (Gowers 2002, 51).

23.8 From Enlightenment to Understanding

As already pointed out, a limitation of Rota's approach to mathematical beauty is that Rota stops short of explaining what he means by 'enlightening'. Actually, 'enlightening' is not the most appropriate word to use in this context, because 'to enlighten' means 'to give insight', and 'insight' suggests that mathematical beauty is appreciated with the instantaneousness of a light bulb being lit. This conflicts with Rota's warning that we must avoid the light bulb mistake.

It seems more appropriate to say that, when mathematicians assert that a piece of mathematics is beautiful, they mean to say that it gives understanding. In fact, Kosso and Breitenbach establish a relation between beauty in natural science and understanding. Kosso maintains that there is "a link between beauty and scientific understanding" (Kosso 2002, 40). Breitenbach maintains that, "in searching for beauty, scientists aim for theories that provide understanding" (Breitenbach 2013, 85). The same can be said about mathematics.

It may seem odd to say that, when mathematicians assert that a piece of mathematics is beautiful, they mean to say that it gives understanding. For, understanding is not commonly associated with beauty. But, as argued in Chap. 22, understanding is the recognition of the fitness of the parts to each other and to the whole. Understanding thus meant is an aesthetic property because, according to a tradition going back to antiquity, one aspect of the appreciation of beauty in a work of art is just the recognition of the fitness of the parts to each other and to the whole.

As Heisenberg points out, in antiquity "there were two definitions of beauty, which stood in a certain opposition to one another" (Heisenberg 2001, 57). The "one describes beauty as the proper conformity of the parts to one another, and to the whole. The other, stemming from Plotinus, describes it, without any reference to the parts, as the translucence of the eternal splendor of the 'one' through the material phenomenon" (ibid.). Heisenberg supports "the first and more sober definition of beauty" (ibid., 69). He also establishes a relation between beauty and understanding. For, he states that, "if the beautiful is conceived as a conformity of the parts to one another and to the whole, and if, on the other hand, all understanding is first made possible by means of this formal connection, the experience of the beautiful becomes virtually identical with the experience of connections either understood or, at least, guessed at" (ibid., 59).

Understanding, meant as the recognition of the fitness of the parts to each other and to the whole, is context-dependent. In particular, it is dependent upon the mathematician's background knowledge, so it is usually recognizable only to the well trained. Appreciating mathematical beauty requires apprehending the ideas involved, in a way that reveals the fitness of the parts to each other and to the whole, and this is something that a poor background knowledge cannot provide.

Of course, the fitness of the parts to each other and to the whole means different things in art, music, literature or mathematics, and, within mathematics, in demonstrations or theorems. Therefore, each case requires special consideration.

23.9 Beauty in Works of Art

In works of art, a kind of fitness of the parts to each other and to the whole is expressed by the golden ratio. Two quantities are said to be in the golden ratio (φ) if the sum of the two quantities is to the larger quantity as the larger quantity is to the smaller. Namely, a and b, with $a>b$, are in the golden ratio if $\frac{a+b}{a}=\frac{a}{b}=\varphi$. Since $\frac{a+b}{a}=1+\frac{b}{a}=1+\frac{1}{\varphi}$, we have $1+\frac{1}{\varphi}=\varphi$, so $\varphi+1=\varphi^2$, hence $\varphi^2-\varphi-1=0$. The positive solution of this equation is $\varphi=\frac{1+\sqrt{5}}{2}=1.61805$, which gives the value of the golden ratio to the first five decimal places.

The golden ratio is observed in many works of art. For example, in Leonardo da Vinci's *Mona Lisa*, the ratio of the height to the width of the canvas is the golden ratio.

Moreover, the proportions of Mona Lisa's body exhibit several golden ratios. For example, the ratio of the height to the width of Mona Lisa's face is the golden ratio.

As another example, in Botticelli's *The Birth of Venus*, the ratio of the width to the height of the canvas is very nearly the golden ratio.

Moreover, the proportions of Venus' body exhibit several golden ratios. For example, the ratio of Venus' height overall to the height of her navel is the golden ratio.

23.10 Beauty in Demonstrations

Of course, the beauty of a demonstration cannot be expressed in terms of the golden ratio, it must involve another kind of relationship between the parts of the demonstration. Specifically, there is fitness of the parts of a demonstration to each other and to the whole when it is clear what the whole idea of the demonstration is, what the contribution of each part of the demonstration to the whole idea is, and why such contribution is essential. Then the demonstration provides understanding. For, it shows why the proposition that is being demonstrated holds, and so, as Rota states, it leads us to perceive the actual, not the logical inevitability of the proposition that is being demonstrated.

This concept of fitness is significant, as it is apparent from the statements of several mathematicians, who point out that, when there is no fitness of the parts of a demonstration to each other and to the whole, a demonstration may not provide understanding.

Thus Poincaré states that, even when the mathematician has cut up "each demonstration into a very great number of elementary operations" and has "ascertained that each is correct," he may not "have grasped the real meaning of the demonstration" since he may not have seized that "which makes the unity of the demonstration" (Poincaré 2013, 217–218).

Mordell states that, "even when a proof has been mastered" and "may be strictly logical and convincing," the "reader may feel that something is missing. The argument may have been presented in such a way as to throw no light on the why and wherefore of the procedure or on the origin of the proof or why it succeeds" (Mordell 1959, 11).

Moreover, as we have already seen, Rota states that, even when a mathematician has been able to follow the proof of a statement and hence to formally verify its truth in the logical sense of the term, he is still missing something.

What the mathematician is missing is the fitness of the parts of the demonstration to each other and to the whole. Only a demonstration which shows such fitness provides understanding.

23.11 An Example of a Beautiful Demonstration

To give a simple example of a beautiful demonstration, we consider the problem: Construct a square which is double the size of a given square. In Plato's *Meno* a solution to this problem is obtained by using the following diagram, where the given square is one of the four smallest squares of which the larger square consists, and the middle square – the one looking like an equilateral kite – is the square which is double the size of the given square.

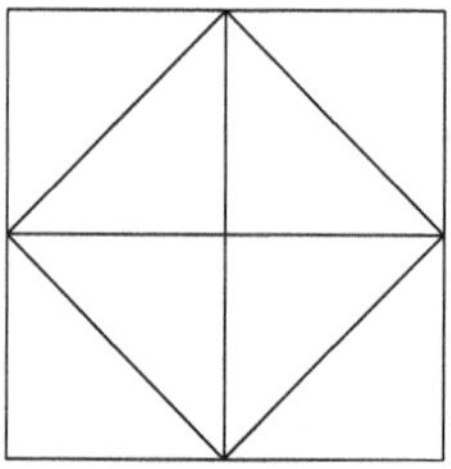

From the diagram it is apparent that the given square consists of two equal right-angled triangles, and the middle square consists of four equal right-angled triangles. This gives a demonstration that the middle square is double the size of the given square. The demonstration is beautiful because it provides understanding, in the sense explained in the previous section.

It might be objected that demonstrations based on diagrams are not genuine demonstrations. But, as argued in Chap. 19, this objection is unjustified.

23.12 Differences in Beauty Between Geometrical Demonstrations

There are significant differences between geometrical demonstrations in terms of understanding, and hence of beauty. For example, let us consider the problem: Show that the square on the hypotenuse of a right-angled triangle is equal to the sum of the squares on the other two sides (the Pythagorean Theorem).

Demonstration 1 (Euclid, *Elements*, I.47). Describe the square *BDEC* on *BC*, and the squares *GB* and *HC* on *BA* and *AC*, respectively. Draw *AL* parallel to either *BD* or *CE*, and draw *AD* and *FC*.

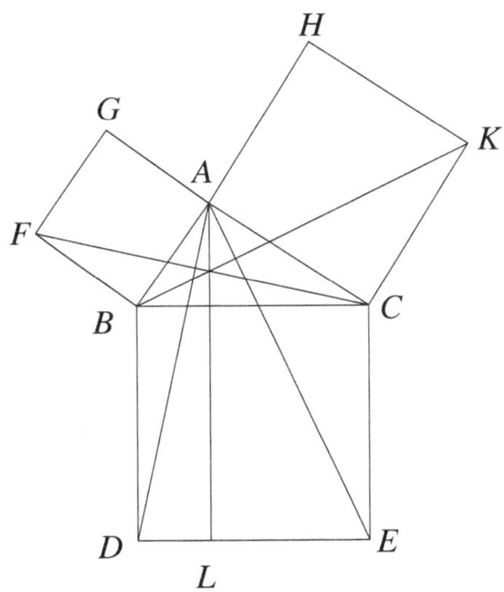

Then triangle *ABD* is equal to the triangle *FBC* because *DB*=*BC*, *FB*=*BA* and angle *ABD* = angle *FBC* (since both angles consist of angle *ABC* plus a right angle).

On the other hand, the triangle *ABD* is half the parallelogram *BL* because they have the same base *BD* and are in the same parallels *BD* and *AL*. In the same way, the triangle *FBC* is half the square *GB* because they have the same base *FB* and are in the same parallels *FB* and *GC*.

Since the triangle *ABD* is equal to the triangle *FBC*, it follows that the parallelogram *BL* is equal to the square *GB*.

Similarly, it can be shown that the parallelogram *CL* is equal to the square *HC*.

Therefore, the whole square *BDEC* is equal to the sum of the two squares *GB* and *HC*.

Demonstration 2. (Euclid, *Elements*, VI.31). Let *AD* be drawn perpendicular to *BC*.

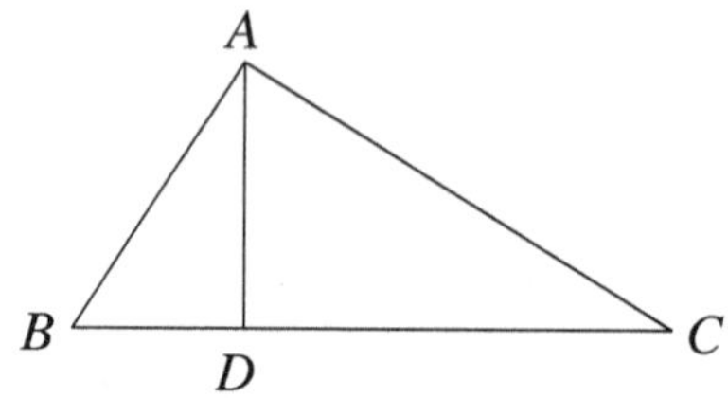

Then the right-angled triangle *ABC* is divided into two right-angled triangles *ABD* and *ADC*. The angles of the triangles *ABD* and *ADC* are equal to the angles of the triangle *ABC*, so the triangles *ABD* and *ADC* are similar to the triangle *ABC*. Now, in similar triangles the corresponding sides are proportional. So, since the triangles *ABD* and *ABC* are similar, we have

$$\frac{BD}{AB}=\frac{AB}{BC},$$

and, since the triangles *ADC* and *ABC* are similar, we have

$$\frac{DC}{AC}=\frac{AC}{BC}.$$

Cross multiplying, we obtain $BD \cdot BC = AB^2$ and $DC \cdot BC = AC^2$. Since $DC = BC - BD$, from the latter we obtain $(BC - BD) \cdot BC = AC^2$, so $BC^2 - BD \cdot BC = AC^2$. Hence $BC^2 - AB^2 = AC^2$, therefore $BC^2 = AB^2 + AC^2$.

Demonstration 1 is rather clumsy. Indeed, Schopenhauer states: “Euclid’s stilted, indeed underhand, proof leaves us without an explanation of why” (Schopenhauer 2010, 98). In that proof, “lines are often drawn without any indication of why,” and the reader “must admit in astonishment what remains completely incomprehensible in its inner workings” (ibid., 96). Similarly, Rav states: “Euclid’s proof is a tour de force to fit a preset methodology of a purely geometric, deductive argument,” so it “hides the heuristic path to the discovery and a more intuitive proof of the Pythagorean theorem” (Rav 2005, 52).

Demonstration 2 is beautiful, because it provides understanding. It shows why the square on the hypotenuse of a right triangle is equal to the sum of the squares on the other two sides. The reason is that any right-angled triangle can be divided into two right-angled triangles which are similar to it, and in similar triangles the corresponding sides are proportional.

23.13 Differences in Beauty Between Arithmetical Demonstrations

There are significant differences also between non-geometrical demonstrations – for example, arithmetical demonstrations – in terms of understanding, and hence of beauty. For example, let us consider the problem: Show that $\sqrt{2}$ is not a fraction.

Demonstration 1 (Euclid, *Elements*, X, Appendix 27). Suppose that $\sqrt{2} = p / q$ where p and q are integers. We may assume that p and q have no common factor – if they do, we cancel it. Now, from our assumption it follows that $2 = p^2/q^2$, so $p^2 = 2q^2$. Then p^2 is even, and therefore p itself is even, since the square of an odd number is odd. Thus $p = 2r$ for some r, so $(2r)^2 = 2q^2$, that is, $4r^2 = 2q^2$, hence $q^2 = 2r^2$. Then q^2 is even, and therefore, as above, q itself is even. So p and q are both even, hence they have a common factor. But by our assumption p and q have no common factor. Contradiction. Therefore $\sqrt{2}$ cannot be a fraction.

Demonstration 2. Suppose that $\sqrt{2} = p / q$ for two integers p and q. Then $2 = p^2/q^2$, so $p^2 = 2q^2$. Since every integer >1 can be represented as a product of primes, and this representation is unique apart from the order of the factors, we may assume that p and q have been so represented. Since $p^2 = p \cdot p$, in the representation of p^2 every prime will occur an even number of times. Similarly in the representation of q^2. Then in the representation of $2q^2$ the prime number 2 will occur an odd number of times. Since $p^2 = 2q^2$ and the representation is unique, this is impossible. Therefore $\sqrt{2}$ cannot be a fraction.

Demonstration 2 is beautiful. Indeed, as Davis and Hersh state, it "exhibits a higher level of aesthetic delight" than Demonstration 1, because it "seems to reveal the heart of the matter," while Demonstration 1 "conceals it" (Davis and Hersh 1981, 299–300). Demonstration 2 "exposes the 'real' reason" (ibid., 333). It shows that $\sqrt{2}$ cannot be a fraction because the parities of the exponent of 2 in the representations of p^2 and q^2 as a product of primes will be different. Thus, Demonstration 2 provides understanding.

23.14 An Example of a Beautiful Theorem

The beauty of a theorem involves another kind of relationship between the parts of the theorem. There is fitness of the parts of a theorem to each other and to the whole when it is clear what the content of the theorem is, what the contribution of each concept to the content is, and what relations the theorem establishes between such concepts.

To give an example of a beautiful theorem, let us consider Euler's identity, $e^{i\pi} + 1 = 0$, or equivalently $e^{i\pi} = -1$. The poll of readers of *The Mathematical Intelligencer* mentioned above, ranked Euler's identity the most beautiful theorem in mathematics (see Wells 1990, 37). In the functional magnetic resonance imaging by which the activity in the brains of a number of mathematicians has been imaged, "the formula most consistently rated as beautiful," both "before and during the scans, was Leonhard Euler's identity" (Zeki et al. 2014, 4). Nahin even states that Euler's identity sets "the gold standard for mathematical beauty," and "will still appear, to the arbitrarily advanced mathematicians ten thousand years hence, to be beautiful and stunning and untarnished by time" (Nahin 2006, xxxii).

Euler's identity is beautiful because it provides understanding. For, it involves five of the most important numbers in mathematics, 0, 1, i, π, e, along with the fundamental concepts of addition, multiplication and exponentiation, and establishes a basic relation between them in a transparent way. More generally, it establishes a relation between what Halmos calls "the three major parts of mathematics," that is, "algebra (a symbolic outgrowth of arithmetic), geometry (the study of shape), and analysis (the abstract version of calculus)" (Halmos 1992, 1325). For, 0, 1, i come from algebra, π comes from geometry, and e comes from analysis, since $e = \lim_{n\to\infty}\left(1+\frac{1}{n}\right)^n$.

To see how Euler's identity comes about, observe that we can represent powers of the complex number $1+\frac{i\pi}{n}$, $\left(1+\frac{i\pi}{n}\right)^2$, $\left(1+\frac{i\pi}{n}\right)^3$, …, $\left(1+\frac{i\pi}{n}\right)^n$ by drawing similar right-angled triangles that share an edge, where 1 is the length of the first spoke, and $\frac{\neq}{n}$ is the length of first edge. For example, for $n=6$, this yields the spiral:

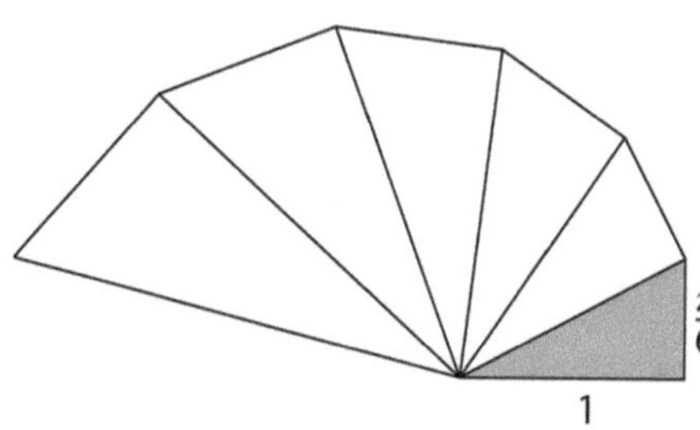

As we increase the number n of similar right-angled triangles, the length of the hypotenuse of each right-angled triangle becomes very close to the length of the previous spoke, so the spiral becomes closer to circular. For example, for $n=24$, we have:

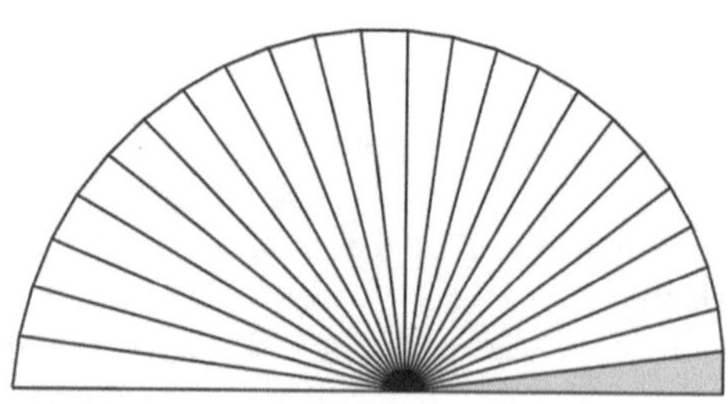

Taking n to the limit, we get an infinite number of spokes that are all the same size, taking us around the unit circle to the opposite side, -1. Therefore, $\lim_{n\to\infty}\left(1+\frac{i\pi}{n}\right)^n = -1$.

On the other hand, from the definition of e it follows that $e^{i\pi} = \lim_{n\to\infty}\left(1+\frac{i\pi}{n}\right)^n$. Therefore, $e^{i\pi} = -1$. This gives away the secret of Euler's identity. (For a similar explanation of Euler's identity, see Conway and Guy 1996, 254–256).

23.15 Mathematical Beauty and Discovery

Even though the quest for beauty may sometimes be an obstacle to the development of science, nevertheless it may also play a positive role in the development of mathematics, not only as a motivation for finding better demonstrations of known propositions, but, more important, as a part of the process of discovery.

As we have seen in Chap. 11, in the past century it has been almost universally held that scientific discovery escapes logical analysis. But discovery does not escape logical analysis, it can be analysed in terms of the analytic method.

Now, in the analytic method, hypotheses are obtained by non-deductive rules. Such rules, however, may yield so many hypotheses that it could be unfeasible to check all of them for plausibility. The sense of beauty may guide us in selecting which hypothesis to check and which to disregard, thus acting as a shortcut.

Avigad states that "mathematics presents us with a combinatorial explosion of options; understanding helps us sift through them, and pick out the ones that are worth pursuing" (Avigad 2008, 320). Indeed, "when we do mathematics, we are like Melville's sailors, swimming in a vast expanse. Just as the sailors cling to sides of their ship, we rely on our understanding to guide us and support us" (ibid.). So far so good. But then Avigad says that understanding is based on a "moment of insight" which occurs when "I am working through a difficult proof" and "I struggle to make sense of it," then "all of a sudden, something clicks, and everything falls into place – now I understand" (ibid., 322). Now, to say that understanding is based on a moment of insight is to commit the light bulb mistake denounced by Rota.

If we agree that understanding is the non-instantaneous recognition of the fitness of the parts to each other and to the whole, then understanding, hence beauty, helps us sift through the hypotheses generated by non-deductive rules, and pick out the ones that are worth pursuing. As Breitenbach states, "because of the link to our capacities of understanding, following beauty can provide a heuristic means for choosing between theories" (Breitenbach 2013, 94). Thus beauty may have a role in the context of discovery. This is an aspect of the role emotion may have in that context (see Chapters 2 and 14).

That beauty may have a role in the context of discovery was already pointed out by Poincaré. According to him, mathematical discovery consists in making new combinations with concepts that are already known, and in selecting those that are useful. The selection is made on the basis of the feeling of mathematical beauty. As already argued in Chap. 11, Poincaré's proposal is subject to limitations, nevertheless the importance of his remark that beauty may have a role in the context of discovery remains.

That the sense of beauty may guide us in selecting which hypothesis to check and which to disregard, is because the hypotheses thus selected provide understanding. On the other hand, that they provide understanding does not guarantee that they are plausible. The hypotheses may let us recognize the fitness of the concepts involved to each other and to the whole, but the concepts involved may overlook or confuse some aspect of the problem the hypotheses are intended to solve. Then the hypotheses may not be plausible. Such was the case of Frege's hypothesis that two functions have equal extensions if and only if they map every object to the same value. The hypothesis was beautiful because, as Frege stated, it provided understanding of what one means when one speaks of extensions of concepts (see Chap. 10). But the hypothesis was not plausible, as it is clear from the fact that it led to Russell's paradox.

Thus, selecting hypotheses is only one part of the analytic method. Another essential part is ascertaining that the hypotheses thus selected are plausible. This requires a separate argument. Dirac claims that "it is more important to have beauty in one's equations that to have them fit experiment" because, "if one is working from the point of view of getting beauty in one's equations, and if one has really a sound insight, one is on a sure line of progress" (Dirac 1963, 47). But it is not so. For, neither beauty nor a sound insight can guarantee that the hypotheses selected on the basis of them are plausible. Establishing that they are plausible essentially involves a comparison with the existing data. As we have seen in Chap. 11, Poincaré also recognized that, once useful combinations have been selected, it is necessary to submit them to verification.

Dirac himself provides evidence that neither beauty nor a sound insight can guarantee that the hypotheses selected on the basis of them are plausible. As already mentioned in Chap. 11, he stuck to his own version of quantum electrodynamics, which made predictions that were often infinite and hence unacceptable, rejecting renormalization, which led to accurate predictions. He declared that he "might have thought that the new ideas were correct if they had not been so ugly" (Dyson 1992, 306). Even if renormalization "has been very successful in setting up rules for handling the infinities and subtracting them away," the "resulting theory is an ugly and incomplete one," and hence "cannot be considered as a satisfactory solution of the problem of the electron" (Dirac 1951, 291).

To assume that, if one is working from the point of view of getting beauty in one's equations, one is on a sure line of progress, amounts to assuming that the world conforms to our aesthetic preferences. But there is no evidence that the laws governing the world should be expressible in beautiful equations. As Dyson asks: "Why should nature care about our feelings of beauty? Why should the electron prefer a beautiful equation to an ugly one?" (Dyson 1992, 305–306).

This by no means diminishes the role of the sense of beauty in selecting hypotheses. As already stated, this role depends on the fact that the non-deductive rules by which hypotheses are obtained in the analytic method may yield so many hypotheses that it would be unfeasible to check all of them for plausibility, and the sense of beauty may guide us in selecting which hypothesis to check and which to disregard, thus acting as a shortcut.

23.16 An Example of the Role of Beauty in Discovery

To give a simple example of the role of mathematical beauty in the context of discovery, we consider a likely story of how an unknown Egyptian mathematician found that the area of a triangle is half the base times the height. Only a likely story, because, as it is generally the case with discoveries, there is no report as to how discovery did happen. Nevertheless, the story seems to be authorized by the description of Problem 51 in the Rhind Mathematical Papyrus.

The Egyptian mathematician observes that, from a right-angled triangle, he can "get its rectangle" (Chace 1927, 92). That is, from a right-angled triangle, he can get a rectangle with the same base and the same height.

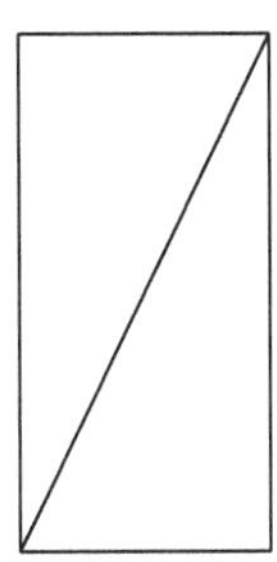

Then the Egyptian mathematician halves the area of the rectangle. Since the area of the rectangle is the base times the height, the area of the right-angled triangle will be half the base times the height.

From this, the Egyptian mathematician might infer several hypotheses, for example, that the area of the isosceles triangle will be half the base times the height.

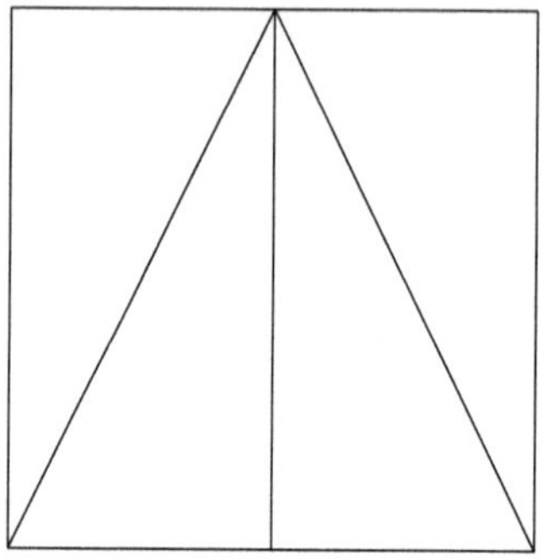

But the Egyptian mathematician selects the more general hypothesis – obtained by induction from a single case – that the area of any triangle is half the base times the height. He selects this hypothesis because it is the aesthetically most appealing one, since it establishes a relation between three basic parameters of any triangle, base, height, and area in a transparent way, and thus provides understanding.

Not only this hypothesis is the aesthetically most appealing one, but it is easy to see that it is plausible. To this purpose the Egyptian mathematician observes that, dropping a perpendicular, he can divide any triangle into two right-angled triangles. On the other hand, as he has already observed, from each of these two right-angled triangles, he can get a rectangle with the same base and the same height. Therefore, from the whole triangle itself he can get a rectangle with the same base and the same height.

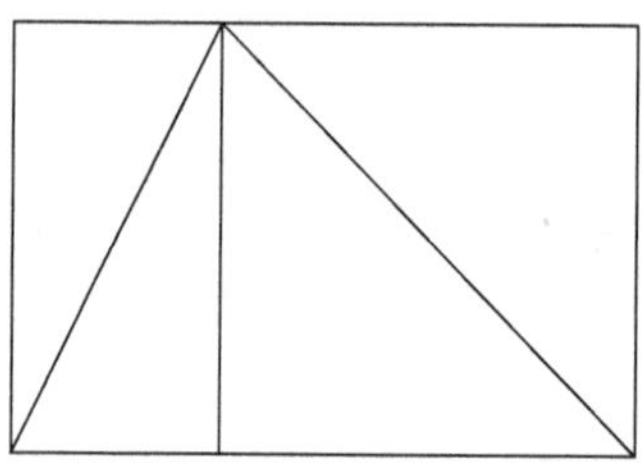

Then the Egyptian mathematician halves the area of such rectangle. Since the area of the rectangle is the base times the height, the area of the whole triangle will be half the base times the height.

This gives a demonstration of the hypothesis. The demonstration is beautiful because it provides understanding. For, it shows why the area of a triangle is half the base times the height.

23.17 Epistemic Role of the Aesthetic Factors

The quest for beauty has often been a motivation for doing research in mathematics. Indeed, von Neumann even states that the mathematician's "criteria of selection, and also those of success, are mainly aesthetical" (von Neumann 1961, 8).

This does not mean that in mathematics the quest for beauty is an end in itself. On the contrary, it is instrumental to the development of mathematics. This fact is often overlooked or denied. For example, Todd states that "aesthetic judgments and the evaluation of scientific theories are odd bedfellows, and their conjunction a just object of suspicion" (Todd 2008, 62). In fact, "aesthetic appreciation and epistemic satisfaction are distinct," and hence one must "avoid collapsing them into each other" (ibid., 75).

But this amounts to confining epistemology to the evaluation of scientific theories, hence to the context of justification, assuming that aesthetic factors cannot have any epistemic role *qua* aesthetic factors. Indeed, if epistemology is confined to the context of justification, then aesthetic factors cannot have any epistemic role at all, not only *qua* aesthetic factors. But, as argued in Chap. 2, epistemology need not be confined to the context of justification. Mathematical beauty can have a role in the context of discovery, because it may guide us in selecting which hypothesis to consider and which to disregard. Therefore, the aesthetic factors can have an epistemic role *qua* aesthetic factors.

References

Avigad, Jeremy. 2008. Understanding proofs. In *The philosophy of mathematical practice*, ed. Paolo Mancosu, 317–353. Oxford: Oxford University Press.

Breitenbach, Angela. 2013. Aesthetics in science: A Kantian proposal. *Proceedings of the Aristotelian Society* 113 : 83–100.

Chace, Arnold Buffum, ed. 1927. *The Rhind mathematical papyrus*. Vol. 1. Oberlin: Mathematical Association of America.

Conway, John, and Richard Guy. 1996. *The book of numbers*. New York: Copernicus.

Davis, Philip J., and Reuben Hersh. 1981. *The mathematical experience*. Boston: Birkhäuser.

Dirac, Paul Adrien Maurice. 1951. A new classical theory of electrons. *Proceedings of the Royal Society of London*, Series A, 209: 291–296.

———. 1963. The evolution of the physicist's picture of nature. *Scientific American* 208 : 43–53.

Dyson, Freeman John. 1956. Obituary: Hermann Weyl. *Nature* 177 : 457–458.

———. 1992. *From Eros to Gaia*. New York: Pantheon Books.

Gowers, Timothy. 2002. *Mathematics: A very short introduction*. Oxford: Oxford University Press.

Halmos, Paul Richard. 1992. Mathematics. In *Academic Press dictionary of science and technology*, ed. Christopher Morris, 1325. San Diego: Academic.

Hardy, Godfrey Harold. 1992. *A mathematician's apology*. Cambridge: Cambridge University Press.

Heisenberg, Werner. 2001. Science and the beautiful. In *Quantum questions*, ed. Ken Wilber, 56–69. Boston: Shambhala.

Hersh, Reuben, and Vera John-Steiner. 2011. *Loving + hating mathematics: Challenging the myths of mathematical life*. Princeton: Princeton University Press.

Hofstadter, Douglas. 1999. *Gödel, Escher, Bach: An eternal golden braid*. New York: Basic Books.
Kant, Immanuel. 1996. *Practical philosophy*. Cambridge: Cambridge University Press.
———. 2000. *Critique of the power of judgment*. Cambridge: Cambridge University Press.
Kosso, Peter. 2002. The omniscienter: Beauty and scientific understanding. *International Studies in the Philosophy of Science* 16 : 39–48.
McAllister, James. 2005. Mathematical beauty and the evolution of the standards of mathematical proof. In *The visual mind II*, ed. Michele Emmer, 15–34. Cambridge: The MIT Press.
Mordell, Louis. 1959. *Reflections of a mathematician*. Montreal: Canadian Mathematical Congress, McGill University.
Nahin, Paul. 2006. *Dr. Euler's fabulous formula: Cures many mathematical ills*. Princeton: Princeton University Press.
Poincaré, Henri. 2013. *The foundations of science: Science and hypothesis – The value of science – Science and method*. Cambridge: Cambridge University Press.
Rav, Yehuda. 2005. Reflections on the proliferous growth of mathematical concepts and tools: Some case histories from mathematicians' workshops. In *Mathematical reasoning and heuristics*, ed. Carlo Cellucci and Donald Gillies, 49–69. London: College Publications.
Rota, Gian-Carlo. 1997. *Indiscrete thoughts*. Boston: Birkhäuser.
Schopenhauer, Arthur. 2010. *The world as will and representation*. Vol. 1. Cambridge: Cambridge University Press.
Sinclair, Nathalie. 2011. Aesthetic considerations in mathematics. *Journal of Humanistic Mathematics* 1 : 2–32.
Snow, Charles Percy. 1973. The classical mind. In *The physicist's conception of nature*, ed. Jagdish Mehra, 809–813. Dordrecht: Springer.
Todd, Cain. 2008. Unmasking the truth beneath the beauty: Why the supposed aesthetic judgments made in science may not be aesthetic at all. *International Studies in the Philosophy of Science* 22 : 61–79.
van Gerwen, Rob. 2011. Mathematical beauty and perceptual presence. *Philosophical Investigations* 34 : 249–267.
von Neumann, John. 1961. The mathematician. In John von Neumann, *Collected works*, vol. 1, 1–9. Oxford: Pergamon Press.
Weinberg, Steven. 1993. *Dreams of a final theory: The search for the fundamental laws of nature*. New York: Vintage.
Wells, David. 1990. Are these the most beautiful? *The Mathematical Intelligencer* 12 (3): 37–41.
Zangwill, Nick. 1998. Aesthetic/sensory dependence. *British Journal of Aesthetics* 38 : 66–81.
Zeki, Semir, John Romaya, Dionigi Benincasa, and Michael Atiyah. 2014. The experience of mathematical beauty and its neural correlates. *Frontiers in Human Neuroscience* 8 : 1–12.

Chapter 24
Mathematics and the World

Abstract This chapter discusses the relation of mathematics to the world considering two questions: What is the relation of mathematical objects to the world? Why is mathematics applicable to the world? As to the first question, the chapter maintains that mathematical objects are not obtained by abstraction from sensible things, or by idealization from our operations of collecting objects. They are hypotheses we make to solve mathematical problems by the analytic method, several of which have an extra-mathematical origin. As to the second question, the chapter maintains, on the one hand, that the applicability of natural mathematics to the world is due to the fact that natural mathematics fits in certain mathematical properties of the world. On the other hand, the applicability of artificial mathematics to the world is due to several factors, starting with the decision of modern science to confine itself to dealing only with some phenomenal properties of the world, mathematical in kind.

24.1 Mathematical Objects and Abstraction

So far we have not specifically discussed the relation of mathematics to the world. This is the aim of this chapter. There are several questions about the relation of mathematics to the world, but the most important ones are: What is the relation of mathematical objects to the world? Why is mathematics applicable to the world?

We first consider the question of the relation of mathematical objects to the world. There are several views of this relation. A first view is that mathematical objects are obtained by abstraction from sensible things, where abstraction is the process by which one considers certain properties of sensible things, leaving out all the other ones. According to this view, although mathematical objects are thought of as separate from sensible things, they are not really separate from them, and exist only in sensible things.

Thus, Aristotle states that "the mathematician carries out his investigation about things reached by abstraction. For, he investigates things leaving out all sensible qualities, such as weight, lightness, hardness," and "considering only quantity and continuity, in one, two or three dimensions, and the properties of these as quantitative and continuous; and he does not consider them in any other respect" (Aristotle, *Metaphysica*, K 3, 1061 a 28–35). The mathematician "thinks of mathematical objects, which are not separate, as being separate, when he thinks of them" (Aristotle,

© Springer International Publishing AG 2017
C. Cellucci, *Rethinking Knowledge*, European Studies in Philosophy of Science 4,
DOI 10.1007/978-3-319-53237-0_24

De Anima, Γ 7, 431 b 15–16). But "since, as it appears, there is nothing that exists as separate from the sensible magnitudes, it is in the sensible forms that" one finds the things "that are spoken of in abstraction" (ibid., Γ 8, 432 a 3–6).

The view that mathematical objects are obtained by abstraction from sensible things, however, is problematic. As Kant points out, "abstraction is only the negative condition under which universal representations can be generated" (Kant 1992, 593). Through it, "nothing is produced, but rather left out" (ibid., 487). In fact, "abstraction does not add anything," but "rather cuts off everything that does not belong to the concept" (ibid., 351).

Since abstraction does not add anything, by abstraction from sensible things we only obtain objects which exist in sensible things, so we do not obtain several kinds of mathematical objects. An obvious example are the objects of transfinite set theory, which clearly do not exist in sensible things.

Moreover, by abstraction from sensible things we do not obtain objects having that exactness which is proper to mathematical objects. For example, from an edge of a table, we do not obtain a straight line, but only a line which is slightly zigzag and with some, however small, breadth.

Furthermore, by abstraction from sensible things we do not necessarily obtain mathematical objects. As Frege points out, if, "in considering a white cat and a black cat, I disregard the properties which serve to distinguish them," then I do not obtain the concept 'two' but rather "the concept 'cat'" (Frege 1960, 45). The "concept 'cat', no doubt, which we have arrived at by abstraction, no longer includes the special characteristics of either, but of it, for just that reason, there is only one" (ibid., 45–46).

That abstraction is a purely negative condition prompts Frege's ironical comment that abstraction "is especially effective. We attend less to a property, and it disappears. By thus making one characteristic mark after another disappear, we obtain more and more abstract concepts," so "inattention is a most effective logical power; this is presumably why professors are absent-minded" (Frege 1984, 197). By continued application of abstraction, "each object is transformed into a more and more bloodless phantom," and ultimately "we obtain from each object a something emptied of all content; but the something obtained from one object differs nevertheless from the something obtained from another object, even though it is not easy to say how" (ibid., 198).

This confirms that the view that mathematical objects are obtained by abstraction from sensible things is problematic.

24.2 Mathematical Objects and Idealization

Another view of the relation of mathematical objects to the world is that mathematical objects are obtained by idealization from our operations of collecting objects. Or, personifying the idealization, mathematical objects are obtained by the ideal operations of collecting of an ideal subject who is an idealization of ourselves.

Thus Kitcher states that mathematical objects are obtained by idealization from the "operations of collecting" which "we are able to perform with respect to any objects" (Kitcher 1983, 12). Or, "we may personify the idealization, by thinking" of mathematical objects as obtained by the ideal operations "of an ideal subject, whose status as an ideal subject resides in her freedom from certain accidental limitations imposed on us" (ibid., 109). To talk of an ideal subject "is not to suppose that there is a mysterious being with superhuman powers" (ibid., 110). Simply, "the ideal subject is an idealization of ourselves" (ibid., 111).

The view that mathematical objects are obtained by idealization from our operations of collecting objects, however, is problematic. For example, let us consider the iterative hierarchy of sets, which is formed in successive stages, starting from the integers, by transfinite iteration of the power-set operation. Kitcher states that we may think of the iterative hierarchy of sets as obtained "from the iterated performance of one fundamental operation, that of collecting," by "the ideal mathematical subject" (ibid., 133). But, since the iterative hierarchy of sets is highly non-denumerable, this would require that the succession of stages be a succession in a highly non-denumerable supertime. Kitcher claims that "the view of the ideal subject as an idealization of ourselves does not lapse when we release the subject from the constraints of our time" (ibid., 147). This, however, conflicts with the fact that a highly non-denumerable supertime cannot be viewed as an idealization of time, since it has a completely different structure. Therefore, the idea of an ideal subject performing superoperations in a highly non-denumerable supertime can hardly be considered an explanation of the relation of mathematical objects to the world.

Indeed, as Parsons points out, the mind of the ideal mathematical subject would "differ not only from finite minds but also from the divine mind as conceived in philosophical theology, for either the latter is thought of as in time, and therefore as doing things in an order with the same structure as that in which finite beings operate, or its eternity is interpreted as complete liberation from succession" (Parsons 1983, 273). Then, contrary to Kitcher's claim, the ideal mathematical subject is not an idealization of ourselves, but really a mysterious being with superhuman powers.

This confirms that the view that mathematical objects are obtained by idealization from our operations of collecting objects is problematic.

24.3 Mathematical Objects and Hypotheses

As argued in Chap. 19, rather than being obtained by abstraction from sensible things or by idealization from our operations of collecting objects, mathematical objects are hypotheses we make to solve mathematical problems by the analytic method.

This does not mean that mathematical objects have no relation to the world, but only that they are not obtained by abstraction from sensible things, or by idealization from our operations of collecting objects. As all other hypotheses, they are introduced to solve problems, several of which have an extra-mathematical origin.

For example, as already mentioned in Chap. 19, Menaechmus introduced conic sections as hypotheses to solve the problem of doubling the cube, a problem which had an extra-mathematical origin. According to a legend, it originated from the fact that, "when the god announced" to the inhabitants of the island of Delos, "through the oracle, that to get rid of a plague" which affected them, "they should construct an altar double of the one that existed," which had a form of a cube, "their craftsmen fell into great perplexity in trying to find how a solid could be made double of another solid, and they went to ask Plato about it. He told them that the god had given this oracle" because he wished "to reproach the Greeks for their neglect of mathematics and their contempt for geometry" (Theon of Smyrna, *Expositio Rerum Mathematicarum ad Legendum Platonem Utilium*, 2.4–12). In fact, some mathematicians of Plato's Academy applied themselves to the problem, and Menaechmus was one of them.

That several mathematical problems have an extra-mathematical origin is essential for the fruitfulness of mathematics. As von Neumann states, "mathematical ideas originate in empirics" and, once they are so conceived, mathematics "begins to live a peculiar life of its own" (von Neumann 1961, 9). Conversely, if mathematics "travels far from its empirical source, or still more, if it is a second and third generation only indirectly inspired by ideas coming from 'reality'," it is "in danger of degeneration" (ibid.).

24.4 Mathematics and Applicability

After considering the question of the relation of mathematical objects to the world, we now consider the question of the applicability of mathematics to the world. From its very beginning, mathematics has been held to be applicable to the world. Already the Rhind Mathematical Papyrus asserts that mathematics provides "accurate reckoning. The entrance into the knowledge of all existing things and all obscure secrets" (Chace 1927, 49).

But why is mathematics applicable to the world? This question was put on the philosophical agenda by the Pythagoreans, it remained on the philosophical agenda up to the end of the nineteenth century, but in the first half of the twentieth century disappeared, because of the foundational interest that dominated the philosophy of mathematics in that period. An example of this is the standard Benacerraf and Putnam anthology in the philosophy of mathematics, which included no article on the applicability of mathematics to the world, and the "lack of material was the reason" (Steiner 1998, 14, footnote 8).

In 1960 Wigner gave a sort of theoretical justification for the lack of material, by advancing the thesis of "the unreasonable effectiveness of mathematics in natural science" (Wigner 1960, 1). According to Wigner, the "enormous usefulness of mathematics in the natural sciences is something bordering on the mysterious," and "there is no rational explanation for it" (ibid., 2). It is "difficult to avoid the impression that a miracle confronts us here" (ibid., 7). The "miracle of the appropriateness of the language of mathematics for the formulation of the laws of physics is a wonderful gift which we neither understand nor deserve. We should be grateful for it and hope that it will remain valid in future research" (ibid., 14).

This theoretical justification, however, is hardly satisfactory, because it is simply a declaration of incapability to provide a rational explanation. Despite Wigner's claims, in what follows we try to give an answer to the question of the applicability of mathematics to the world, starting from the distinction between natural and artificial mathematics, made in Chap. 18.

24.5 Natural Mathematics and the World

In the case of natural mathematics, an account of the applicability of mathematics to the world is straightforward. Natural mathematics is applicable to the world because the systems of core knowledge of which it consists, being a result of biological evolution, fit in certain mathematical properties of the world. They fit in such properties much in the same sense in which the movements of the fishes and the shape of their fins fit in certain hydrodynamic properties of water, or the eyes fit in certain physical properties of light. If the systems of core knowledge in question did not fit in certain mathematical properties of the world, we could not survive.

That the systems of core knowledge of which natural mathematics consists fit in certain mathematical properties of the world, is no more a miracle than we can breathe because the earth has oxygen – just what we need to breathe. Saying that it is a miracle that we can breathe because the earth has oxygen, would be to forget that the human species evolved to breathe what is here, not the other way around. Much in the same way, saying that it is a miracle that the systems of core knowledge fit in certain mathematical properties of the world, would be to forget that the human species evolved to survive in a world having those mathematical properties, not the other way around.

24.6 Artificial Mathematics and the World

In the case of artificial mathematics, an account of the applicability of mathematics to the world is not so straightforward.

Since antiquity, several accounts have been given of the applicability of artificial mathematics to the world. In Chap. 7 we have already discussed one such account and found it wanting: the account put forward by mathematical structural realism. In what follows, we discuss some other accounts and find them also wanting: the theistic account, the parallelism account, the friendly universe account, and the mapping account. Then we propose an alternative account.

Since the question of the applicability of natural mathematics to the world has already been considered above, in what follows 'mathematics' will stand for 'artificial mathematics', unless stated otherwise.

24.7 The Theistic Account

An account of the applicability of mathematics to the world that has had large following since antiquity is the theistic account, according to which mathematics is applicable to the world because the world has been formed by a mathematician-God. Thus Plato states that, when God undertook the work of setting in order the universe, he gave shape to fire and water and earth "by means of forms and numbers" (Plato, *Timaeus*, 53 B 4–5).

The theistic account has been reasserted several times since Plato. Thus Kepler states that "God, in creating the universe and regulating the order of the cosmos, had in view the five regular solids of geometry, as known since the days of Pythagoras and Plato, and fixed, according to those dimensions, the number of heavens, their proportions, and the relations of their movements" (Kepler 1937–, I, 9).

Leibniz states that, "at the very origin of things there is a certain divine mathematics or metaphysical mechanism" (Leibniz 1965, VII, 304). Already the ancients said that "God made everything by weight, measure, number" (ibid., VII, 184). Indeed, "when God calculates and exercises his thought, the world is made" (ibid., VII, 191, footnote).

Dirac states that "God is a mathematician of a very high order, and he used very advanced mathematics in constructing the universe" (Dirac 1963, 53).

The theistic account, however, essentially depends on the assumption that there exists a creator-God. This assumption is an act of faith, so the theistic account can hardly be considered a rational account of the applicability of mathematics to the world.

24.8 The Parallelism Account

Another account of the applicability of mathematics to the world is Hilbert's parallelism account, according to which mathematics is applicable to the world because there is a "parallelism between nature and thought" (Hilbert 1996f, 1160). According to Hilbert, the evidence for this is that several mathematical theories, originally formulated for some other purpose, find application to the world. The "older examples for this are conic sections, which one studied long before one suspected that our planets or even electrons move in such a course" (ibid.). This fact may be called, "in a sense different from that in Leibniz, pre-established harmony" (ibid.). Only by admitting that there is a parallelism between nature and thought, we can account for "the numerous and surprising analogies and that apparently harmony which the mathematician so often perceives" (Hilbert 2000, 243).

The parallelism account, however, is inadequate because, that several mathematical theories, originally formulated for some other purpose, find application to nature, is not evidence for a parallelism between nature and thought, on the contrary, it is the phenomenon to be explained. Moreover, the parallelism account does not explain why only certain parts of mathematics have applications to nature. Furthermore, the parallelism account implies that the mathematics used to formulate a physical theory should have a physical counterpart. This conflicts with the fact that mathematics often uses concepts and results, such as the concepts and results of transfinite set theory, to which no physical meaning can be assigned. Also, the parallelism account implies that we can advance in the knowledge of nature only using correct mathematics. This conflicts with the fact that, from the seventeenth to the nineteenth century, the developments in physics took place using incorrect mathematics, Newton's and Leibniz's infinitesimal calculus, which was based on inconsistent principles. In addition, the parallelism account implies that the mathematics used to treat certain physical phenomena should be uniquely determined by them. This contrasts with the fact that often different mathematical theories can be used to treat the same physical phenomena. For example, we can treat quantum mechanics using either group theory or Hilbert space theory. We can treat special relativity using either the tensor calculus or the vector calculus. We can treat optics using either geometry or the matrix calculus.

24.9 The Friendly Universe Account

Another account of the applicability of mathematics to the world is Steiner's friendly universe account, according to which mathematics is applicable to the world because there is a "true 'correspondence'" between "the human brain and the physical world as a whole" (Steiner 1998, 176). Therefore, the universe is "an intellectually 'user friendly' universe, a universe which allows our species to discover things about it" (ibid., 8). This has been "a necessary factor (not the only factor) in discovering today's fundamental physics" (ibid.). That there is a true correspondence between the human brain and the physical world as a whole depends on the fact that "the human race is in some way privileged, central to the scheme of things" (ibid., 55). It means that "Somebody cares enough about us to put us in a special place" (ibid., 56, footnote 20). This is "consistent with natural theology" (ibid., 10). Moreover, it shows "the importance of the enterprise of scientific inquiry from a religious point of view" (ibid., 10).

The friendly universe account, however, is inadequate because it is a combination of the theistic account and Hilbert's parallelism explanation, and inherits their limits. Indeed, on the one hand, like the theistic account, the friendly universe account essentially depends on the assumption that there exists Somebody who put us in a special place, and this assumption is an act of faith. On the other hand, like the parallelism account, the friendly universe account depends on the assumption that there is a true correspondence between the human brain and the physical world as a whole, but this assumption is begging the question. For, it is exactly what is in question.

24.10 The Mapping Account

Another account of the applicability of mathematics to the world is Bueno and Colyvan's mapping account, according to which mathematics is applicable to the world because we can "establish certain mappings between the empirical set up and appropriate mathematical structures" (Bueno and Colyvan 2011, 353). Specifically, first we establish "a mapping from the empirical setup to a convenient mathematical structure. We call this step immersion" (ibid.). Then we draw mathematical "consequences from the mathematical formalism, using the mathematical structure obtained in the immersion step. We call this step derivation" (ibid.). Finally, we interpret the mathematical consequences thus drawn "in terms of the initial empirical setup. We call this step interpretation" (ibid., 353–354).

The mapping account, however, is inadequate because the immersion and the interpretation mappings are mathematical objects, hence so must be their domain and codomain. This contrasts with the fact that the empirical setup is not a mathematical object. Bueno and Colyvan think of solving this problem by saying that the immersion step must map a mathematically represented empirical setup to a mathematical object, because "it will be hard to even talk about the empirical setup in question without leaning heavily on the mathematical structure, prior to the immersion step" (ibid., 354). But then the immersion step is a mapping, not from the world, but from a mathematical representation of the world, to a mathematical object. Similarly, the interpretation step is a mapping from a mathematical object, not to the world, but to a mathematical representation of the world.

Bueno and French claim that, with the mapping account, "mathematics is applied by bringing structure from a mathematical domain" into "a physical, but mathematized, domain" (Bueno and French 2012, 88). Then, however, the immersion and the interpretation mappings do not account for the applicability of mathematics to the world, but only for the applicability of mathematics to an already mathematized domain, a mathematical representation of the world.

24.11 Galileo's Philosophical Revolution and Mathematics

Given the inadequacy of all the above accounts of the applicability of mathematics to the world, two alternatives are possible.

One may agree with Wigner that the unreasonable effectiveness of mathematics in natural science is something bordering on the mysterious, and there is no rational explanation for it.

Thus, Hamming asks: "Is it not remarkable that 6 sheep plus 7 sheep make 13 sheep; that 6 stones plus 7 stones make 13 stones? Is it not a miracle that the universe is so constructed that such a simple abstraction as a number is possible? To me this is one of the strongest examples of the unreasonable effectiveness of mathematics. Indeed, I find it both strange and unexplainable" (Hamming 1980, 84).

However, as stated above, this view seems hardly satisfactory, because it is simply a declaration of incapability to provide a rational explanation.

Alternatively, one may try to give an account of the applicability of mathematics to the world, more satisfactory than those considered above. To this aim, it must be noted that, since mathematics is a result of cultural evolution, which is complex, the applicability of mathematics to the world cannot be expected to be due to a single factor, it may be due to several factors. To investigate these factors, it is useful to ask: Why was there no essential use of mathematics in Greek science, while its use is so essential in modern science?

Clearly, this must be related to Galileo's philosophical revolution from which modern science originated: the decision to renounce Aristotle's aim to penetrate the essence of natural substances, dealing only with some of their phenomenal properties which are mathematical in kind, such as location, motion, shape, or size (see Chap. 8). As long as, following Aristotle, science aimed at penetrating the essence of natural substances, mathematics had no chance of being applicable to the natural world, because essence is not mathematical in kind. For this reason Aristotle stated that the method of mathematics is not well suited to physics. Only with Galileo's philosophical revolution, a mathematical treatment of nature became possible and even necessary.

That, with Galileo's philosophical revolution, a mathematical treatment of nature became possible and even necessary, shows that it is unwarranted to claim, as Wigner does, that the enormous usefulness of mathematics in the natural sciences is something bordering on the mysterious, and there is no rational explanation for it. Wigner's claimWigner's claim is as unwarranted as that of the ichthyologist, described by Eddington, who "casts a net into the water and brings up a fishy assortment," then, "surveying his catch," concludes that "no sea-creature is less than two inches long" (Eddington 1939, 16). Like Wigner, the ichthyologist could say that this is something bordering on the mysterious and that there is no rational explanation for it. But the ichthyologist's conclusion, that no sea-creature is less than two inches long, only holds of his catch, it depends on the structure of his net. In fact, there are sea-creatures less than two inches long, but the ichthyologist's net is not adapted to catch them.

A main reason of the enormous usefulness of mathematics in the natural sciences is then that, as a result of Galileo's philosophical revolution, modern science contents itself with dealing with some phenomenal properties of the world which are mathematical in kind. This means, however, that just as the ichthyologist's net is adapted only to catch sea-creatures not less than two inches long, mathematics is adapted to catch only those aspects of the world which are mathematical in kind. For this reason, Russell states that "physics is mathematical not because we know so much about the physical world, but because we know so little; it is only its mathematical properties that we can discover" (Russell 1995a, 125).

Anyhow, the decision to renounce Aristotle's aim to penetrate the essence of natural substances, dealing only with some of their phenomenal properties mathematical in kind, is a main factor of the applicability of mathematics to the world.

24.12 The Fusion of Mathematics and Physics

To investigate the factors of the applicability of mathematics to the world, it is also useful to observe that, until the nineteenth century, there was a fusion of mathematics and physics. This means that, until then, the question was not that of the applicability of mathematics to the world, but rather that of the validity of scientific theories as a whole.

The separation of mathematics and physics began in the nineteenth century. As Kline says, "the Greeks, Descartes, Newton, Euler, and many other believed mathematics to be the accurate description of real phenomena" and "regarded their work as the uncovering of the mathematical design of the universe" (Kline 1972, 1028). But, in the nineteenth century, "mathematicians began to introduce concepts that had little or no direct physical meaning" (ibid., 1029). So "mathematics was progressing beyond concepts suggested by experience," and was no longer "a reading of nature" (ibid., 1030). Then, "after about 1850, the view that mathematics can introduce and deal with rather arbitrary concepts and theories that do not have immediate physical interpretation" gained "acceptance," even geometry "cut its bonds to physical reality" (ibid., 1031). Finally, in the twentieth century, notwithstanding some single exception such as von Neumann or Hermann Weyl, the separation of mathematics and physics was fully accomplished.

At the origin of the separation of mathematics and physics there was the emergence of the formal axiomatic method (see Chap. 12) and the rise of the axiomatic ideology (see Chap. 20).

Thus Bourbaki states that, before the twentieth century "a great deal of effort was spent on trying to derive mathematics from experimental truths," but "the axiomatic method has shown that the 'truths' from which it was hoped to develop mathematics, were but special aspects of general concepts" (Bourbaki 1950, 231). This gave rise to the separation of mathematics and physics, and to the development of "monster-structures, entirely without applications," whose "sole merit was that of showing the exact bearing of each axiom, by observing what happened if one omitted or changed it" (ibid., 230, footnote). The separation of mathematics and physics led to conclude that the "intimate connection" between mathematics and physics, "of which we were asked to admire the harmonious inner necessity, was nothing more than a fortuitous contact of two disciplines," and that "we are completely ignorant as to the underlying reasons for" the connection between mathematics and physics, and "we shall perhaps always remain ignorant of them" (ibid., 231).

Anyway, at least until the nineteenth century, the fusion of mathematics and physics is another factor of the applicability of mathematics to the world.

24.13 Limits to the Applicability of Mathematics

Wigner's claim of the enormous usefulness of mathematics in the natural sciences needs qualification. Several people believe that, although mathematics is adapted to catch only those aspects of the world which are mathematical in kind, mathematics is adapted to catch all those aspects. The following two examples, however, show that this belief is unjustified.

1) *Deterministic chaos.* A very simple physical system like the solar system is deterministic, in the sense that each state is completely determined by the previous one. But even a small error in the measurement of the initial state is multiplied several times at each next state, until, from a certain point onwards, the behaviour of the

system becomes absolutely unpredictable. Since the initial state of the system can be known with a precision that, however great, is always limited to a finite number of decimals, from a certain point onwards mathematics will be unable to predict the state of the system. For this reason, mathematics is unable to accurately calculate the orbits of planets beyond a certain time. For example, an error of less than 15 metres in the measurement of the earth's position would make it impossible to predict the orbit of the earth 100 million years into the future.

2) *Renormalization*. In quantum field theory, and also elsewhere, renormalization is a technique for dealing with infinities arising in the calculation of physical quantities. It is, however, a mathematically incorrect technique. Thus, Dirac states that renormalization involves "neglecting infinities which appear in its equations, neglecting them in an arbitrary way. This is just not sensible mathematics. Sensible mathematics involves neglecting a quantity when it is small – not neglecting it just because it is infinitely great and you do not want it!" (Dirac 1978, 36). And yet, thanks to renormalization, in quantum electrodynamics the agreement between theory and experiment achieves a very high accuracy. For example, in the calculation of the magnetic moment of the electron, it has a precision of four parts in thousand billions. This means that, if we were to measure the distance from Los Angeles to New York to this accuracy, our measure would be exact to the thickness of a human hair. No correct mathematics is known capable of dealing as successfully with quantum electrodynamics phenomena.

These examples, and others that could be made, show that there are phenomena, mathematical in kind, which mathematics cannot catch. These phenomena range from a ball bouncing in a roulette wheel or the smoke rising from a cigarette, to the orbits of the planets in the solar system or the evolution of the whole universe.

As Steiner points out, there are "instances in which scientists fail to find appropriate mathematical descriptions of natural phenomena," and such instances "outnumber the successes by far" (Steiner 1998, 9). Then, rather than saying that mathematics is unreasonably effective in natural science, we should rather say that mathematics is reasonably ineffective in natural science. (That mathematics is reasonably ineffective in natural science is also maintained in Abbott 2013, Grattan-Guinness 2008).

24.14 Mathematics and Simplicity

The reason why there are properties of the world, mathematical in kind, which mathematics cannot catch, is that mathematics is able to deal successfully only with the simplest properties of the world which are mathematical in kind. As Schwartz states, mathematics "is able to deal successfully only with the simplest of situations, more precisely, with a complex situation only to the extent that rare good fortune makes this complex situation hinge upon a few dominant simple factors" (Schwartz 1992, 21–22).

Since mathematics is able to deal successfully only with the simplest of situations, with understandable opportunism scientists select such situations. Namely, they select what problems to work on based on the fact that those problems are simple enough to be amenable to a mathematical treatment. In other words, they select what problems to work on largely for their ability to allow mathematical treatment.

But this does not cancel the fact that there are several complex situations which do not hinge upon a few dominant simple factors, and which scientists are unable to treat by mathematics. Confronted with these situations, scientists leave mathematical models aside and take refuge in empirical modelling, namely, in computer modelling based on empirical observations, rather than on mathematically describable relationships of the system being modelled.

In particular, this is the case of biological and cognitive systems, which consist of a huge number of parts that do not interact linearly, and whose macroscopic, observable parts, behave in ways not derivable from simple microlevel principles. This makes it very difficult to deal with such systems by means of mathematical models. But difficulties also arise with certain physical phenomena, such as superconductivity, convection of complex fluids, or oscillations in reaction-diffusion systems (see, for example, Hooker 2011).

Balaguer claims that "the mathematical realm is so robust that it provides an apparatus for all situations. That is, no matter how the physical world worked, there would be a mathematical theory that truly described part of the mathematical realm and that could be used to help us do empirical science" (Balaguer 1998, 143). Or "perhaps this is a bit strong," but "we don't need anything nearly as strong as the claim about all physical setups in order to" account for "the applicability of mathematics. All we need is this: for most physical setups, there is a mathematical apparatus that could be used to help us do empirical science" (ibid., 202, footnote 17).

But, since there are several complex situations which do not hinge upon a few dominant simple factors and which scientists are unable to treat by mathematics, it is hardly justified even to claim that, for most physical setups, there is a mathematical apparatus that could be used to help us do empirical science.

That scientists select what problems to work on largely for their ability to allow mathematical treatment, is another factor of the applicability of mathematics to the world.

24.15 Mathematics and Simplification

Even when mathematics is applicable to the world, the mathematics required cannot be expected to be found ready made in textbooks or in mathematics journals, but must be adapted to the phenomena under investigation.

To this aim, the phenomena must be simplified. For example, particles must be treated as point masses, molecules as perfectly elastic and spherical, and surfaces as frictionless. Also, some aspects of the phenomena under investigation must be ignored before mathematics can even begin to be applied to them.

For this reason, as we have seen in Chap. 21, Hilbert states that the application of a theory to the world of appearances always requires a certain measure of good will and tactfulness. For example, it requires "that we substitute the smallest possible bodies for points and the longest possible ones, e.g. light-rays, for lines. We also must not be too exact in testing the propositions, for these are only theoretical propositions" (Hilbert 1980a, 41).

Moreover, only a small part of mathematics finds an application to the world. Some mathematicians even take pride of this. Thus Hardy states that what finds an application are only "the dull and elementary parts of applied mathematics" and "the dull and elementary parts of pure mathematics" (Hardy 1992, 132). Dieudonné states that "of all the striking progress" in recent mathematics, almost "not a single one, with the possible exception of distribution theory, had anything to do with physical applications; and even in the theory of partial differential equations, the emphasis is now much more on 'internal' and structural problems than on questions having a direct physical significance" (Dieudonné 1964, 248).

Furthermore, often what finds application are not deep theorems. For example, what is used in quantum theory is not any deep theorem about Hilbert space, but only the definition of a Hilbert space and little more. What is used in the so-called Grand Unified Theory is only some fairly elementary group theory. Admittedly, some advanced mathematics is used in areas of physics such as string theory, but those areas are still mostly empirically unconfirmed. In addition, the mathematics used in such areas has been developed to a significant extent by physicists themselves. Even Wigner admits that "the concepts which were chosen were not selected arbitrarily from a listing of mathematical terms but were developed, in many if not most cases, independently by the physicist and recognized then as having been conceived before by the mathematician" (Wigner 1960, 7).

In addition, sometimes what finds application is some non-rigorous mathematics. An example of this is renormalization, which has been already mentioned above.

That, even when mathematics is applicable to the world, the mathematics required cannot be expected to be found ready made in textbooks or in mathematics journals but must be adapted to the the phenomena under investigation, and what finds application are not deep theorems, and even non-rigorous mathematics, is another factor of the applicability of mathematics to the world.

24.16 Mathematics and Approximations

In applying mathematics to the world, allowance must always be made for the fact that all physical laws are only approximations to reality.

Thus Galileo states that, "just as an accountant must make allowances for containers, straps, and other packing items if he wants his calculations to come out right in regard to sugar, silk, and wool, so, when the geometrical philosopher wants to find in the concrete the effects demonstrated in the abstract, he must deduct the impediments of matter, if he knows how to do this" (Galilei 1968, VII, 234). Errors arise when the geometrical philosopher does not deduct the impediments of matter correctly, so errors "derive neither from the abstract nor from the concrete, neither from geometry nor from physics, but from the calculator, who does not know how to do the accounting correctly" (ibid.).

Since all physical laws are only approximations to reality, one might always find some other mathematical means that provide better approximations. All approximations, however, while sufficient for certain purposes, will not be adequate generally, since they will not permit long-range predictions. For example, in the case of Kepler's laws, the orbits of the planets are only roughly elliptic, and are not easily calculated due to the confusing influence of Keplerian, Newtonian, Einsteinian, thermodynamic, and chaotic factors. Even when they can be calculated, the calculations permit accurate predictions only over a comparatively short period of time. This is sufficient for certain purposes, but the calculations do not permit accurate predictions over time scales of millions of years.

That, in applying mathematics to the world, allowance must always be made for the fact that all physical laws are only approximations to reality, is another factor of the applicability of mathematics to the world.

24.17 The Applicability of Mathematics and Evolution

A prerequisite for the applicability of mathematics to the world is the ability of human beings to develop a mathematics applicable to the world.

This prerequisite can be satisfied since, as Atiyah points out, "if one views the brain in its evolutionary context, then the mysterious success of mathematics in the physical sciences is at last partially explained. The brain evolved in order to deal with the physical world, so it should not be too surprising that it has developed a language, mathematics, that is well suited for the purpose" (Atiyah 1995).

As already stated in Chap. 18, mathematics is the product of brains that are a result of biological evolution, and is based on them. The mathematical abilities with which biological evolution has endowed us and which are the foundation of natural mathematics are also, indirectly, the foundation of artificial mathematics, since the latter is developed on the basis of those abilities. Moreover, artificial mathematics is based on the same method as natural mathematics, namely the analytic method.

Even the role of the role of the sense of beauty in selecting hypotheses within the analytic method is a result of biological evolution, because our sense of beauty is shaped by evolution, and hence is sensitive to the environment. Pincock even maintains that "our tendency to make aesthetic judgments is an adaptation precisely because these judgments track objective features of our environment" (Pincock 2012, 184–185).

That the brain evolved in order to deal with the physical world and has developed a mathematics well suited for the purpose, is another factor of the applicability of mathematics to the world.

24.18 Explaining the Pre-established Harmony

A question related to that of the applicability of mathematics to the world is Hilbert's pre-established harmony question: How could it be that mathematical theories, originally formulated for some other purpose, find application to the world? Several people believe that, like the question of the applicability of mathematics to the world, Hilbert's pre-established harmony question is something bordering on the mysterious, and that there is no rational explanation for it.

Thus Weinberg states that "it is very strange that mathematicians are led by their sense of mathematical beauty to develop formal structures" that later physicists find "useful, even where the mathematician had no such goal in mind" (Weinberg 1993, 125). It is "positively spooky how the physicist finds the mathematician has been there before him or her" (Weinberg 1986, 725). It is "as if Neil Armstrong in 1969 when he first set foot on the surface of the moon had found in the lunar dust the footsteps of Jules Verne" (Weinberg 1993, 125).

Dyson states that "one of the central themes of science" is "the mysterious power of mathematical concepts to prepare the ground for physical discoveries which could not have been foreseen or even imagined by the mathematicians who gave the concepts birth" (Dyson 1999, ix).

Bourbaki states that "mathematics appears" as "a storehouse of abstract forms – the mathematical structures; and it so happens – without our knowing why – that certain aspects of empirical reality fit themselves into these forms, as if through a kind of pre-adaptation" (Bourbaki 1950, 231). One may ask why "some of the most intricate theories in mathematics become an indispensable tool to the modem physicist," but "fortunately for us, the mathematician does not feel called upon to answer such questions" (Bourbaki 1949, 2).

Against the belief that Hilbert's pre-established harmony question is something bordering on the mysterious and that there is no rational explanation for it, Bangu argues that to "wonder how could it be that a mathematician was 'there' first is like wondering how could it be that all the people we find in a hospital are sick" (Bangu 2016, 28). For, it is "a matter of sociological/professional scientific fact" that, when a scientist has a major insight, "such a scientist has to embed his insight into a mathematical formalism," otherwise "the theory is extremely unlikely to draw any interest from the scientific community" (ibid., 27). Therefore, "there can't be any cases of (major) achievements in physics in which the mathematician hasn't been 'there' before. The very fact that there is an achievement" recognized as such "is already a guarantee that there was a mathematician 'there' first" (ibid., 28). So, "there is no anticipation, no pre-adaptation, no pre-established harmony, no miracle – only a filtering ('ecological') effect operating in the scientific environment" (ibid., 27).

Bangu's argument, however, does not explain why, as a matter of sociological/professional scientific fact, the scientific community is unlikely to have any interest in physical theories not embedded into mathematical formalisms. What is more important, it does not explain how could it be that mathematical theories, originally formulated for some other purpose, find application to the world. For, Bangu's argument would be perfectly compatible with the assumption that the mathematical formalisms into which physical theories are embedded are developed to that purpose. Indeed, such was the case of the infinitesimal calculus, which was first developed for the purposes of physics and astronomy in the seventeenth century.

A better explanation is implicit in the above analysis of the factors of the applicability of mathematics to the world, in particular that, until the nineteenth century, there was a fusion of mathematics and physics; that scientists select what problems to work on largely for their ability to allow mathematical treatment by means of mathematical theories already available; that the mathematics required cannot be expected to be found ready made in textbooks or in mathematics journals, but must be adapted to the phenomena under investigation, and what finds application are not deep theorems, and even non-rigorous mathematics; that all physical laws are only approximations to reality, so one might always find some other mathematical means that provide better approximations, and in any case physical laws do not permit long-range predictions.

References

Abbott, Derek. 2013. The reasonable ineffectiveness of mathematics. *Proceedings of the IEEE* 101: 2147–2153.

Atiyah, Michael. 1995. Creation v. discovery. *Times Higher Education Supplement*, 29 September.

Balaguer, Mark. 1998. *Platonism and anti-platonism in mathematics*. Oxford: Oxford University Press.

Bangu, Sorin. 2016. On 'The unreasonable effectiveness of mathematics in the natural sciences'. In *Models and inferences in science*, ed. Emiliano Ippoliti, Fabio Sterpetti, and Thomas Nickles, 11–29. Cham: Springer.

Bourbaki, Nicholas. 1949. Foundations of mathematics for the working mathematician. *The Journal of Symbolic Logic* 14 : 1–8.

———. 1950. The architecture of mathematics. *The American Mathematical Monthly* 57 : 221–232.

Bueno, Otávio, and Mark Colyvan. 2011. An inferential conception of the application of mathematics. *Noûs* 45 : 345–374.

Bueno, Otávio, and Steven French. 2012. Can mathematics explain physical phenomena? *The British Journal for the Philosophy of Science* 63 : 85–113.

Chace, Arnold Buffum, ed. 1927. *The Rhind mathematical papyrus*. Vol. 1. Oberlin: Mathematical Association of America.

Dieudonné, Jean. 1964. Recent developments in mathematics. *The American Mathematical Monthly* 71 : 239–248.

Dirac, Paul Adrien Maurice. 1963. The evolution of the physicist's picture of nature. *Scientific American* 208 : 43–53.

———. 1978. *Directions in physics*. New York: Wiley.

Dyson, Freeman John. 1999. Foreword. In Michael Monastyrsky, *Riemann, topology, and physics*, ix–ix, Boston: Birkhäuser.

Eddington, Arthur. 1939. *The philosophy of physical science*. Cambridge: Cambridge University Press.

Frege, Gottlob. 1960. *The foundations of arithmetic: A logico-mathematical enquiry into the concept of number*. New York: Harper.

———. 1984. *Collected papers on mathematics, logic, and philosophy*. Oxford: Blackwell.

Galilei, Galileo. 1968. *Opere*. Florence: Barbera.
Grattan-Guinness, Ivor. 2008. Solving Wigner's mystery: The reasonable (though perhaps limited) effectiveness of mathematics in the natural sciences. *The Mathematical Intelligencer* 30 (3): 7–17.
Hamming, Richard. 1980. The unreasonable effectiveness of mathematics. *The American Mathematical Monthly* 87 : 81–90.
Hardy, Godfrey Harold. 1992. *A mathematician's apology*. Cambridge: Cambridge University Press.
Hilbert, David. 1980a. Letter to Frege, 29 December 1899. In Gottlob Frege, *Philosophical and mathematical correspondence*, 38–41. Oxford: Blackwell.
———. 1996f. Logic and the knowledge of nature. In *From Kant to Hilbert: A source book in the foundations of mathematics*, II, ed. William Ewald, 1157–1165. Oxford: Oxford University Press.
———. 2000. Mathematical problems. Appendix to Jeremy Gray, *The Hilbert challenge*, 240–282. Oxford: Oxford University Press.
Hooker, Cliff, ed. 2011. *Philosophy of complex systems*. Amsterdam: North Holland.
Kant, Immanuel. 1992. *Lectures on logic*. Cambridge: Cambridge University Press.
Kepler, Johannes. 1937–. *Gesammelte Werke*. München: Beck.
Kitcher, Philip. 1983. *The nature of mathematical knowledge*. Oxford: Oxford University.
Kline, Morris. 1972. *Mathematical thought from ancient to modern times*. Oxford: Oxford University Press.
Leibniz, Gottfried Wilhelm. 1965. *Die Philosophischen Schriften*. Hildesheim: Olms.
Parsons, Charles. 1983. *Mathematics in philosophy: Selected essays*. Ithaca: Cornell University Press.
Pincock, Christopher. 2012. *Mathematics and scientific representation*. Oxford: Oxford University Press.
Russell, Bertrand. 1995a. *An outline of philosophy*. London: Routledge.
Schwartz, Jack. 1992. The pernicious influence of mathematics on science. In *Discrete thoughts*, ed. Mark Kac, Gian-Carlo Rota, and Jack Schwartz, 19–25. Boston: Birkhäuser.
Steiner, Mark. 1998. *The applicability of mathematics as a philosophical problem*. Cambridge: Harvard University Press.
von Neumann, John. 1961. The mathematician. In John von Neumann, *Collected works*, vol. 1, 1–9. Oxford: Pergamon Press.
Weinberg, Steven. 1986. Lecture on the applicability of mathematics. *Notices of the American Mathematical Society* 33 : 725–728.
———. 1993. *Dreams of a final theory: The search for the fundamental laws of nature*. New York: Vintage.
Wigner, Eugene Paul. 1960. The unreasonable effectiveness of mathematics in the natural sciences. *Communications on Pure and Applied Mathematics* 13 : 1–14.

Part V
Coda

Chapter 25
Knowledge and the Meaning of Human Life

Abstract Not only knowledge plays a central role in human life at all levels, from survival to improving the quality of life, but, since antiquity, several people have claimed that knowledge is the purpose and meaning of human life. This chapter maintains that this claim is unjustified, and that all attempts to show that human life has a purpose and meaning from an external and higher point of view are bound to be unsuccessful. Human life has indeed a purpose and meaning, but only from an internal point of view. As it has been widely held since antiquity, this purpose and meaning is happiness, which however does not consist in knowledge. On the other hand, even if happiness does not consist in knowledge, knowledge is a precondition of happiness.

25.1 Knowledge and Purpose and Meaning of Human Life

In Chap. 6 it has been argued that knowledge plays a central role in human life at all levels, from survival to improving the quality of life. Since antiquity, however, it has even been maintained that there is a connection between knowledge and the purpose and meaning of human life. This raises question: What is the nature of the connection between knowledge and the purpose and meaning of human life?

Since antiquity, several people have answered this question by saying that knowledge is the purpose and meaning of human life.

Thus Aristotle states that theoretical wisdom, namely knowledge of the first principles and what follows from them, "is by nature our purpose" and is "the ultimate thing for the sake of which we have come to be" (Aristotle, *Protrepticus* Düring, 17). Theoretical wisdom is "knowledge of a more perfect kind, and the supreme end of this is contemplation" (ibid., 66). Indeed, contemplation "is of all things the most desirable for men, comparable, I think, to eyesight, which everyone would choose to have even if nothing other than sight itself were to result from it" (ibid., 70). Since the supreme end of theoretical wisdom is contemplation, theoretical wisdom makes our life free, for "thought and contemplation" are what "we describe as the free life" (ibid., 43). Being knowledge of the first principles, theoretical wisdom is knowledge of God, because "God is thought to be among the causes of all things and to be a first principle" (Aristotle, *Metaphysica*, A 2, 983 a 8–9). Indeed, "on such a principle does the heaven and the world of nature depend" (ibid., Λ 7, 1072 b 13–14).

© Springer International Publishing AG 2017

C. Cellucci, *Rethinking Knowledge*, European Studies in Philosophy of Science 4,
DOI 10.1007/978-3-319-53237-0_25

Russell states that "the life of the instinctive man is shut up within the circle of his private interests" (Russell 1997, 157). But, "if our life is to be great and free, we must escape this prison," and the main way of doing so is through knowledge, since "all acquisition of knowledge is an enlargement of the Self" (ibid., 158). Through knowledge, our mind "becomes capable of that union with the universe which constitutes its highest good" (ibid., 161). Knowledge "makes us citizens of the universe," and "in this citizenship of the universe consists man's true freedom, and his liberation from the thraldom of narrow hopes and fears" (ibid.). The union with the universe is the purpose of religion, since "the essence of religion is the union with the universe" (Russell 1985a, 105). Religion "derives its power from the sense of union with the universe which it is able of to give" (Russell 1985b, 121).

What Aristotle and Russell propose as the purpose and meaning of human life is the ideal of self-liberation through knowledge: only through knowledge human beings can become really free. This ideal is religious in kind, as it is clear from the fact that Aristotle identifies the liberation of human beings with the knowledge of God, and Russell identifies it with that union with the universe in which, in his opinion, religion consists.

However, the view that knowledge is the purpose and meaning of human life is hardly defensible.

Indeed, suppose that, as Aristotle maintains, the purpose and meaning of human life is knowledge, since it is through theoretical wisdom that we can have knowledge of the principle of all things, namely God. This raises the question: Does God exist? Several proofs of the existence of God have been proposed, because, as Dummett says, God's existence "is held by most believers to be a truth attainable by purely rational thought" (Dummett 2010, 42). However, none of such proofs is generally accepted as proving what it purports to prove.

Also, suppose that, as Russell maintains, the purpose and meaning of human life is knowledge, since it is through knowledge that our mind becomes capable of that union with the universe which is its highest good. This raises the question: Why such union? Russell's answer, that through knowledge man reaches his liberation from the thraldom of narrow hopes and fears, is inadequate because to have knowledge of the universe is not necessarily a liberation. It can make us even more aware of the constraints to which we are subjected. For example, to know that the solar system will eventually cease to exist, or that the universe itself will eventually die, can hardly be said to have a liberating effect.

25.2 Purpose and Meaning of Human Life and Evolution

The inadequacy of the view that knowledge is the purpose and meaning of human life does not mean, however, that knowledge has nothing to say about this question. In fact, an important part of our knowledge, the theory of evolution, has something to say about it.

One of the most basic questions human beings are inclined to ask is: Why human life? The answer of the theory of evolution is: Human life exists in virtue of the fact that it is the result of an adaptation. But, it will be objected, this answer is based on a misunderstanding. What one means by the question 'Why human life?' is not: In virtue of what does human life exist? It is rather: What is the purpose and meaning of human life?

Actually, the theory of evolution gives an answer also to this question: Human life has no purpose and meaning, in the sense of an ideal goal to which it tends. This, of course, is a negative answer, but a clear and neat one. Thus the theory of evolution gives an answer to both senses of the question: Why human life? Human life exists in virtue of the fact that it is the result of an adaptation, and has no purpose and meaning, in the sense of an ideal goal to which it tends.

In particular, from the viewpoint of the theory of evolution, the purpose and meaning of human life cannot consist in progress. Evolution favours those characteristics which promote the propagation of the genes that originate them, but this is not a progress by any of the moral, aesthetic or emotional standards by which we consider something to be a progress. Even if one is willing to consider the propagation of genes as a progress in itself, natural selection cannot be considered as productive of progress, since an adaptation can be judged good or bad only relative to the environment in which it takes place. It cannot be considered a progress in an absolute sense.

Nor, from the viewpoint of the theory of evolution, can the purpose and meaning of human life consist in an increase in complexity. Although evolution favours the most complex organisms when they produce greater genetic adequacy, the same holds of the least complex ones. In a relatively stable environment, a species can progressively become better adapted, but this is simply local progress, not the global one that would allow one to speak of progress in an absolute sense. In any case, it is not progress in the sense of an approximation to an ideal goal to which the species would tend.

25.3 Purpose and Meaning of Human Life and God

But, even if the theory of evolution gives an answer to both senses in which one can mean the question 'Why human life?', this answer would not satisfy several people. Presumably, they would want an answer like that of theistic religions, for which human life exists in virtue of the fact that it has been created by God, and its purpose and meaning is to contribute to God's design. This answer, however, presupposes that God exists and, as already stated, none of the proofs of the existence of God that have been proposed is generally accepted as proving what it purports to prove.

Against this, it might be objected that God must exist, because only the belief in God can endow our life with a purpose and meaning.

Thus Wittgenstein maintains that "to believe in God means to understand the question about the meaning of life" and "to see that life has a meaning" (Wittgenstein

1979, 74). For, "the meaning of life, i.e., the meaning of the world, we can call God," hence "to pray is to think about the meaning of life" (ibid., 73).

Dummett maintains that only "religious belief endows our lives with a point" (Dummett 2010, 45).

But this objection implicitly assumes that human life has a purpose and meaning only if it is part of God's design. This assumption raises the question: In what sense has God's design a purpose and meaning? This leads to the dilemma: Either God's design has a purpose and meaning as it is part of a higher design, or it has a purpose and meaning in itself. Both horns of the dilemma are impossible. The first horn raises the question: In what sense has the higher design a purpose and meaning? This leads to an infinite regress, in virtue of which the world might exist without presupposing the existence of God. The second horn raises the question: If something can have a purpose and meaning in itself, why could not that something be human life itself? That is, why could not the purpose and meaning of human life reside in human life itself?

Moreover, even admitting that human life has a purpose and meaning only if it contributes to God's design, a further question would arise: Once God's design would have been realized, what purpose and meaning would remain to human life? There would remain none, because then there would be no further purposes to realize.

25.4 Why God?

Furthermore, even admitting that God must exist because only the belief in God can endow our life with a purpose and meaning, the question would arise: Why God? Like the question 'Why human life?' one can understand this question in two senses, namely, 'In virtue of what does God exist?' or 'What is the purpose and meaning of God?'

It might be objected that, while the question 'In virtue of what does human life exist?' is legitimate, the question 'In virtue of what does God exist?' is illegitimate. For, there must be a first principle in virtue of which every other thing exists, and that first principle is God, otherwise there would be an infinite regress. But this objection is not valid because, even admitting that there must be a first principle in virtue of which everything else exists, there is no evidence that such first principle is God rather than, say, as Russell maintains, the universe. Moreover, as John Stuart Mill's father impressed upon his son, "the question, 'Who made me?' cannot be answered," because "any answer only throws the difficulty a step further back, since the question immediately presents itself, 'Who made God?'" (Mill 1963–1986, I, 44).

It might be also objected that, while the question 'What is the purpose and meaning of human life?' is legitimate, the question 'What is the purpose and meaning of God?' is illegitimate. For, there must be a purpose to which all other purposes must be subordinate, otherwise there would be an infinite regress, and God is that purpose. But this objection is not valid, once again because, even admitting that there must be a purpose to which all other purposes must be subordinate, there is no evidence that this purpose is God rather than, say, the universe. Indeed, Spinoza maintains that the purpose and meaning of human life is "the knowledge of the union which the mind has with the whole of nature" (Spinoza 2002, 6).

Human life has indeed a purpose and meaning for us, in the sense that each of us, in one way or the other, manages to give some purpose and meaning to his own life. But there is no proof that God exists, nor that there exists a God's design to which human beings are called to contribute.

25.5 Belief in God and Rationality

Of course, even if none of the proofs of the existence of God is generally accepted as proving what it purports to prove, this does not mean that one cannot believe that God exists, that human life exists because it has been created by God, and that its purpose and meaning is to contribute to God's design. It only means that there is no proof to support this belief. What is in question here is not the belief, but the ground of the belief.

The belief in God is essentially different from the belief in things that eventually prove to be implausible. For example, although today we no longer believe in the Ptolemaic theory because the arguments against it are stronger than those for it, human beings believed in it for centuries. And, at that time, it was rational to believe in it, because the arguments for it were stronger than those against it. But, in the case of the belief in God, arguments for it stronger than those against it have never existed. This has been recognized from the beginnings of Christianity. Thus Tertullian stated: "Certum est, quia impossibile [It is certain, because it is impossible]" (Tertullian, *De Carne Christi*, V, 25–26). And an assertion of uncertain origin stated: "Credo quia absurdum [I believe because it is absurd]."

Dummett claims that there is "every reason to think" that philosophy can settle the question "whether there are rational grounds for believing in the existence of God," that this question "can be resolved positively," and that philosophy "will even do so in the lifetime of our great-grandchildren" (Dummett 2010, 151). But this claim has no rational basis, only an emotional one. As Pascal says, when confronted with the question whether God exists or does not exist, "reason cannot make you choose one way or the other," it "cannot make you defend either of the two choices" (Pascal 1995, 153).

This, however, cannot be used, as Pascal does, "to humble reason, which would like to be the judge of everything" (ibid., 36). According to Pascal, the inability of reason to prove God's existence "proves only the weakness of our reason" (ibid., 35). There is an alternative way of knowing the existence of God, namely by means of the heart, and "it is on this knowledge by means of the heart" that "reason has to rely, and must base all its arguments" (ibid., 36).

But the heart, hence emotion, by itself gives us no knowledge. To rely on emotion with regard to questions such as the existence of God, is to project our wishes outside us, taking them for reality. Only reason would be entitled to give us knowledge about such questions, everything else is emotional projection, consolatory fabulation.

25.6 Morality and God

There is no better ground for the argument that God must exist, otherwise there would be no morality and everything would be permitted. This is the 'Karamazov thesis': "If there's no infinite God, then there's no virtue either" (Dostoevsky 1992, 536). And, if "there is no virtue," then "everything is permitted" (ibid., 68).

This argument overlooks that religions without God, such as Buddhism, have been as effective in favouring morally good behaviour as theistic religions, and many non-believers behave in ways which are morally superior to those of many believers in theistic religions. Not to speak of the innumerable people tortured, women burnt, slaughters perpetrated in the name of some God. As Dawkins points out, "religious wars really are fought in the name of religion," while "I cannot think of any war that has been fought in the name of atheism," since one of the motives "for war is an unshakeable faith that one's own religion is the only true one" (Dawkins 2006, 278).

The argument that God must exist, otherwise there would be no morality and everything would be permitted, is based on the assumption that 'morally good' is what God wants. This assumption leads to the dilemma: Either what is morally good is wanted by God because it is morally good in itself, or it is morally good because it is wanted by God. This is a variant of the *Euthyphro* dilemma: "Is the holy being loved by the gods because it is holy, or is it holy because it is being loved by the gods?" (Plato, *Euthyphro*, 10 a 2–3).

Both horns of the dilemma are impossible. The first horn of the dilemma implies that the fact that something is morally good does not depend of God, there exists a higher moral standard to which God must conform. But then it is unjustified to say that God must exist, otherwise there would be no morality and everything would be permitted. The second horn of the dilemma implies that what is morally good is the result of an act of God's will. But then it is arbitrary, God could want something which is not morally good.

Against this, it cannot be objected that God could never want anything of the kind, because he only wants what is morally good. This would imply that what is morally good is not so because God wants it, but because it is morally good in itself, thus one would be taken back to the first horn of the dilemma. Therefore, it is unjustified to say that God must exist otherwise there would be no morality.

25.7 Intelligibility of the World and Naturalism

From what has been said above, it is clear that the answer of theistic religions to the question 'Why human life?' is inadequate. This brings us back to the answer of the theory of evolution: Human life exists in virtue of the fact that it is the result of an adaptation, and has no purpose and meaning, in the sense of an ideal goal to which it tends. Since there seem to be no alternatives, such an answer appears to be the only plausible one.

Dembski criticizes Darwin because "toward the end of his life Darwin would deny design any epistemic force" (Dembski 2002, 80). It was "Darwin's expulsion of design from biology that made possible the triumph of naturalism in Western culture" (ibid., 14). And "within naturalism the intelligibility of the world must always remain a mystery" (ibid., 230).

Actually, quite the opposite is true. To begin with, only certain features of the world are intelligible to human beings, not all. Such features are intelligible to them because human beings have cognitive architectures suitable to comprehend them, and those architectures are a result of evolution. Human beings may comprehend only those features of the world which are accessible to the cognitive architectures with which evolution has endowed them. Therefore, only naturalism can explain the intelligibility of the world for us.

25.8 Purpose and Meaning of Human Life from an External Point of View

That human life has no purpose and meaning, in the sense of an ideal goal to which it tends, implies that there exists no justification of human life from an external and higher point of view. We all find ourselves with a life to live by the very fact that we were born, but, from an external and higher point of view, our life has no purpose and meaning.

This is implicit in the origin itself of our individual life. It arises from the fertilization of an egg by a spermatozoon, but several hundred millions spermatozoa intervene in it. Two factors determine which spermatozoon will fertilize the egg. The first is that some spermatozoa find themselves in an optimal position at the optimal moment. The second one is that, among them, a more fitted one reaches the egg before the less fitted. Thus, in the generation of an individual human life, only two factors intervene: chance and the survival of the fittest. This implies that an individual life cannot be said to have a predesigned purpose and meaning.

Moreover, from an external and higher point of view, not only human life has no purpose and meaning, but has no intrinsic importance. However much we may feel that a world without us would lack a crucial piece of it, with respect to the history of the world our existence is a contingent and negligible fact. As Hume says, "the life of a man is of no greater importance to the universe than that of an oyster" (Hume 1998, 100). Of course, our existence may be very important to the persons to whom we are dear but, with respect to the history of the world, this is a contingent and negligible fact. Our existence, as well as its importance to the persons to whom we are dear, are an inessential aspect of the history of the world.

This gives us a definitely unpleasant feeling, since the idea that our existence is an inessential aspect of the history of the world is difficult to endure. If we could forget ourselves, we could perhaps imagine a world without us. But we are incapable of doing so. What we can imagine is only our world, namely, the world as we conceive it, without us in it. But this is, once again, our world, because the subject who imagines the world without us is inevitably us. Thus to exclude us from our world would mean for us to exclude the world itself, not merely to exclude us within it. Therefore, for us, to suppose that we had never existed would amount to supposing that the world had never existed.

And yet, however unbearable it may appear to us, we must accept the idea that our existence is an inessential aspect of the history of the world, and hence that, from an external and higher point of view, our existence is a contingent and negligible fact.

Actually, not only our existence is a contingent and negligible fact, but so is the existence of the world itself. Not only it would have made no difference if we had never existed, but it would have made no difference if the world itself had never existed. To think that the world has a purpose and meaning is an illusion. All the more so, it is an illusion to think that human life has a purpose and meaning. If the world has no purpose and meaning, *a fortiori* human life has no purpose and meaning, in the sense of an ideal goal to which human life tends and which can be discovered and revealed by some philosopher, religious leader or scientist. Those philosophers, religious leaders or scientists who pretend to have discovered such an ideal goal and to reveal it, deceive us, and perhaps even deceive themselves.

25.9 Purpose and Meaning of Human Life from an Internal Point of View

That our life has no purpose and meaning from an external and higher point of view does not mean, however, that it has no purpose and meaning from an internal point of view. In fact, our life has a purpose and meaning to us, and to the persons to whom we are dear. Even if, from an external and higher point of view, it would not matter if we, and the persons who are dear to us and to whom we are dear, did not exist at all, it does matter to us and to them, and that is all there is to it.

This should induce a form of humility in us: the recognition that our life has a purpose and meaning, not in an absolute sense, but only in this relative sense. But this is no limitation, because for us there is only our world, namely the world as we conceive and perceive it, and the most important place in it is occupied by us and the persons who are dear to us and to whom we are dear.

Actually, not only this is no limitation, but is the necessary precondition of a life lived to the full.

Russell maintains that theistic religions are based "primarily and mainly upon fear," which "is partly the terror of the unknown," and partly "the wish to feel that you have a kind of elder brother who will stand by you in all your troubles and disputes. Fear is the basis of the whole thing – fear of the mysterious, fear of defeat, fear of death" (Russell 2005, 18).

But even those who do not share this view will recognize that theistic religions consider individual life as something that is not complete in itself, but is lived in view of something else – an elsewhere, an afterlife, a reward in the kingdom of God. Theistic religions do so, although they are sometimes ambiguous on the subject, as it appears from Luke's statement: "Behold, the kingdom of God is within you" (Luke 17: 21).

By considering individual life as something which is to be lived in view of something else, theistic religions prevent us from living our life to the full. As Law states, "belief in a god or gods can actually be an impediment to our living full and meaningful lives," in particular, by "wasting our lives promoting false beliefs because of a mistaken expectation of a life to come" (Law 2012, 38). Only those who do not place the value of individual life in an elsewhere, can live a life which is not a wait for something else. Therefore, they will try to fill their life with meaning in each moment, being aware that their life can find its accomplishment only in itself, not in a chimerical elsewhere.

25.10 Happiness and the Purpose and Meaning of Human Life

But what is the purpose and meaning of human life from an internal point of view? The most convincing answer remains Aristotle's: the purpose and meaning of human life is happiness.

Indeed, Aristotle claims that "there is a very general agreement among the majority of men" concerning the purpose of human life, in fact, "both the common people and people of superior refinement say that it is happiness" (Aristotle, *Ethica Nicomachea*, A 4, 1095 a 17–19). Happiness is "the purpose of human actions" (ibid., K 6, 1176 a 31–32). We choose "honour, pleasure, reason, and every virtue" only "for the sake of happiness, judging that it is by means of them that we shall be happy. Happiness, on the other hand, no one chooses for the sake of these, nor, in general, for anything other than itself" (ibid., A 7, 1097 b 2–6). We "always choose it in itself" (ibid., A 7, 1097 b 1–2). For, "happiness is clearly something perfect and self-sufficient, being the end to which our actions are directed" (ibid., A 7, 1097 b 20–21). By 'happiness', it is meant here happiness "in a complete life. For, as one swallow does not make a spring, nor does one day, so one day, or a short time, does not make a man blessed and happy" (ibid., A 7, 1098 a 18–20).

The answer that the purpose and meaning of human life is happiness, partially compensates us for the unpleasantness of the conclusion that, from an external and higher point of view, our existence is a contingent and negligible fact. We know that everything will come to an end, that our life has no absolute justification, that it has no purpose and meaning from an external and higher point of view. But, that something can make us happy, is for us a sufficient purpose and reason to live. Admittedly, since everything will come to an end, we will eventually lose the game, but, if something will have made us happy, we will have earned our life.

It is for this reason that, as Rilke says, "again and again, however we know the landscape of love | and the little churchyard there, with its sorrowing names | and the frighteningly silent abyss into which the others | fall: again and again the two of us walk out together | under the ancient trees, lie down again and again | among the flowers, face to face with the sky" (Rilke 1998, 881).

But is the the answer that the purpose and meaning of human life is happiness justified? Some people deny it, because they identify happiness with pleasure, and assume that pleasure and the satisfaction of desire constitute the traditional conception of happiness.

Thus, Thomson states that the "activities and experiences that in part constitute a meaningful life cannot be adequately explained in terms of pleasure and the satisfaction of desire, which constitute the traditional conception of happiness. Consequently, happiness as traditionally conceived is not the meaning of life" (Thomson 2003, 78).

But this is unwarranted, because pleasure and the satisfaction of desire do not constitute the traditional conception of happiness.

Thus, Aristotle states that "men of the most vulgar type" seem "to identify the good, or happiness, with pleasure" because "the mass of mankind are evidently quite slavish in their tastes, preferring a life suitable to beasts" (Aristotle, *Ethica Nicomachea*, A 5, 1095 b 16–20). On the contrary, happiness is "activity of the soul in accordance with virtue, and if there are more than one virtue, in accordance with the best and most perfect" (ibid., A 7, 1098 a 17–18).

Kant states that "virtue (as worthiness to be happy) is the supreme condition of whatever can even seem to us desirable and hence of all our pursuit of happiness and that it is therefore the supreme good" (Kant 1996, 228). Indeed, "when a thoughtful human being has overcome incentives to vice and is aware of having done his often bitter duty, he finds himself in a state that could well be called happiness, a state of contentment and peace of soul in which virtue is its own reward" (ibid., 510–511).

Even independently of whether pleasure and the satisfaction of desire constitute the traditional conception of happiness, identifying happiness with pleasure is untenable. On the one hand, we can take pleasure staring at a sunset, and yet not necessarily be happy. On the other hand, we can be happy because we are satisfied with our life, and yet not feel pleasure.

So, we are faced once again with the question: Is the answer that the purpose and meaning of human life is happiness justified?

Clearly, the answer is unjustified from the viewpoint of theistic religions, which do not guarantee that the purpose and meaning of human life is happiness. Human beings can be unhappy having faith in God, and happy not having faith in God, or vice versa. It is not the faith or lack of faith in God that gives happiness. Even if faith in God gave happiness, one might be reminded of Bernard Shaw's remark: "The fact that a believer is happier than a sceptic is no more to the point than the fact that a drunken man is happier than a sober one" (Shaw 2008, 92).

The answer is also unjustified from the viewpoint of the theory of evolution. The latter does not guarantee that the purpose and meaning of human life is happiness.

And yet, even if the answer is unjustified both from the viewpoint of religions and the theory of evolution, the pursuit of happiness is a powerful impulse which urges human beings to desire their own survival and live life to the full. Without that impulse, it would be difficult to explain the tenacious will of so many human beings to preserve their life beyond any reasonableness, even in the most appalling life conditions.

25.11 Happiness and Knowledge

The view that the purpose and meaning of human life is happiness raises the question: What is happiness? As it was to be expected, philosophers for whom the purpose and meaning of human life is knowledge, answer that happiness consists in knowledge.

Thus, Aristotle states that "the theoretically wise is happy in the highest degree" (Aristotle, *Ethica Nicomachea*, K 8, 1179 a 32). He lives more happily "who most attains truth. This is the one who is theoretically wise and speculates according to the most exact knowledge" (Aristotle, *Protrepticus* Düring, 85). Since the supreme end of the most exact knowledge is contemplation, "happiness extends as far as contemplation, and the more contemplation there is in one's life, the happier one is, not incidentally, but in virtue of the contemplation, since this is honourable in itself. Happiness, therefore, will be some form of contemplation" (Aristotle, *Ethica Nicomachea*, K 8, 1178 b 28–32).

Russell states that "all the conditions of happiness are realized in the life of the man of science" (Russell 1930, 146). A "life devoted to science is therefore a happy life" (Russell 1994, 60). The "desire for a larger life and wider affairs, for an escape from private circumstances, and even from the whole recurring human cycle of birth and death, is fulfilled by the impersonal cosmic outlook of science as by nothing else" (ibid.).

But the claim that happiness consists in knowledge is hardly convincing, both in Aristotle's and Russell's version.

For, even admitting that, as Aristotle maintains, the theoretically wise is happy and even happy in the highest degree, he is not necessarily the only one to be happy. There are also those who are made happy by other things, such as work, love, friendship, the feeling that their life is important to their family and friends.

Also, even admitting that, as Russell maintains, a life devoted to science is a happy life, it is not necessarily the only happy life. Russell contrasts it with "the life of the instinctive man," who is all "shut up within the circle of his private interests," in which "family and friends may be included, but the outer world is not regarded except as it may help or hinder what comes within the circle of instinctive wishes" (Russell 1997, 157–158). But there are also those who are made happy by things which come within the circle of their private interests.

Indeed, Russell himself implicitly admits this when, in addition to knowledge, as a means to be happy, he indicates to help one's fellows so as "to shed sunshine on their path, to lighten their sorrows by the balm of sympathy, to give them the pure joy of a never-tiring affection, to strengthen failing courage, to instil faith in hours of despair" (Russell 1994, 18). We will be rewarded for this with the feeling that, "wherever a spark of the divine fire kindled in their hearts, we were ready with encouragement, with sympathy, with brave words in which high courage glowed" (ibid.).

25.12 The Nature of Happiness

That happiness cannot be said to consist in knowledge, is just one aspect of the general fact that the question 'What is happiness?' does not admit a unique answer. What makes one happy differs from person to person, since it depends on what one wants, which in turn depends on what one is. Moreover, it changes according to the different ages and conditions of life. For children, what makes them happy depends on their parents. For lovers, the object of their love is everything to them. Therefore children and lovers have a feeling that everything is within easy reach, that for them it is enough to stretch out their hand and seize it. For, parents or the beloved one are the main things that matter to them, and they are a friendly and benevolent will.

Also, for the young, what makes them happy is to expand in the world. This seems within easy reach to them, because they have a still very limited experience of the human condition. As Russell points out, it is for this reason that, "for the young, there is nothing unattainable; a good thing desired with the whole force of a passioned will, and yet impossible, is to them not credible" (Russell 1994, 14). But "to every man comes, sooner or later, the great renunciation" (ibid.). Indeed, "by death, by illness, by poverty, or by the voice of duty, we must learn, each one of us, that the world was not made for us, and that, however beautiful may be the things we crave, Fate may nevertheless forbid them" (ibid.).

That the question 'What is happiness?' does not admit a unique answer, does not mean, however, that it is impossible to state some minimal conditions for happiness.

First, happiness is the will to live. An essential condition for a happy life is to have a life, and to desire to continue to have one. So, happiness is first of all that strong attachment to life, that flame that catches again and again and takes deeper roots after each sorrow. Of course, this is minimal happiness, but is the mother of all happiness.

Moreover, happiness is to have something: to have some interests, affections, something to do and someone to love. Having them, we expand in the world and multiply ourselves in it. It is true that, for some people, the supreme form of happiness consists in renouncing everything. But they do so only to possess what they desire more than everything else: to be fully themselves, without ties. An extreme form of happiness as renunciation is Paul of Tarsus' "cupio dissolvi [I desire to be dissolved]" (Paul of Tarsus, *Epistle to the Philippians*, 1.23–24). This is shared by the great mystics, for whom self-dissolution is the means to get rid of the ties of the body and reach the supreme good of the union with God. But, without going to such extremes, happiness is also to give something to others, since this is a way of feeling that one's existence is useful to someone else.

Furthermore, happiness is the hope that we will have tomorrow what is denied to us today. Happiness and hope are strictly connected. Kant says that "all hope concerns happiness" (Kant 1998, A805/B833). But one may also assert the converse: all happiness concerns hope. Even if the experience of life tells us that hope is often a fable, we go on telling ourselves that fable and believing it, like children. All children's fables are similar, as all lovers' discourses are similar, since they have the colour of hope. Hope is what gives us the strength to face difficulties, and our human condition becomes unbearable when hope fails.

25.13 Seeking Happiness in One's Individual Life

In any case, instead of seeking a chimerical purpose and meaning of our life from an external and higher point of view, we can only seek happiness within our individual life. The universe does not care whether our individual life is miserable or marvellous, and no outer and higher entity worries about our happiness. To seek it is entirely up to us.

That human life has no purpose and meaning, in the sense of an ideal goal to which it tends, and that the purpose and meaning of human life consists only in what makes us individually happier, can be depressing to those who draw impulse and justification for what they do from the belief that it is important not only to them, but absolutely. Moreover, it may appear cruel that, if we do not have a happy life, a life worth living, we cannot even hope, as theistic religions promise, to find a compensation for suffering in an afterlife. But the very fact that we look for a compensation when we do not have a happy life, shows that a happy life is in itself something that is worth living. We may feel defrauded by life if we do not have the goods we desire, or the joys we long for, but we cannot deny that those goods and joys exist. They exist for the very fact that we would like to have them.

It may also appear cruel that life is so short. When we are born, we suddenly enter a marvellous garden, we walk in its alleys for a short time, and then we go out of it in the same sudden way in which we entered it. Indeed, as Quasimodo says, "each of us is alone on the heart of the earth | pierced by a ray of sun: | and suddenly it's evening" (Quasimodo 1984, 29).

Then it is true, as Russell says, that "brief and powerless is man's life; on him and all his race the slow, sure doom falls pitiless and dark" (Russell 1994, 18). His life is a "march through the night, surrounded by invisible foes, tortured by weariness and pain, towards a goal that few can hope to reach, and where none may tarry long" (ibid.). But the very fact that we regret that our time in the world is so brief, indicates that life, at least a happy life, is worth living. Otherwise we would have no reason to regret its brevity and desire its prolongation.

25.14 Brevity and Value of Human Life

It is just the brevity of human life that gives it its value and makes it a precious good. Many human activities gain their meaning from the fact that human life is so short.

The brevity of human life does not mean lack of reality or value, on the contrary it shows how real and valuable life is. Seneca states: "Quam stultum est aetatem disponere ne crastini quidem dominum! [How foolish it is to arrange one's life, when one is not even master of tomorrow!]" (Seneca, *Epistulae ad Lucilium*, 101: 4). But almost the opposite is the case, because the point of the brevity of human life is that it advises us to make the most of our life.

Just because human life is so short, we must not waste it. We must make the most of it because there is not much of it, and what there is of it may be an invaluable gift for those who have it. The brevity of human life must make us value the present because we might not have a future, it must warn us to avoid wasting our life, to make the best we can of it.

25.15 Knowledge as a Precondition of Happiness

That the question 'What is happiness?' does not admit a unique answer, and hence the answer is not necessarily 'knowledge', does not mean that knowledge is irrelevant to happiness. On the contrary, knowledge is an important precondition of it.

Indeed, an essential condition for a happy life is to have a life, and without knowledge life, biological life itself, would not have existed nor would continue to exist. For, as maintained in Chap. 6, knowledge is a response to the environment that is essential for survival. It is through knowledge that life, from the first unicellular organisms to human beings, has been able to exist and preserve itself. Those primordial unicellular organisms which developed rudimentary sense organs through which they solved the problem of their survival, were the first to discover that knowledge is indispensable to life. And, as Sagan says, "our future depends on how well we know this cosmos in which we float like a mote of dust in the morning sky," so "knowledge is prerequisite to survival" (Sagan 1980, 4).

Moreover, an essential condition for a happy life is to know who we are and what we want, and to a large extent this depends on what we know. We reflect reality, and reality is for us what we have access to and we know. Generally, our aspirations, desires and hopes essentially depend on what we know.

Furthermore, a condition for a happy life is not to be paralyzed by the fear that originates from prejudice, and many prejudices arise from lack of knowledge. Lack of knowledge also generates superstition, and prejudice and superstition are the causes of so many fears and human sufferings. This has negative consequences even from a biological point of view, because those who act on the basis of prejudices, or moved by superstition, diminish their ability to interact with the environment in an effective and optimal way.

This permits us to give an answer to the question posed at the beginning of this chapter: What is the nature of the connection between knowledge and the purpose and meaning of human life? As argued in Chap. 6, knowledge has a biological role, it is sought as a means to satisfy that basic necessity of life that is survival, and has a cultural role. But, in addition, knowledge is sought as being a precondition of that state of emotional well-being that we call 'happiness'. In these three roles – biological, cultural and as a precondition of happiness – knowledge shows its nature and finds its reason, its purpose and its accomplishment.

References

Dawkins, Richard. 2006. *The God delusion*. London: Bantam Press.

Dembski, William. 2002. *Intelligent design: The bridge between science & theology*. Westmont: InterVarsity Press.

Dostoevsky, Fyodor. 1992. *The brothers Karamazov*. London: Everyman's Library.

Dummett, Michael. 2010. *The nature and future of philosophy*. New York: Columbia University Press.

Hume, David. 1998. *Dialogues concerning natural religion – Of the immortality of the soul – Of suicide – Of miracles*. Indianapolis: Hackett.

Kant, Immanuel. 1996. *Practical philosophy*. Cambridge: Cambridge University Press.

———. 1998. *Critique of pure reason*. Cambridge: Cambridge University Press.

Law, Stephen. 2012. The meaning of life. *Think* 11 : 25–38.

Mill, John Stuart. 1963–1986. *Collected works*. Toronto: University of Toronto Press.

Pascal, Blaise. 1995. *Pensées and other writings*. Oxford: Oxford University Press.

Quasimodo, Salvatore. 1984. *Complete poems*. New York: Schocken Books.

Rilke, Rainer Maria. 1998. *Die Gedichte*. Frankfurt am Main: Insel.

Russell, Bertrand. 1930. *The conquest of happiness*. London: Allen & Unwin.

———. 1985a. Prisons. In Bertrand Russell, *Collected papers*, vol. 12, 97–109. London: Allen & Unwin.

———. 1985b. The essence of religion. In Bertrand Russell, *Collected papers*, vol. 12, 110–122. London: Allen & Unwin.

———. 1994. *Mysticism and logic*. London: Routledge.

———. 1997. *The problems of philosophy*. Oxford: Oxford University Press.

———. 2005. *Why I am not a Christian and other essays on religion and related subjects*. London: Routledge.

Sagan, Carl. 1980. *Cosmos*. New York: Random House.

Shaw, Bernard. 2008. *Preface to Androcles and the lion*. London: Icon.

Spinoza, Benedictus. 2002. *Complete works*. Indianapolis: Hackett.

Thomson, Garrett. 2003. *On the meaning of life*. Belmont: Wadsworth.

Wittgenstein, Ludwig. 1979. *Notebooks 1914–1916*. Oxford: Blackwell.

Chapter 26
Conclusion

Abstract The conclusion summarizes some of the main theses of the book. In the past century, the view that philosophy aims at knowledge and methods to acquire knowledge has been abandoned, and this has contributed to the increasing irrelevance of the subject. The book attempts to revive this view. This requires a rethinking of knowledge. According to it, knowledge is a natural process, continuous with the biological process by which life is sustained and evolved. Specifically, knowledge is problem solving by the analytic method, so it does not consist of truths but of plausible hypotheses, and is never definitive, hence it is an ongoing process. That knowledge is problem solving by the analytic method applies to all kinds of knowledge, from perceptual knowledge to scientific and mathematical knowledge. The importance of a rethinking of knowledge depends on the fact that knowledge plays a central role in human life at all levels, from survival to improving the quality of life, and is also a precondition of happiness.

26.1 The Challenge to Philosophy

In the seventeenth century, philosophy suffered a trauma from which it has not recovered yet, the birth of modern science. The latter has invaded many areas traditionally covered by philosophy, thus making philosophy increasingly irrelevant and presenting it with the challenge to legitimate itself.

Since the seventeenth century, a great deal of philosophy has been an attempt to provide an answer to this challenge. Nevertheless, the challenge has not been successfully met, so much so that several scientists, and even some philosophers, have concluded that philosophy is dead and has dissolved into the sciences.

The question then arises whether philosophy can still be fruitful and what kind of philosophy can be such. Asking these questions is nothing really new, because philosophy has always called into question everything, including itself. But, with the birth of modern science, such questions have become more pressing, as well as more difficult and embarrassing.

Giving an answer to these questions requires a rethinking of philosophy, and its relationship with scientific knowledge and knowledge generally.

© Springer International Publishing AG 2017

C. Cellucci, *Rethinking Knowledge*, European Studies in Philosophy of Science 4,
DOI 10.1007/978-3-319-53237-0_26

26.2 The Nature of Philosophy

In the past century, the prevailing view of philosophy has been that philosophy does not aim at knowledge and methods to acquire knowledge, but only clarifies what we already know; in philosophy there are no great essential problems in the sense of science; questions about knowledge no longer lie at the heart of the philosophical enterprise; the objectives of philosophy are essentially different from those of the sciences; philosophy cannot be a matrix out of which new sciences could develop; and the solution of philosophical problems cannot advance knowledge.

By assigning philosophy only a marginal role, this view has contributed to the increasing irrelevance of the subject. Indeed, if philosophy does not aim at knowledge, if in philosophy there are no great essential problems in the sense of science, then philosophy is led to focus on minute, inconsequential questions, of no importance for our understanding of the world, including ourselves.

On the contrary, this book maintains that philosophy aims at knowledge and methods to acquire knowledge; it concerns itself with great essential problems in the sense of science; questions about knowledge are central to philosophy; the objectives of philosophy are not essentially different from those of the sciences, the only difference being that philosophy deals with questions which are beyond the present sciences; philosophy can be a matrix out of which new sciences can develop; and, although solutions to philosophical problems are always temporary, they advance knowledge.

26.3 The Nature of Knowledge

The view that philosophy aims at knowledge and methods to acquire knowledge requires a rethinking of knowledge. In the past century, the prevailing view has been that knowledge is justified true belief, hence it is a state of mind which matches the external world.

On the contrary, this book maintains that knowledge is a natural process, continuous with the biological processes by which life is sustained and evolved. Human beings make hypotheses about the environment and acquire knowledge on it, on the basis of which they assume behaviours that ensure their survival. It is through knowledge that life can preserve itself, so knowledge is essential for survival.

Being essential for survival, knowledge has a biological role. In addition, it has a cultural role. Knowledge is a means not only for survival, but also for improving the quality of life, from the routine workings of everyday life to global issues.

The cultural role of knowledge does not reduce to the biological one, since a culture is a body of cognitions that human beings transmit from generation to generation non-genetically, namely not through DNA, but through what they do and communicate. On the other hand, the cultural role of knowledge is not opposed to the biological role, since it is the product of organisms which are the outcome of biological evolution. Rather, the cultural role of knowledge is an enhancement of the biological role, and cannot be separated from it.

26.4 The Relation of Knowledge to Reality

There are several questions about knowledge. A first question is the relation of knowledge to reality. An approach to this question, which has had a large following in the past century, is scientific realism. According to it, the world is an objective reality that exists independently of human thought; the aim of science is to have true theories about the world, where 'true' is understood in the sense of the concept of truth as correspondence; scientific inquiry yields knowledge of the truth about the objective reality investigated by scientists; and the truth about the world, of which scientific inquiry yields knowledge, is definitive.

On the contrary, this book maintains that the aim of science is to have plausible hypotheses about the world, namely hypotheses such that the arguments for them are stronger than the arguments against them, on the basis of the existing knowledge; scientific inquiry yields knowledge which does not consist of truths, but instead, of plausible hypotheses; and scientific knowledge is never definitive, it is an ongoing process, since there are no problems conclusively solved but only problems more or less solved.

26.5 The Objectivity of Knowledge

Another question about knowledge is the objectivity of knowledge. According to scientific realism, the truth about the world of which scientific inquiry yields knowledge, is objective in the absolute sense of being totally independent of any subject. The aim of science is to have knowledge which is objective in this absolute sense.

On the contrary, this book maintains that the aim of science is to have knowledge which is objective not in the absolute sense of being totally independent of any subject, but only in the relative sense of being as independent as possible of any particular human subject. More specifically, the aim of science is to have knowledge which is plausible, namely, such that the arguments for it are stronger than the arguments against it. This is the only kind of objectivity we can achieve.

26.6 The Question of Discovery

Another question about knowledge is how knowledge is acquired, hence the question of discovery. In the past century it has been generally believed that this question is vacuous, because there is no method for acquiring knowledge; discovery is a purely subjective, psychological matter, so it cannot be an object of logic but only of psychology; and the basic laws of science can be reached only by intuition, then, from them, conclusions are deduced and compared with experience.

This belief has had a very negative impact on logic, the method of science, and philosophy. It has led to restrict logic to deductive logic, to separate it from the method of science, and to replace the study and development of method with largely irrelevant discussions on the definition of the concept of knowledge.

On the contrary, this book maintains that the belief that the question of discovery is vacuous, has no foundation; since antiquity, a method of discovery has been known and widely used, the analytic method, and there is nothing subjective and psychological about this method; and the basic laws of science are reached by non-deductive reasoning, then they are tested for plausibility by comparison with experience.

26.7 Modelling Scientific Knowledge

Another question about knowledge is how scientific theories are built. An approach to this question, which has had a large following in the past century, is the hypothetico-deductive model of science. It assumes that there is no method for acquiring knowledge; science is based on the axiomatic method; building a scientific theory is a matter of choosing certain hypotheses and deducing consequences from them; the hypotheses can be chosen arbitrarily, but are subject to the condition that they must be consistent with the total system of hypotheses; and the hypotheses can and must be tested by comparing the consequences deduced from them with the observational and experimental data.

On the contrary, this book maintains that science is based on the analytic method. This model assumes that there is a method for acquiring knowledge; science is based on the analytic method; building a scientific theory is a matter of choosing certain problems and finding hypotheses that permit to solve them; the hypotheses are introduced by some non-deductive rule; and the hypotheses must be plausible, namely, the arguments for them must be stronger than the arguments against them, on the basis of the existing knowledge.

26.8 Knowledge as Problem Solving

The analytic method, namely the method by which knowledge is acquired, when solving problems produces knowledge. Conversely, knowledge is produced solving problems by the analytic model. Therefore, knowledge is problem solving by the analytic method.

This implies that problems are the starting points of knowledge. According to an influential tradition going back to antiquity, the starting point of knowledge is sense perception; knowledge begins with sense perception, then from sense perception there arises memory; from memory, when it occurs often in connection with the same thing, there arises experience, since, though there are many memories, they make up a single experience; and from experience there arises the starting point of scientific knowledge.

On the contrary, this book maintains that the starting point of knowledge is not sense perception, but problems; sense perception itself, in particular vision, is problem solving by the analytic method; since the starting point of knowledge are problems, problem posing is an important process, since it suggests that there may be relations between the data; and solving a problem establishes that such relations really exist.

26.9 The Nature of Perceptual Knowledge

That knowledge is problem solving by the analytic method applies to all knowledge. As already mentioned above, this includes perceptual knowledge.

According to an influential tradition, perception is a passive process, determined entirely by the features of the external world. Vision is supposed to be based either on the fact that the eye is like a *camera obscura* and the retina is the screen and the brain watches the screen, or on the fact that we observe mental images resembling objects of the external world.

On the contrary, this book maintains that perception is based on the fact that we form hypotheses about objects of the external world from the stimuli on our sense organs, by means of non-deductive inferences. The non-deductive inferences involved in perception are unconscious, and moreover they are non-propositional, being transformations of data rather than transformations of propositions.

26.10 The Nature of Mathematical Knowledge

That knowledge is problem solving by the analytic method typically applies to scientific knowledge and mathematical knowledge. In particular, with respect to scientific knowledge, the case of mathematical knowledge raises special problems. Therefore, the book focuses on this case.

In the past century, the prevailing view of mathematics has been that mathematics is theorem proving by the axiomatic method; mathematics is independent of experience; and mathematics is absolutely certain.

On the contrary, this book maintains that mathematics is problem solving by the analytic method; mathematics is not independent of experience, because several mathematical problems have an extra-mathematical origin, and the solutions of mathematical problems are only plausible, so they are evaluated in terms of their compatibility with the existing knowledge; and mathematics is not absolutely certain, not only because mathematicians often make mistakes which sometimes remain undetected for a long time, but also because it is impossible to prove that mathematics does not contain mistakes.

26.11 The Nature of Mathematical Objects

That mathematics is problem solving by the analytic method has implications for the nature of mathematical objects. In the past century, a very influential view of mathematical objects has been mathematical platonism, according to which mathematical objects are non-physical things which exist independently of the human mind.

On the contrary, this book maintains that mathematical objects are hypotheses mathematicians make to solve mathematical problems by the analytic method. Mathematical objects exist only in the minds of the mathematicians who hypothesize them, and in those of people who make use of them.

This follows from the fact that mathematics is a human product, it is part of being human. Natural mathematics, namely, the mathematics resulting from biological evolution, is a set of abilities that are essential for the survival of human beings. Artificial mathematics, namely, the mathematics as a discipline resulting from cultural evolution, is a set of concepts and procedures that allow human beings to deal with the world and make it understandable to themselves. Both natural mathematics and artificial mathematics solve problems by the analytic method.

26.12 Mathematics and Intuition

That mathematics is problem solving by the analytic method has implications also for the role of intuition in mathematics. In the past century, the prevailing view has been that intuition has a central role in mathematics, because intuition is the faculty by which we grasp mathematical objects.

On the contrary, this book maintains that, since mathematics is problem solving by the analytic method and in the analytic method hypotheses are obtained by non-deductive rules, intuition plays no role in mathematics.

Indeed, intuition plays no role in the discovery of hypotheses, because they are obtained from the problem, and possibly other data, by non-deductive rules, so not by intuition but by inference. Intuition plays no role in the justification of hypotheses, because their plausibility is established by comparing the arguments for and the arguments against them on the basis of the existing knowledge, so not by intuition but by inference.

What several mathematicians call 'intuition' is simply their feeling of 'almost knowing' some hypothesis without having consciously gone through a step-by-step reasoning process to get there. This feeling can be explained in terms of the fact that they arrived at the hypothesis through some unconscious non-deductive inference, which led them to trust that the hypothesis might be plausible.

26.13 The Nature of Mathematical Demonstration

That mathematics is problem solving by the analytic method has implications also for the nature of mathematical demonstration. In the past century, the prevailing view has been that axiomatic demonstration is the basic concept of demonstration; a demonstration consists in a deduction of a proposition from given axioms which are true, either in the strong sense that there is a system of things, specified in advance, for which the axioms are true, or in the weak sense that they are consistent; and the purpose of axiomatic demonstration is to justify and teach an already acquired proposition.

On the contrary, this book maintains that analytic demonstration is the basic concept of demonstration; a demonstration consists in a non-deductive derivation of a hypothesis from a problem and possibly other data, where the hypothesis is a sufficient condition for the solution of the problem and is plausible, then it consists in a non-deductive derivation of a new hypothesis from the previous hypothesis and possibly from other data, where the new hypothesis is a sufficient condition for the solution of the problem posed by the previous hypothesis and is plausible, and so on, *ad infinitum*; and the purpose of analytic demonstration is to discover hypotheses that are sufficient conditions for the solution of a problem, and are plausible.

26.14 Mathematical Explanations

An important question about mathematics is whether there are mathematical explanations of mathematical facts, and mathematical explanations of empirical facts.

In the past century, the prevailing view about mathematical explanations of mathematical facts has been that there are no such explanations, in particular, there is no objective distinction between explanatory and non-explanatory demonstrations.

On the contrary, this book maintains that there are mathematical explanations of mathematical facts, in particular, there is an objective distinction between explanatory and non-explanatory demonstrations. Moreover, there are two different approaches to explanatory demonstrations: the static approach, according to which the function of explanatory demonstrations is to convince the audience that a proposition should be accepted, and the dynamic approach, according to which the function of explanatory demonstrations is to reveal how the demonstration was discovered.

Also, in the past century, the prevailing view about mathematical explanations of empirical facts has been that there are such explanations, namely, explanations of empirical facts in which mathematical facts play an essential role. Many examples are commonly cited as evidence for this.

On the contrary, this book maintains that there are no mathematical explanations of empirical facts, and all examples commonly cited as evidence for the existence of such explanations do not really prove what they purport to prove.

26.15 The Applicability of Mathematics

Another important question about mathematics is why mathematics is applicable to the world. In the past century, the prevailing view has been that the applicability of mathematics to the world is something bordering on the mysterious, for which there is no rational explanation.

On the contrary, this book maintains that the applicability of mathematics to the world is by no means mysterious and can be rationally accounted for. On the one hand, the applicability of natural mathematics to the world is due to the fact that natural mathematics fits in certain mathematical properties of the world. On the other hand, the applicability of artificial mathematics to the world is due to several factors, starting with the decision of modern science to deal only with some phenomenal properties of the world, mathematical in kind.

26.16 The Role of Knowledge in Human Life

It has been stated above that a philosophy aimed at knowledge and methods to acquire knowledge requires a rethinking of knowledge. The importance of this rethinking depends on the fact that knowledge plays a central role in the life of all living beings.

It is through knowledge that life, from the prokaryotes to human beings, has been able to preserve itself. The prokaryotes which solved their survival problem using data obtained through their rudimentary sense organs, were the first to discover that knowledge is indispensable to life, and indeed a precondition of it.

In particular, knowledge plays a central role in human life at all levels, from survival to improving the quality of life, and is a precondition of happiness. Therefore, a philosophy which wants to be still fruitful must aim at knowledge and improve our methods to acquire it.

Bibliography

Abbott, Derek. 2013. The reasonable ineffectiveness of mathematics. *Proceedings of the IEEE* 101: 2147–2153.

Aliseda, Atocha. 2006. *Abductive reasoning: Logical investigations into discovery and explanations*. Berlin: Springer.

Andersen, Ross. 2012. Has physics made philosophy and religion obsolete? Interview to Lawrence Krauss. *The Atlantic*, April 23.

Anderson, Chris. 2008. The end of theory: The data deluge makes the scientific method obsolete. *Wired Magazine*, 23 June.

Anscombe, Elisabeth. 1981. *The collected philosophical papers*. Oxford: Blackwell.

Appel, Kenneth, and Wolfgang Haken. 1989. *Every planar map is four colorable*. Providence: American Mathematical Society.

Arana, Andrew. 2015. On the depth of Szemerédi's theorem. *Philosophia Mathematica* 23: 163–176.

Archimedes. 2013. *Opera omnia*. Cambridge: Cambridge University Press.

Atiyah, Michael. 1984. Interview by Roberto Minio. *The Mathematical Intelligencer* 6 (1): 9–19.

———. 1995. Creation v. discovery. *Times Higher Education Supplement*, 29 September.

———. 2005. Joint interview with Isadore Singer, by Martin Raussen and Christian Skau. *Notices of the American Mathematical Society* 52: 223–231.

Auden, Wystan. 1950. *Collected shorter poems 1930–1944*. London: Faber & Faber.

Austin, John Langshaw. 1970. *Philosophical papers*. Oxford: Oxford University Press.

Avigad, Jeremy. 2008. Understanding proofs. In *The philosophy of mathematical practice*, ed. Paolo Mancosu, 317–353. Oxford: Oxford University Press.

Ayer, Alfred Jules. 1990. *Language, truth and logic*. London: Penguin Books.

Bacon, Francis. 1961–1986. *Works*. Stuttgart Bad Cannstatt: Frommann Holzboog.

Baker, Alan. 2005. Are there genuine mathematical explanations of physical phenomena? *Mind* 114: 223–238.

———. 2009. Mathematical explanation in science. *The British Journal for the Philosophy of Science* 60: 611–633.

Baker, Alan, and Mark Colyvan. 2011. Indexing and mathematical explanation. *Philosophia Mathematica* 19: 323–334.

Balaguer, Mark. 1998. *Platonism and anti-platonism in mathematics*. Oxford: Oxford University Press.

Bangu, Sorin. 2016. On 'The unreasonable effectiveness of mathematics in the natural sciences'. In *Models and inferences in science*, eds. Emiliano Ippoliti, Fabio Sterpetti, and Thomas Nickles, 11–29. Cham: Springer.

© Springer International Publishing AG 2017

C. Cellucci, *Rethinking Knowledge*, European Studies in Philosophy of Science 4,
DOI 10.1007/978-3-319-53237-0

Bardi, Jason Socrates. 2006. *The calculus wars: Newton, Leibniz, and the greatest mathematical clash of all time*. New York: Thunder's Mouth Press.

Barrow, John David. 2000. *Between inner space and outer space: Essays on science, art, and philosophy*. Oxford: Oxford University Press.

Barwise, Jon, and John Etchemendy. 1996. Visual information and valid reasoning. In *Logical reasoning with diagrams*, eds. Gerard Allwein and Jon Barwise, 3–25. Oxford: Oxford University Press.

Bealer, George. 1996. On the possibility of philosophical knowledge. *Philosophical Perspectives* 10: 1–34.

Bennett, Maxwell Richard, and Peter Michael Stephan Hacker. 2003. *Philosophical foundations of neuroscience*. Oxford: Blackwell.

Berkeley, George. 1948–1957. *Works*. London: Nelson.

Berman, Jules. 2013. *Principles of big data: Preparing, sharing, and analyzing complex information*. Amsterdam: Elsevier.

Blackburn, Simon. 2010. *Practical tortoise raising and other philosophical essays*. Oxford: Oxford University Press.

Boltzmann, Ludwig. 1974. *Theoretical physics and philosophical problems: Selected writings*. Dordrecht: Springer.

Boole, George. 1958. *An investigation of the laws of thought on which are founded the mathematical theories of logic and probabilities*. Mineola: Dover.

Boolos, George. 1998. *Logic, logic, and logic*. Cambridge: Harvard University Press.

Borges, Jorge Luis. 1975. Of exactitude in science. In Jorge Luis Borges, *A universal history of infamy*, 131. London: Penguin Books.

Bostock, David. 2009. *Philosophies of mathematics: An introduction*. Oxford: Wiley-Blackwell.

Bourbaki, Nicholas. 1949. Foundations of mathematics for the working mathematician. *The Journal of Symbolic Logic* 14: 1–8.

———. 1950. The architecture of mathematics. *The American Mathematical Monthly* 57: 221–232.

———. 1994. *Elements of the history of mathematics*. New York: Springer.

Brannon, Elizabeth. 2005. The independence of language and mathematical reasoning. *Proceedings of the National Academy of Sciences of the United States of America* 102: 3177–3178.

Breitenbach, Angela. 2013. Aesthetics in science: A Kantian proposal. *Proceedings of the Aristotelian Society* 113: 83–100.

Brouwer, Luitzen Egbertus Jan. 1975. *Collected works*, vol. 1. Amsterdam: North-Holland.

Bueno, Otávio, and Mark Colyvan. 2011. An inferential conception of the application of mathematics. *Noûs* 45: 345–374.

Bueno, Otávio, and Steven French. 2012. Can mathematics explain physical phenomena? *The British Journal for the Philosophy of Science* 63: 85–113.

Butterworth, Brian. 1999. *The mathematical brain*. London: Macmillan.

Byers, William. 2007. *How mathematicians think*. Princeton: Princeton University Press.

Calvino, Italo. 1974. *Invisible cities*. New York: Harcourt Brace Jovanovich.

———. 1988. *Six memos for the next millennium*. Cambridge: Harvard University Press.

———. 2010. *Palomar*. New York: Random House.

———. 2014. *Collection of sand*. New York: Houghton Mifflin Harcourt.

Cambiano, Giuseppe. 1967. Il metodo ipotetico e le origini della sistemazione euclidea della geometria. *Rivista di Filosofia* 58: 115–149.

Cantor, Georg. 1996. Foundations of a general theory of manifolds: A mathematico-philosophical investigation into the theory of the infinite. In *From Kant to Hilbert: A source book in the foundations of mathematics*, ed. William Ewald, vol. 2, 881–920. Oxford: Oxford University Press.

Capozzi, Mirella. 2013. *Kant e la logica I*. Naples: Bibliopolis.

Carl, Wolfgang. 1994. *Frege's theory of sense and reference: Its origin and scope*. Cambridge: Cambridge University Press.

Carnap, Rudolf. 1935. *Philosophy and logical syntax*. London: Kegan Paul.

———. 1939. *Foundations of logic and mathematics*. Chicago: Chicago University Press.
Carnap, Rudolph. 1966. *Philosophical foundations of physics: An introduction to the philosophy of science*. New York: Basic Books.
Carnap, Rudolf. 2001. *Logical syntax of language*. London: Routledge.
———. 2003. *The logical structure of the world – Pseudoproblems in philosophy*. Chicago: Open Court.
Carroll, Lewis. 1977. *Symbolic logic*. New York: Clarkson N. Potter.
———. 1996. *Sylvie and Bruno concluded*. In *The complete illustrated Lewis Carroll*, 457–674. Ware: Wordsworth.
Cellucci, Carlo. 1995. Frege e le origini della logica matematica. Introduction to Gottlob Frege, *Leggi fondamentali dell'aritmetica*, ix–lxvi. Rome: Teknos.
———. 1998. *Le ragioni della logica*. Rome: Laterza.
———. 2007. *La filosofia della matematica del Novecento*. Rome: Laterza.
———. 2013a. *Rethinking logic: Logic in relation to mathematics, evolution, and method*. Dordrecht: Springer.
———. 2013b. Top-down and bottom-up philosophy of mathematics. *Foundations of Science* 18: 93–106.
———. 2017. Varieties of maverick philosophy of mathematics. In *Humanizing mathematics and its philosophy*, ed. Bharath Sriraman, forthcoming. Basel: Birkhäuser.
Chace, Arnold Buffum (ed.). 1927. *The Rhind mathematical papyrus*, vol. 1. Oberlin: Mathematical Association of America.
Chakrabarti, Kisor Kumar. 2010. *Classical Indian philosophy of induction: The Nyāya viewpoint*. Lanham: Lexington Books.
Chang, Hasok. 2012. *Is water H_2O? Evidence, realism and pluralism*. Dordrecht: Springer.
Chisholm, Roderick. 1989. *Theory of knowledge*. London: Prentice-Hall.
Church, Alonzo. 2011. Letter to Stephen Kleene, 29 November 1935. In Appendix B to Rod Adams, *An early history of recursive functions and computability: From Gödel to Turing*, 223–227. Boston: Docent Press.
Clark, Andy. 2003. *Natural-born cyborgs: Minds, technologies, and the future of human intelligence*. Oxford: Oxford University Press.
Clavius, Christophorus. 1601. *In sphaeram Ioannis de Sacro Bosco commentarius*. Venice: Sub signo Solis.
Colyvan, Mark. 2013. *An introduction to the philosophy of mathematics*. Cambridge: Cambridge University Press.
Conway, John, and Richard Guy. 1996. *The book of numbers*. New York: Copernicus.
Cooper, William. 2001. *The evolution of reason: Logic as a branch of biology*. Cambridge: Cambridge University Press.
Crick, Francis. 1982. *Life itself: Its origin and nature*. New York: Simon & Schuster.
Curry, Haskell Brooks. 1951. *Outlines of a formalist philosophy of mathematics*. Amsterdam: North-Holland.
———. 1977. *Foundations of mathematical logic*. Mineola: Dover.
Daly, Christopher, and Simon Langford. 2009. Mathematical explanation and indispensability arguments. *The Philosophical Quarterly* 59: 641–658.
Darwin, Charles. 1903. *More letters of Charles Darwin*. London: Murray.
———. 2009a. *The origin of species by means of natural selection*. Cambridge: Cambridge University Press.
———. 2009b. *The life and letters of Charles Darwin, including an autobiographical chapter*. Cambridge: Cambridge University Press.
David, Marian. 2004. Theories of truth. In *Handbook of epistemology*, eds. Ilkka Niiniluoto, Matti Sintonen, and Jan Woleński, 331–414. Dordrecht: Springer.
Davies, Edward Brian. 2008. Interview. In *Philosophy of mathematics: 5 questions*, eds. Vincent Hendricks and Hannes Leitgeb, 87–99. New York: Automatic Press/VIP.
Davis, Martin. 2005. What did Gödel believe and when did he believe it? *The Bulletin of Symbolic Logic* 11: 194–206.

Davis, Philip. 1972. Fidelity in mathematical discourse: Is one and one really two? *The American Mathematical Monthly* 79: 252–263.
Davis, Philip J., and Reuben Hersh. 1981. *The mathematical experience*. Boston: Birkhäuser.
Dawkins, Richard. 2006. *The God delusion*. London: Bantam Press.
Dawson, John W. 2015. *Why prove it again? Alternative proofs in mathematical practice*. Basel: Birkhäuser.
Dedekind, Richard. 1967. Letter to Keferstein, 27 February 1890. In *From Frege to Gödel: A source book in mathematical logic, 1879–1931*, ed. Jean van Heijenoort, 99–103. Cambridge: Harvard University Press.
———. 1996a. Was sind und was sollen die Zahlen? In *From Kant to Hilbert: A source book in the foundations of mathematics*, ed. William Ewald, vol. 2, 790–833. Oxford: Oxford University Press.
———. 1996b. Letter to Heinrich Weber, 24 January 1888. In *From Kant to Hilbert: A source book in the foundations of mathematics*, ed. William Ewald, vol. 2, 834–835. Oxford: Oxford University Press.
Dehaene, Stanislas. 1997. *The number sense: How the mind creates mathematics*. Oxford: Oxford University Press.
———. 2009. Origins of mathematical intuitions: The case of arithmetic. *Annals of the New York Academy of Sciences* 1156: 232–259.
Dehaene, Stanislas, and Laurent Cohen. 2007. Cultural recycling of cortical maps. *Neuron* 56: 384–398.
Dehaene, Stanislas, Véronique Izard, Pierre Pica, and Elizabeth Spelke. 2006. Core knowledge of geometry in an Amazonian indigene group. *Science* 311: 381–384.
de l'Hospital, Guillaume. 1696. *Analyse des infiniment petits*. Paris: Imprimerie Royale.
Dembski, William. 2002. *Intelligent design: The bridge between science & theology*. Westmont: InterVarsity Press.
De Morgan, Augustus. 1835. Review of George Peacock, *A treatise on algebra*. *Quarterly Journal of Education* 9: 91–110, 293–311.
Dennett, Daniel. 1996. *Darwin's dangerous idea: Evolution and the meaning of life*. London: Penguin Books.
———. 2013. *Intuition pumps and other tools for thinking*. New York: Norton.
Descartes, René. 1996. *Oeuvres*. Paris: Vrin.
Detlefsen, Michael. 1992. On an alleged refutation of Hilbert's program using Gödel's first incompleteness theorem. In *Proof, logic and formalization*, ed. Michael Detlefsen, 199–235. London: Routledge.
Devitt, Michael. 2013. Hilary and me: Tracking down Putnam on the realism issue. In *Reading Putnam*, ed. Maria Baghramian, 101–120. London: Routledge.
Devlin, Keith. 1990. *Mathematics: The new golden age*. London: Penguin Books.
———. 2005. *The math instinct: Why you're a mathematical genius (along with lobsters, birds, cats, and dogs)*. New York: Thunder's Mouth Press.
Dewey, John. 1911. Brief studies in realism. I. *The Journal of Philosophy. Psychology and Scientific Method* 8: 393–400.
———. 1938. *Logic: The theory of inquiry*. New York: Holt, Rinehart & Winston.
———. 2004. *Reconstruction in philosophy*. Mineola: Dover.
Dieter, Kevin, Bo Hu, David Knill, Randolph Bike, and Duje Tadin. 2013. Kinestesis can make an invisible hand visible. *Psychological Science* 25: 66–75.
Dieudonné, Jean. 1964. Recent developments in mathematics. *The American Mathematical Monthly* 71: 239–248.
———. 1969. *Foundations of modern analysis*. New York: Academic Press.
———. 1975. L'abstraction et l'intuition mathématique. *Dialectica* 29: 39–54.
———. 2013. *Mathematics – the music of reason*. Berlin: Springer.
Dilthey, Wilhelm. 1989. *Introduction to the human sciences*. Princeton: Princeton University Press.

Dirac, Paul Adrien Maurice. 1951. A new classical theory of electrons. *Proceedings of the Royal Society of London*, Series A, 209: 291–296.
———. 1963. The evolution of the physicist's picture of nature. *Scientific American* 208: 43–53.
———. 1978. *Directions in physics*. New York: Wiley.
Dostoevsky, Fyodor. 1992. *The brothers Karamazov*. London: Everyman's Library.
Doyle, Arthur Conan. 2003. *Sherlock Holmes: The complete novels and stories*, vol. I. New York: Bantham.
Dummett, Michael. 1973. *Frege: Philosophy of language*. New York: Harper & Row.
———. 1978. *Truth and other enigmas*. Cambridge: Harvard University Press.
———. 1991. *The logical basis of metaphysics*. Cambridge: Harvard University Press.
———. 2006. *Thought and reality*. Oxford: Oxford University Press.
———. 2007. The place of philosophy in European culture. *European Journal of Analytic Philosophy* 3: 21–30.
———. 2010. *The nature and future of philosophy*. New York: Columbia University Press.
Dyson, Freeman John. 1956. Obituary: Hermann Weyl. *Nature* 177: 457–458.
———. 1992. *From Eros to Gaia*. New York: Pantheon Books.
———. 1999. Foreword. In Michael Monastyrsky, *Riemann, topology, and physics*, ix–x. Boston: Birkhäuser.
Dyson, Freeman. 2012. What can you really know? *The New York Review of Books* 59. *No.* 17 (November 8): 18–20.
Eddington, Arthur. 1939. *The philosophy of physical science*. Cambridge: Cambridge University Press.
Einstein, Albert. 1954. A testimonial. In Jacques Hadamard, *The psychology of invention in the mathematical field*, 142–143. Mineola: Dover.
———. 1998. Letter to Eduard Study, 25 September 1918. In Albert Einstein, *Collected papers*, vol. 8, 651–653. Princeton: Princeton University Press.
———. 2002. Induction and deduction in physics. In Albert Einstein, *Collected papers*, vol. 7, 108–109. Princeton: Princeton University Press.
———. 2010. *Ideas and opinions*. New York: Crown.
Einstein, Albert, and Leopold Infeld. 1966. *The evolution of physics*. New York: Simon & Schuster.
Ellis, Brian. 2002. *The philosophy of nature: A guide to the new essentialism*. Chesham: Acumen.
———. 2008. Essentialism and natural kinds. In *The Routledge companion to philosophy of science*, eds. Martin Curd and Stathis Psillos, 139–148. London: Routledge.
Engel, Pascal. 2002. *Truth*. Chesham: Acumen.
Etcoff, Nancy. 1999. *Survival of the prettiest: The science of beauty*. New York: Anchor.
Feferman, Solomon. 1998. *In the light of logic*. Oxford: Oxford University Press.
Feyerabend, Paul K. 1975. Machamer on Galileo. *Studies in History and Philosophy of Science* 5: 297–304.
———. 1981–2016. *Philosophical papers*. Cambridge: Cambridge University Press.
———. 1993. *Against method*. London: Verso.
Field, Hartree. 1980. *Science without numbers*. Oxford: Blackwell.
———. 1989. *Realism, mathematics and modality*. Oxford: Blackwell.
Finocchiaro, Maurice A. 2010. *Defending Copernicus and Galileo: Critical reasoning in the two affairs*. Dordrecht: Springer.
Fodor, Jerry. 1974. Special sciences (or: the disunity of science as a working hypothesis). *Synthese* 28: 97–115.
———. 1981. *Representation: Philosophical essays on the foundations of cognitive science*. Brighton: Harvester Press.
———. 1986. Why paramecia don't have mental representations. *Midwest Studies in Philosophy* 10: 3–23.
———. 2004. Water's water everywhere. *London Review of Books* 26 (20): 17–19.
Frankfurt, Harry. 1958. Peirce's notion of abduction. *The Journal of Philosophy* 55: 593–597.
Frege, Gottlob. 1960. *The foundations of arithmetic: A logico-mathematical enquiry into the concept of number*. New York: Harper.

———. 1964. On the scientific justification of a concept-script. *Mind* 73: 155–160.
———. 1967. *Begriffsschrift*, a formula language, modeled upon that of arithmetic, for pure thought. In *From Frege to Gödel: A source book in mathematical logic, 1879–1931*, ed. Jean van Heijenoort, 5–82. Cambridge: Harvard University Press.
———. 1979. *Posthumous writings*. Oxford: Blackwell.
———. 1980. *Philosophical and mathematical correspondence*. Oxford: Blackwell.
———. 1984. *Collected papers on mathematics, logic, and philosophy*. Oxford: Blackwell.
———. 2013. *Basic laws of arithmetic*. Oxford: Oxford University Press.
Friend, Michèle. 2014. *Pluralism in mathematics: A new position in philosophy of mathematics*. Dordrecht: Springer.
Gadamer, Hans Georg. 1998. *Reason in the age of science*. Cambridge: The MIT Press.
Galavotti, Maria Carla. 2014. For a bottom-up approach to the philosophy of science. In *From a heuristic point of view*, ed. Cesare Cozzo, and Emiliano Ippoliti, 195–211. Newcastle upon Tyne: Cambridge Scholars Publishing.
Galilei, Galileo. 1968. *Opere*. Florence: Barbera.
Gentzen, Gerhard. 1969. Investigations into logical deduction. In *The collected papers of Gerhard Gentzen*, ed. Manfred E. Szabo, 68–131. Amsterdam: North-Holland.
George, Alexander, and Daniel J. Velleman. 2002. *Philosophies of mathematics*. Oxford: Blackwell.
Gettier, Edmund L. 1963. Is justified true belief knowledge? *Analysis* 23: 121–123.
Giaquinto, Marcus. 2008. Visualizing in mathematics. In *The philosophy of mathematical practice*, ed. Paolo Mancosu, 22–42. Oxford: Oxford University Press.
Gibson, James. 1976. The myth of passive perception: A reply to Richards. *Philosophy and Phenomenological Research* 37: 234–238.
Gillies, Donald. 2014. Serendipity and mathematical logic. In *From a heuristic point of view*, ed. Cesare Cozzo, and Emiliano Ippoliti, 23–39. Newcastle upon Tyne: Cambridge Scholars Publishing.
Giusti, Enrico. 1980. *Bonaventura Cavalieri and the theory of indivisibles*. Rome: Cremonese.
Glock, Hans-Johann. 2008. *What is analytic philosophy?* Cambridge: Cambridge University Press.
Gödel, Kurt. 1986–2002. *Collected works*. Oxford: Oxford University Press.
Goodman, Nelson. 1976. *Languages of art: An approach to a theory of symbols*. Indianapolis: Hackett.
Goodman, Nicolas D. 1991. Modernizing the philosophy of mathematics. *Synthese* 88: 119–126.
Gowers, Timothy. 2002. *Mathematics: A very short introduction*. Oxford: Oxford University Press.
———. 2006. Does mathematics need a philosophy? In *18 unconventional essays on the nature of mathematics*, ed. Reuben Hersh, 182–200. New York: Springer.
———. 2007. Mathematics, memory and mental arithmetic. In *Mathematical knowledge*, ed. by Mary Leng, Alexander Paseau, and Michael Potter, 33–58. Oxford: Oxford University Press.
Gowers, Timothy, and Michael Nielsen. 2009. Massively collaborative mathematics. *Nature* 461(15 October): 879–881.
Grabiner, Judith. 1983. The changing concept of change: The derivative from Fermat to Weierstrass. *Mathematics Magazine* 56: 195–206.
———. 2005. *The origins of Cauchy's rigorous calculus*. Mineola: Dover.
Grattan-Guinness, Ivor. 2008. Solving Wigner's mystery: The reasonable (though perhaps limited) effectiveness of mathematics in the natural sciences. *The Mathematical Intelligencer* 30 (3): 7–17.
Grosholz, Emily. 1992. Was Leibniz a mathematical revolutionary? In *Revolutions in mathematics*, ed. Donald Gillies, 117–133. Oxford: Oxford University Press.
———. 2000. The partial unification of domains, hybrids, and the growth of mathematical knowledge. In *The growth of mathematical knowledge*, eds. Emily Grosholz and Herbert Breger, 81–91. Dordrecht: Springer.
———. 2007. *Representation and productive ambiguity in mathematics and the sciences*. Oxford: Oxford University Press.
———. 2015. Review of Carlo Cellucci, *Rethinking logic: Logic in relation to mathematics, evolution and method. Philosophia Mathematica* 23: 136–140.

Grothendieck, Alexander. 1985. *Récoltes et semailles. Réflexions et témoignage sur un passé de mathématicien*. Montpellier: Université des Sciences et Techniques du Languedoc.
Hadamard, Jacques. 1954. *The psychology of invention in the mathematical field*. Mineola: Dover.
Halmos, Paul Richard. 1968. Mathematics as a creative art. *American Scientist* 56: 375–389.
———. 1980. The heart of mathematics. *The American Mathematical Monthly* 87: 519–524.
———. 1992. Mathematics. In *Academic Press dictionary of science and technology*, ed. Christopher Morris, 1325. San Diego: Academic Press.
Hamming, Richard. 1980. The unreasonable effectiveness of mathematics. *The American Mathematical Monthly* 87: 81–90.
———. 1998. Mathematics on a distant planet. *The American Mathematical Monthly* 105: 640–650.
Hanna, Robert. 2006. *Kant, science, and human nature*. Oxford: Oxford University Press.
Hanson, Norwood Russell. 1965. *Patterns of discovery*. Cambridge: Cambridge University Press.
Hardy, Godfrey Harold. 1929. Mathematical proof. *Mind* 38: 1–25.
———. 1992. *A mathematician's apology*. Cambridge: Cambridge University Press.
Hart, Wilbur Dyre. 1996. Introduction. In *The philosophy of mathematics*, ed. Wilbur Dyre Hart, 1–13. Oxford: Oxford University Press.
Hawking, Stephen. 1988. *A brief history of time: From the big bang to black holes*. New York: Bantam Books.
Hawking, Stephen, and Leonard Mlodinow. 2010. *The grand design*. New York: Bantam Books.
Heidegger, Martin. 1968. *What is called thinking?* New York: Harper & Row.
———. 1972. *On time and being*. Chicago: The University of Chicago Press.
———. 1994. *Basic questions of philosophy: Selected 'problems' of 'logic'*. Bloomington: Indiana University Pres.
———. 1998. *Pathmarks*. Cambridge: Cambridge University Press.
Heisenberg, Werner. 2001. Science and the beautiful. In *Quantum questions*, ed. Ken Wilber, 56–69. Boston: Shambhala.
Hempel, Carl Gustav. 1966. *Philosophy of natural science*. Englewood Cliffs: Prentice-Hall.
———. 2000. *Selected philosophical essays*. Cambridge: Cambridge University Press.
———. 2001. *The philosophy of Carl G. Hempel*: Oxford University Press, Oxford.
Hempel, Carl Gustav, and Paul Oppenheim. 1948. Studies in the logic of explanation. *Philosophy of Science* 15: 135–175.
Hersh, Reuben. 1979. Some proposals for reviving the philosophy of mathematics. *Advances in Mathematics* 31: 31–50.
———. 1997. *What is mathematic, really?* Oxford: Oxford University Press.
———. 2014. *Experiencing mathematics: What do we do, when we do mathematics?* Providence: American Mathematical Society.
Hersh, Reuben, and Vera John-Steiner. 2011. *Loving + hating mathematics: Challenging the myths of mathematical life*. Princeton: Princeton University Press.
Hertz, Heinrich Rudolf. 1956. *The principles of mechanics presented in a new form*. Mineola: Dover.
Hespos, Susan J., and Kristy vanMarle. 2012. Physics for infants: Characterizing the origins of knowledge about objects, substances, and number. *WIREs Cognitive Science* 3: 19–27.
Hesseling, Dennis E. 2003. *Gnomes in the fog: The reception of Brouwer's intuitionism in 1920s*. Basel: Springer.
Hilbert, David. 1902–1903. Über den Satz von der Gleichheit der Basiswinkel im gleichschenkligen Dreieck. *Proceedings of the London Mathematical Society* 35: 50–68.
———. 1905. *Logische Prinzipien des mathematisches Denkens*. Unpublished manuscript. Göttingen: Bibliothek, Mathematischen Institut, Universität Göttingen.
———. 1967a. On the infinite. In *From Frege to Gödel: A source book in mathematical logic, 1879–1931*, ed. Jean van Heijenoort, 369–392. Cambridge: Harvard University Press.
———. 1967b. The foundations of mathematics. In *From Frege to Gödel: A source book in mathematical logic, 1879–1931*, ed. Jean van Heijenoort, 464–479. Cambridge: Harvard University Press.

———. 1980a. Letter to Frege, 29 December 1899. In Gottlob Frege, *Philosophical and mathematical correspondence*, 38–41. Oxford: Blackwell.
———. 1980b. Letter to Frege, 22 September 1900. In Gottlob Frege, *Philosophical and mathematical correspondence*, 50–51. Oxford: Blackwell.
———. 1987. *Grundlagen der Geometrie*. Stuttgart: Teubner.
———. 1992. *Natur und mathematischen Erkennen: Vorlesungen, gehalten 1919–1920 in Göttingen*. Basel: Birkhäuer.
———. 1996a. On the concept of number. In *From Kant to Hilbert: A source book in the foundations of mathematics, II*, ed. William Ewald, 1092–1095. Oxford: Oxford University Press.
———. 1996b. Axiomatic thought. In *From Kant to Hilbert: A source book in the foundations of mathematics*, II, ed. William Ewald, 1107–1115. Oxford: Oxford University Press.
———. 1996c. The new grounding of mathematics: First report. In *From Brouwer to Hilbert*, II, ed. William Ewald, 1117–1134. Oxford: Oxford University Press.
———. 1996d. The logical foundations of mathematics. In *From Kant to Hilbert: A source book in the foundations of mathematics*, II, ed. William Ewald, 1134–1148. Oxford: Oxford University Press.
———. 1996e. The grounding of elementary number theory. In *From Kant to Hilbert: A source book in the foundations of mathematics*, II, ed. William Ewald, 1149–1157. Oxford: Oxford University Press.
———. 1996f. Logic and the knowledge of nature. In *From Kant to Hilbert: A source book in the foundations of mathematics, II*, ed. William Ewald, 1157–1165. Oxford: Oxford University Press.
———. 2000. Mathematical problems. Appendix to Jeremy Gray, *The Hilbert challenge*, 240–282. Oxford: Oxford University Press.
———. 2004a. Die Grundlagen der Geometrie. In *David Hilbert's lectures on the foundations of geometry 1891–1902*, eds. Michael Hallett and Ulrich Majer, 72–123. Berlin: Springer.
———. 2004b. Grundlagen der Euklidischen Geometrie. In *David Hilbert's lectures on the foundations of geometry 1891–1902*, eds. Michael Hallett and Ulrich Majer, 221–286. Berlin: Springer.
———. 2004c. Grundlagen der Geometrie. In *David Hilbert's lectures on the foundations of geometry 1891–1902*, eds. Michael Hallett and Ulrich Majer, 540–602. Berlin: Springer.
———. 2013. Beweis des Tertium non datur. In David Hilbert, *Lectures on the foundations of arithmetic and logic 1917–1933*, eds. William Ewald and Wilfried Sieg, 985–989. Berlin: Springer.
Hilbert, David, and Wilhelm Ackermann. 1928. *Grundzüge der theoretischen Logik*. Berlin: Springer.
Hilbert, David, and Paul Bernays. 1968–1970. *Grundlagen der Mathematik*. Berlin: Springer.
Hintikka, Jaakko. 1996. *The principles of mathematics revisited*. Cambridge: Cambridge University Press.
———. 2012. Method of analysis: A paradigm of mathematical reasoning? *History and Philosophy of Logic* 33: 49–67.
Hintikka, Jaakko, and Unto Remes. 1976. Ancient geometrical analysis and modern logic. In *Essays in memory of Imre Lakatos*, eds. Robert Cohen, Paul Feyerabend, and Max Wartofsky, 253–276. Dordrecht: Springer.
Hippocrates. 2005. *On ancient medicine*. Leiden: Brill.
Hoffmann, Michael. 1999. Problems with Peirce's concept of abduction. *Foundations of Science* 4: 271–305.
Hofstadter, Douglas. 1999. *Gödel, Escher, Bach: An eternal golden braid*. New York: Basic Books.
Hooker, Cliff, ed. 2011. *Philosophy of complex systems*. Amsterdam: North Holland.
Hume, David. 1975. *Enquiries concerning human understanding and concerning the principles of morals*. Oxford: Oxford University Press.
———. 1978. *A treatise of human nature*. Oxford: Oxford University Press.
———. 1998. *Dialogues concerning natural religion – Of the immortality of the soul – Of suicide – Of miracles*. Indianapolis: Hackett.

Hunt, Gavin R., and Russell D. Gray. 2006. Tool manufacture by New Caledonian crows: Chipping away at human uniqueness. *Acta Zoologica Sinica* 52(Supplement): 622–625.
Huntley, H.E. 1970. *The divine proportion: A study of mathematical beauty*. Mineola: Dover.
Husserl, Edmund. 1960. *Cartesian meditation: An introduction to phenomenology*. The Hague: Martinus Nijhoff.
———. 1970. *The crisis of European sciences and transcendental phenomenology*. Evanston: Northwestern University Press.
———. 2001. *Logical investigations*. London: Routledge.
———. 2002. Philosophy as rigorous science. *The New Yearbook for Phenomenology and Phenomenological Philosophy* 2: 249–295.
Jaynes, Edwin, and George Bretthorst. 2003. *Probability theory*. Cambridge: Cambridge University Press.
Joachim, Harold. 1906. *The nature of truth*. Oxford: Oxford University Press.
Kac, Mark, and Stanislaw M. Ulam. 1992. *Mathematics and logic*. Mineola: Dover.
Kant, Immanuel. 1900–. *Gesammelte Schriften (Akademie Aausgabe)*. Berlin: Königlich Preußischen Akademie der Wissenschaften.
———. 1992. *Lectures on logic*. Cambridge: Cambridge University Press.
———. 1996. *Practical philosophy*. Cambridge: Cambridge University Press.
———. 1998. *Critique of pure reason*. Cambridge: Cambridge University Press.
———. 1999. *Correspondence*. Cambridge: Cambridge University Press.
———. 2000. *Critique of the power of judgment*. Cambridge: Cambridge University Press.
———. 2002. *Theoretical philosophy after 1781*. Cambridge: Cambridge University Press.
———. 2007. *Anthropology, history, and education*. Cambridge: Cambridge University Press.
Kantorovich, Aharon. 1993. *Scientific discovery: Logic and tinkering*. Albany: State University of New York Press.
Karihaloo, Bhushan, K. Zhang, and J. Wang. 2013. Honeybee combs: How the circular cells transform into rounded hexagons. *Journal of the Royal Society Interface* 10 (86). doi:10.1098/rsif.2013.0299.
Kekes, John. 2015. Philosophy: Scientific or humanistic? *Philosophy* 90: 241–262.
Kepler, Johannes. 1937– . *Gesammelte Werke*. München: Beck.
Khalidi, Muhammad. 2009. How scientific is scientific essentialism? *Journal for General Philosophy of Science* 40: 85–101.
Kitcher, Philip. 1983. *The nature of mathematical knowledge*. Oxford: Oxford University.
———. 1991. Persuasion. In *Persuading science: The art of scientific rhetoric*, eds. Marcello Pera and William Shea, 3–27. Sagamore Beach: Science History Publications.
———. 2011. Epistemology without history is blind. *Erkenntnis* 75: 505–524.
Kline, Morris. 1964. *Mathematics in Western culture*. Oxford: Oxford University Press.
———. 1972. *Mathematical thought from ancient to modern times*. Oxford: Oxford University Press.
———. 1985. *Mathematics for the nonmathematician*. Mineola: Dover.
Knorr, Wilbur Richard. 1993. *The ancient tradition of geometric problems*. Mineola: Dover.
Körner, Stephan. 1986. *The philosophy of mathematics: An introductory essay*. Mineola: Dover.
Kosso, Peter. 2002. The omniscienter: Beauty and scientific understanding. *International Studies in the Philosophy of Science* 16: 39–48.
Krantz, Steven G. 2011. *The proof is in the pudding: The changing nature of mathematical proof*. New York: Springer.
Kripke, Saul. 1980. *Naming and necessity*. Cambridge: Harvard University Press.
Kuhn, Thomas Samuel. 1957. *The Copernican revolution*. Cambridge: Harvard University Press.
———. 1977. *The essential tension: Selected studies in scientific tradition and change*. Chicago: The University of Chicago Press.
———. 1996. *The structure of scientific revolutions*. Chicago: The University of Chicago Press.
Lakatos, Imre. 1961. *Essays in the logic of mathematical discovery*. Cambridge: University of Cambridge.

———. 1971. Reply to critics. In *PSA 1970: In memory of Rudolf Carnap*, eds. Roger C. Buck and Robert S. Cohen, 174–182. Dordrecht: Springer.

———. 1976. *Proofs and refutations: The logic of mathematical discovery*. Cambridge: Cambridge University Press.

———. 1978. *Philosophical papers*. Cambridge: Cambridge University Press.

Lam, Clement. 1990. How reliable is a computer-based proof? *The Mathematical Intelligencer* 12 (1): 8–12.

Lange, Marc. 2013. What makes a scientific explanation distinctively mathematical? *The British Journal for the Philosophy of Science* 64: 485–511.

Laudan, Larry. 1977. *Progress and its problems*. Berkeley: University of California Press.

———. 1981. *Science and hypothesis: Historical essays on scientific methodology*. Dordrecht: Springer.

Laugwitz, Detlef. 2000. Controversies about numbers and functions. In *The growth of mathematical knowledge*, eds. Emily Grosholz and Herbert Breger, 177–198. Dordrecht: Springer.

Law, Stephen. 2012. The meaning of life. *Think* 11: 25–38.

Lecat, Maurice. 1935. *Erreurs de mathématiciens des origines à nos jours*. Bruxelles: Castaigne.

Lehman, Hugh. 1979. *Introduction to the philosophy of mathematics*. Oxford: Blackwell.

Leibniz, Gottfried Wilhelm. 1965. *Die Philosophischen Schriften*. Hildesheim: Olms.

Lewis, David. 1996. Elusive knowledge. *Australasian Journal of Philosophy* 74: 549–567.

Lobachevsky, Nikolai. 1898–1899. *Zwei geometrische Abhandlungen*. Leipzig: Teubner.

Locke, John. 1975. *An essay concerning human understanding*. Oxford: Oxford University Press.

Lorenz, Konrad. 1977. *Behind the mirror*. London: Methuen.

———. 2009. Kant's doctrine of the *a priori* in the light of contemporary biology. In *Philosophy after Darwin: Classic and contemporary readings*, ed. Michael Ruse, 231–247. Princeton: Princeton University Press.

Lyon, Aidan, and Mark Colyvan. 2008. The explanatory power of phase spaces. *Philosophia Mathematica* 16: 227–243.

Lyons, Timothy D. 2006. Scientific realism and the stratagema de Divide et Impera. *The British Journal for the Philosophy of Science* 57: 537–560.

Mach, Ernst. 1976. *Knowledge and error: Sketches on the psychology of enquiry*. Dordrecht: Springer.

Macintyre, Angus. 2005. The mathematical significance of proof theory. *Philosophical Transactions of the Royal Society* 363: 2419–2435.

———. 2011. The impact of Gödel's incompleteness theorems on mathematics. In *Kurt Gödel and the foundations of mathematics: Horizons of truth*, eds. Matthias Baaz et al., 3–25. Cambridge: Cambridge University Press.

Mac Lane, Saunders. 1986. *Mathematics: Form and function*. New York: Springer.

———. 1994. Response. In Michael Atiyah et al., Responses to 'Theoretical mathematics: Toward a cultural synthesis of mathematics and theoretical physics', by A. Jaffe and F. Quinn, 190–193. *Bulletin of the Americal Mathematical Society* 30: 178–207.

Mäenpää, Petri. 1997. From backward reduction to configurational analysis. In *Analysis and synthesis in mathematics: History and philosophy*, eds. Michael Otte and Marco Panza, 201–226. Dordrecht: Springer.

Maloney, Christopher. 1989. *The mundane matter of the mental language*. Cambridge: Cambridge University Press.

Mancosu, Paolo. 2008. Introduction. In *The philosophy of mathematical practice*, ed. Paolo Mancosu, 1–21. Oxford: Oxford University Press.

Maslow, Abraham H. 1966. *The psychology of science: A reconnaissance*. New York: Harper & Row.

McAllister, James. 2005. Mathematical beauty and the evolution of the standards of mathematical proof. In *The visual mind II*, ed. Michele Emmer, 15–34. Cambridge: The MIT Press.

Meikle, Laura I., and Jacques D. Fleuriot. 2003. Formalizing Hilbert's *Grundlagen* in Isabelle/Isar. In *Theorem proving in higher order logics*, eds. D. Basin and B. Wolff, 319–334. Berlin: Springer.

Meinong, Alexius. 1968–1978. *Gesamtausgabe*. Graz: Akademische Druck- und Verlagsanstalt.
Menn, Stephen. 2002. Plato and the method of analysis. *Phronesis* 47: 193–223.
Merton, Robert K., and Elinor Barber. 2004. *The travels and adventures of serendipity*. Princeton: Princeton University Press.
Mill, John Stuart. 1963–1986. *Collected works*. Toronto: University of Toronto Press.
Moore, George Edward. 1942. An Autobiography. In *The Philosophy of G. E. Moore*, ed. Paul Arthur Schilpp, 3–39. La Salle: Open Court.
Mordell, Louis. 1959. *Reflections of a mathematician*. Montreal: Canadian Mathematical Congress, McGill University.
Mosconi, Jean. 2014. The development of the concept of machine computability from 1936 to the 1960s. In *Constructivity and computability in historical and philosophical perspective*, eds. Jacques Dubucs and Michel Bourdeau, 37–56. Dordrecht: Springer.
Musgrave, Alan. 1988. The ultimate argument for scientific realism. In *Relativism and realism in science*, ed. Robert Nola, 229–252. Dordrecht: Springer.
———. 1996. Realism, truth and objectivity. In *Realism and anti-realism in the philosophy of science*, eds. Robert S. Cohen, Risto Hilpinen, and Qiu Renzong, 19–44. Dordrecht: Springer.
———. 1999. How to do without inductive logic. *Science & Education* 8: 395–412.
———. 2006. Responses. In *Rationality and reality: Conversations with Alan Musgrave*, eds. Colin Cheyne and John Worrall, 293–333. Dordrecht: Springer.
———. 2011. Popper and hypothetico-deductivism. In *Inductive logic*, eds. Dov Gabbay, Stephan Hartmann, and John Woods, 205–234. Amsterdam: North-Holland.
Nagel, Thomas. 1986. *The view from nowhere*. Oxford: Oxford University Press.
———. 1987. *What does it all mean? A very short introduction to philosophy*. Oxford: Oxford University Press.
———. 1997. *The last word*. Oxford: Oxford University Press.
Nahin, Paul. 2006. *Dr. Euler's fabulous formula: Cures many mathematical ills*. Princeton: Princeton University Press.
Nelsen, Roger B. 1993. *Proofs without words: Exercises in visual thinking*. Washington: The Mathematical Association of America.
Netz, Reviel. 2000. Why did Greek mathematicians publish their analyses? In *Ancient & medieval traditions in the exact sciences*, ed. Patrick Suppes, Julis M. Moravcsik, and Henry Mendell, 139–157. Stanford: CSLI Publications.
Newton, Isaac. 1714–1716. An account of the book entituled 'Commercium Epistolicum Collinii et aliorum, de analysi promota'. *Philosophical Transactions of the Royal Society of London* 29: 173–224.
———. 1720. *Universal arithmetick: or, a treatise of arithmetical composition and resolution. To which is added, Dr. Halley's method of finding the roots of equations arithmetically*. London: J. Senex, W. Taylor, T. Warner, and J. Osborn.
———. 1952. *Opticks, or a treatise of the reflections, refractions, inflections & colours of light*. Mineola: Dover.
———. 1962. *Mathematical principles of natural philosophy – The system of the world*. Berkeley: University of California Press.
———. 1967–1981. *The mathematical papers*. Cambridge: Cambridge University Press.
Novalis. 1996. Christianity or Europe. A fragment. In *The early political writings of the German romantics*, ed. Frederick Beiser, 59–79. Cambridge: Cambridge University Press.
———. 2007. *Notes for a romantic encyclopedia: Das Allgemeine Brouillon*. Albany: State University of New York Press.
Nozick, Robert. 1997. *Socratic puzzles*. Cambridge: Harvard University Press.
Ó Cairbre, Fiacre. 2009. The importance of being beautiful in mathematics. *IMTA Newsletter* 109: 29–45.
Ortega y Gasset, José. 1957. *On love: Aspects of a single theme*. New York: Grenwich Editions/ Meridian Books.
Ostrovsky, Yuri, Aaaron Andalman, and Pawan Sinha. 2006. Vision following extended congenital blindness. *Psychological Science* 17: 1009–1014.

Pappus. 1876–1878. *Collectio*. Berlin: Weidmann.

Parsons, Charles. 1983. *Mathematics in philosophy: Selected essays*. Ithaca: Cornell University Press.

Pascal, Blaise. 1904–1914. *Oeuvres*. Paris: Hachette.

———. 1995. *Pensées and other writings*. Oxford: Oxford University Press.

Pasteur, Luis. 1922–1939. *Oeuvres*. Paris: Masson.

Patokorpi, Erkki. 2007. Logic of Sherlock Holmes in technology enhanced learning. *Educational Technology & Society* 10 (1): 171–185.

Peano, Giuseppe. 1973. Definitions in mathematics. In Giuseppe Peano, *Selected works*, 235–246. London: Allen & Unwin.

Peckhaus, Volker. 2002. Regressive analysis. *Logical Analysis and History of Philosophy* 5: 97–110.

Peirce, Charles Sanders. 1931–1958. *Collected papers*. Cambridge: Harvard University Press.

Pennings, Timothy. 2003. Do dogs know calculus? *The College Mathematics Journal* 34: 178–182.

Pica, Pierre, Cathy Lemer, Véronique Izard, and Stanislas Dehaene. 2004. Exact and approximate arithmetic in an Amazonian indigene group. *Science* 306: 499–503.

Pincock, Christopher. 2007. A role for mathematics in the physical sciences. *Noûs* 41: 253–275.

———. 2012. *Mathematics and scientific representation*. Oxford: Oxford University Press.

Pinker, Steven. 1995. *The language instinct*. New York: Harper-Collins.

———. 2002. *The blank slate. The modern denial of human nature*. London: Penguin Books.

Plantinga, Alvin. 2011. *Where the conflict really lies: Science, religion, and naturalism*. Oxford: Oxford University Press.

Plotkin, Henry. 1997. *Darwin machines and the nature of knowledge*. Cambridge: Harvard University Press.

Poggiolesi, Francesca. 2011. *Gentzen calculi for modal propositional logic*. Dordrecht: Springer.

———. 2012. On the importance of being analytic: The paradigmatic case of the logic of proofs. *Logique et Analyse* 55: 443–461.

Poincaré, Henri. 1910. *Savants et écrivains*. Paris: Flammarion.

———. 2013. *The foundations of science: Science and hypothesis – The value of science – Science and method*. Cambridge: Cambridge University Press.

Pólya, George. 1941. Heuristic reasoning and the theory of probability. *The American Mathematical Monthly* 48: 450–465.

———. 1954. *Mathematics and plausible reasoning*. Princeton: Princeton University Press.

———. 1971. *How to solve it: A new aspect of mathematical method*. Princeton: Princeton University Press.

Popper, Karl Raimund. 1945. *The open society and its enemies*. London: Routledge.

———. 1959. *The logic of scientific discovery*. London: Hutchinson.

———. 1972. *Objective knowledge*. Oxford: Oxford University Press.

———. 1974. *Conjectures and refutations: The growth of scientific knowledge*. London: Routledge.

———. 1985. Evolutionary epistemology. In *Open questions in quantum physics*, eds. Gino Tarozzi and Alwyn van der Merwe, 395–413. Dordrecht: Springer.

———. 1996. *In search of a better world*. London: Routledge.

———. 1999. *All life is problem solving*. London: Routledge.

———. 2000. *Realism and the aim of science*. London: Routledge.

———. 2002. *Unended quest: An intellectual autobiography*. London: Routledge.

———. 2005. *The logic of scientific discovery*. London: Routledge.

Post, Emil Leon. 1965. Absolutely unsolvable problems and relatively undecidable propositions: Account of an anticipation. In *The undecidable*, ed. Martin Davis, 340–433. New York: Raven Press.

Prawitz, Dag. 2014. The status of mathematical knowledge. In *From a heuristic point of view*, ed. Cesare Cozzo, and Emiliano Ippoliti, 73–90. Newcastle upon Tyne: Cambridge Scholars Publishing.

Proclus. 1992. *In primum Euclidis Elementorum librum commentarii*. Hildesheim: Olms.
Psillos, Stathis. 1999. *Scientific realism: How science tracks truth*. London: Routledge.
Putnam, Hilary. 1975–1983. *Philosophical papers*. Cambridge: Cambridge University Press.
———. 1978. *Meaning and the moral sciences*. London: Routledge.
———. 1992. *Renewing philosophy*. Cambridge: Harvard University Press.
———. 2001. Reply to Bernard Williams' 'Philosophy as a humanistic discipline'. *Philosophy* 76: 605–614.
Pylyshyn, Zenon. 2003. *Seeing and visualizing: It's not what you think*. Cambridge: The MIT Press.
Quasimodo, Salvatore. 1984. *Complete poems*. New York: Schocken Books.
Quine, Willard Van Orman. 1969. *Ontological relativity and other essays*. New York: Columbia University Press.
———. 1981a. *Mathematical logic*. Cambridge: Harvard University Press.
———. 1981b. *Theories and things*. Cambridge: Harvard University Press.
Raatikainen, Panu. 2003. More on Putnam and Tarski. *Synthese* 135: 37–47.
Raspe, Rudolf Erich. 2005. *The adventures of baron Munchausen*. Mineola: Dover.
Rav, Yehuda. 2005. Reflections on the proliferous growth of mathematical concepts and tools: Some case histories from mathematicians' workshops. In *Mathematical reasoning and heuristics*, eds. Carlo Cellucci and Donald Gillies, 49–69. London: College Publications.
Reid, Thomas. 1827. *Essays on the powers of the human mind*. London: Tegg.
Remes, Pauliina. 2014. Plato: Interaction between the external body and the perceiver in the *Timaeus*. In *Active perception in the history of philosophy, from Plato to modern philosophy*, eds. José Filipe Silva and Mikko Yrjönsuuri, 9–30. Cham: Springer.
Rescher, Nicholas. 1993. American philosophy today. *The Review of Metaphysics* 46: 717–745.
———. 2007. *Dialectics: A classical approach to inquiry*. Frankfurt: Ontos Verlag.
Resnik, Michael David, and David Kushner. 1987. Explanation, independence and realism in mathematics. *British Journal for the Philosophy of Science* 38: 141–158.
Ribet, Kenneth A. 1990. From the Taniyama-Shimura conjecture to Fermat's last theorem. *Annales de la Faculté Mathematiques et des Sciences de Toulouse* 11: 116–139.
Rilke, Rainer Maria. 1998. *Die Gedichte*. Insel: Frankfurt am Main.
Rocke, Alan J. 2011. *Image and reality: Kekulé, Kopp, and the scientific imagination*. Chicago: The University of Chicago Press.
Rodin, Andrei. 2014. *Axiomatic method and category theory*. Berlin: Springer.
Rorty, Richard. 1980. *Philosophy and the mirror of nature*. Princeton: Princeton University Press.
———. 1991–2007. *Philosophical papers*. Cambridge: Cambridge University Press.
———. 1992. Twenty-five years after. In *The linguistic turn: Essays in philosophical method*, ed. Richard Rorty, 371–374. Chicago: The University of Chicago Press.
———. 1999. *Philosophy and social hope*. London: Penguin Books.
Rota, Gian-Carlo. 1981. Introduction. In Philip J. Davis and Reuben Hersh, *The mathematical experience*, xvii–xxix. Boston: Birkhäuser.
———. 1997. *Indiscrete thoughts*. Boston: Birkhäuser.
Rovelli, Carlo. 2014. Science is not about certainty. In *The universe*, ed. John Brockman, 214–228. New York: Harper.
Russell, Bertrand. 1920. *Introduction to mathematical philosophy*. London: Allen & Unwin.
———. 1930. *The conquest of happiness*. London: Allen & Unwin.
———. 1945. *History of Western philosophy*. New York: Simon & Schuster.
———. 1949. Answer to questionnaire. In Eliot Dole Hutchinson, *How to think creatively*, 110–112. New York: Abingdon-Cokesbury Press.
———. 1950. *Unpopular essays*. New York: Simon & Schuster.
———. 1960. Introduction. In Ernest Gellner, *Words and things*, 13–15. Boston: Beacon Press.
———. 1971. *Autobiography*. London: Allen & Unwin.
———. 1973. *Essays in analysis*. London: Allen & Unwin.
———. 1985a. Prisons. In Bertrand Russell, *Collected papers*, vol. 12, 97–109. London: Allen & Unwin.

———. 1985b. The essence of religion. In Bertrand Russell, *Collected papers*, vol. 12, 110–122. London: Allen & Unwin.
———. 1988. Theory of knowledge. In Bertrand Russell, *Collected papers*, vol. 9, 194–202. London: Routledge.
———. 1992a. *Human knowledge: Its scope and limits*. London: Routledge.
———. 1992b. *Theory of knowledge: The 1913 manuscript*. London: Routledge.
———. 1993. *Our knowledge of the external world*. London: Routledge.
———. 1994. *Mysticism and logic*. London: Routledge.
———. 1995a. *An outline of philosophy*. London: Routledge.
———. 1995b. *My philosophical development*. London: Routledge.
———. 1997. *The problems of philosophy*. Oxford: Oxford University Press.
———. 2005. *Why I am not to Christian and other essays on religion and related subjects*. London: Routledge.
———. 2010. *The principles of mathematics*. London: Routledge.
Ryle, Gilbert. 2009. *Collected papers*. London: Routledge.
Sagan, Carl. 1980. *Cosmos*. New York: Random House.
Sankey, Howard. 2008. *Scientific realism and the rationality of science*. Aldershot: Ashgate.
———. 2014. Scientific realism and basic common sense. *Kairos* 10: 11–24.
Santayana, George. 1948. *The life of reason: Reason in common sense*. New York: Scribner.
Schopenhauer, Arthur. 2010. *The world as will and representation*, vol. 1. Cambridge: Cambridge University Press.
Schwartz, Jack. 1992. The pernicious influence of mathematics on science. In *Discrete thoughts*, eds. Mark Kac, Gian-Carlo Rota, and Jack Schwartz, 19–25. Boston: Birkhäuser.
Schwartz, Stephen P. 2012. *A brief history of analytic philosophy: From Russell to Rawls*. New York: Wiley.
Searle, John Rogers. 2008. *Philosophy in a new century*. Cambridge: Cambridge University Press.
Sellars, Wilfrid. 1997. *Empiricism and the philosophy of mind*. Cambridge: Harvard University Press.
Senechal, Marjorie. 1998. The continuing silence of Bourbaki – An interview with Pierre Cartier. *The Mathematical Intelligencer* 20 (1): 22–28.
Shaffer, Dennis, Scott Krauchunas, Marianna Eddy, and Michael McBeath. 2004. How dogs navigate to catch frisbees. *Psychological Science* 15: 437–441.
Shanks, Daniel. 1993. *Solved and unsolved problems in number theory*. New York: Chelsea.
Shapiro, Stewart. 1983. Conservativeness and incompleteness. *The Journal of Philosophy* 80: 521–531.
———. 2000. *Thinking about mathematics. The philosophy of mathematics*. Oxford: Oxford University Press.
Shaw, Bernard. 2008. *Preface to Androcles and the lion*. London: Icon.
Siegfried, Tom. 2013. Rise of big data underscores need for theory. *ScienceNews*, 3 December.
Sierpinska, Anna. 1994. *Understanding in mathematics*. London: The Falmer Press.
Simonton, Dean Keith. 2004. *Creativity in science: Chance, logic, genius, and Zetigeist*. Cambridge: Cambridge University Press.
Sinclair, Nathalie. 2011. Aesthetic considerations in mathematics. *Journal of Humanistic Mathematics* 1: 2–32.
Singh, Simon. 1998. *Fermat's last theorem*. New York: Anchor Books.
Snow, Charles Percy. 1973. The classical mind. In *The physicist's conception of nature*, ed. Jagdish Mehra, 809–813. Dordrecht: Springer.
Spelke, Elizabeth. 2011. Natural number and natural geometry. In *Space, time and number in the brain*, eds. Stanislas Dehaene and Elizabeth Merrit Brannon, 287–317. London: Elsevier.
Spinoza, Benedictus. 2002. *Complete works*. Indianapolis: Hackett.
Spitz, Ellen Handler. 1994. *Museums of the mind: Magritte's labyrinth and other essays in the arts*. New Haven: Yale University Press.
Steiner, Mark. 1998. *The applicability of mathematics as a philosophical problem*. Cambridge: Harvard University Press.

Striker, Gisela. 2009. Commentary. In Aristotle, *Prior Analytics*, Book I, ed. Gisela Striker, 67–246. Oxford: Oxford University Press.
Suppes, Patrick. 1972. *Axiomatic set theory*. Mineola: Dover.
———. 1993. *Models and methods in the philosophy of science: Selected essays*. Dordrecht: Springer.
Tait, William Walker. 2001. Beyond the axioms: The question of objectivity in mathematics. *Philosophia Mathematica* 9: 21–36.
Tarski, Alfred. 1944. The semantic conception of truth and the foundations of semantics. *Philosophy and Phenomenological Research* 4: 341–376.
———. 1969. Truth and proof. *Scientific American* 220: 63–77.
———. 1983. *Logic, semantics, metamathematics*. Indianapolis: Hackett.
———. 1994. *Introduction to logic and to the methodology of deductive sciences*. Oxford: Oxford University Press.
Tegmark, Max. 2008. The mathematical universe. *Foundations of Physics* 38: 101–150.
Tennant, Neil. 1986. The withering away of formal semantics? *Mind and Language* 1: 302–318.
Thom, René. 1998. 'Modern' mathematics: An educational and philosophical error? In *New directions in the philosophy of mathematics*, ed. Thomas Tymoczko, 67–78. Princeton: Princeton University Press.
Thomas, Robert. 2014. Reflections on the objectivity of mathematics. In *From a heuristic point of view*, ed. Cesare Cozzo, and Emiliano Ippoliti, 241–256. Newcastle upon Tyne: Cambridge Scholars Publishing.
Thomson, Garrett. 2003. *On the meaning of life*. Belmont: Wadsworth.
Tichý, Pavel. 1974. On Popper's definitions of verisimilitude. *The British Journal for the Philosophy of Science* 25: 155–160.
Todd, Cain. 2008. Unmasking the truth beneath the beauty: Why the supposed aesthetic judgments made in science may not be aesthetic at all. *International Studies in the Philosophy of Science* 22: 61–79.
Tóth, László Fejes. 1964. What the bees know and what they do not know. *Bulletin of the American Mathematical Society* 70: 468–481.
Turing, Alan Mathison. 2001. *Collected works: Mathematical logic*. Amsterdam: North-Holland.
Vaihinger, Hans. 2008. *The philosophy of 'as if'*. London: Routledge.
van Fraassen, Bas. 1980. *The scientific image*. Oxford: Oxford University Press.
———. 1985. Empiricism in the philosophy of science. In *Images of science: Essays on realism and empiricism*, eds. Paul Churchland and Clifford Hooker, 245–308. Chicago: The University of Chicago Press.
———. 1989. *Laws and symmetry*. Oxford: Oxford University Press.
———. 2000. The false hopes of traditional epistemology. *Philosophy and Phenomenological Research* 60: 253–280.
———. 2008. *Scientific representation: Paradoxes of perspective*. Oxford: Oxford University Press.
van Gerwen, Rob. 2011. Mathematical beauty and perceptual presence. *Philosophical Investigations* 34: 249–267.
Visell, Yvon. 2009. Tactile sensory substitution: Models for enaction in HCI. *Interacting with Computers* 21: 38–53.
von Helmholtz, Hermann. 1867. *Handbuch der Physiologischen Optik*. Leipzig: Voss.
———. 1995. *Science and culture: Popular and philosophical essays*. Chicago: The University of Chicago Press.
von Neumann, John. 1955. *Mathematical foundations of quantum mechanics*. Princeton: Princeton University Press.
———. 1961. The mathematician. In John von Neumann, *Collected works*, vol. 1, 1–9. Oxford: Pergamon Press.
———. 2003. Letter to Gödel, 29 November 1930. In Kurt Gödel, *Collected Works*, vol. 5, 339–340. Oxford: Oxford University Press.

Wang, Hao. 1996. *A logical journey: From Gödel to philosophy*. Cambridge: The MIT Press.
Weinberg, Steven. 1986. Lecture on the applicability of mathematics. *Notices of the American Mathematical Society* 33: 725–728.
———. 1993. *Dreams of a final theory: The search for the fundamental laws of nature*. New York: Vintage.
Wells, David. 1990. Are these the most beautiful? *The Mathematical Intelligencer* 12 (3): 37–41.
Weyl, Hermann. 1949. *Philosophy of mathematics and natural science*. Princeton: Princeton University Press.
———. 1950. *The theory of groups and quantum mechanics*. Mineola: Dover.
———. 1985. Axiomatic versus constructive procedures in mathematics. *The Mathematical Intelligencer* 7 (4): 10–17, 38.
———. 1998. On the new foundational crisis of mathematics. In *From Brouwer to Hilbert: The debate on the foundations of mathematics in the 1920s*, ed. Paolo Mancosu, 86–118. Oxford: Oxford University Press.
———. 2009. *Mind and nature: Selected writings on philosophy, mathematics, and physics*. Princeton: Princeton University Press.
Whewell, William. 1847. *The philosophy of the inductive sciences, founded upon their history*. London: Parker.
———. 1858. *Novum organon renovatum*. London: Parker.
Whitehead, Alfred North, and Bertrand Russell. 1925–1927. *Principia mathematica*. Cambridge: Cambridge University Press.
Wiedijk, Freek. 2008. Formal proof – Getting started. *Notices of the American Mathematical Society* 55: 1408–1414.
Wigner, Eugene Paul. 1960. The unreasonable effectiveness of mathematics in the natural sciences. *Communications on Pure and Applied Mathematics* 13: 1–14.
Williams, Bernard. 1985. *Ethics and the limits of philosophy*. Cambridge: Harvard University Press.
———. 2002. *Truth & truthfulness: An essay in genealogy*. Princeton: Princeton University Press.
———. 2006. *Philosophy as a humanistic discipline*. Princeton: Princeton University Press.
———. 2014. *Essays and reviews 1959–2002*. Princeton: Princeton University Press.
Williams, Michael. 1999. *Groundless belief*. Princeton: Princeton University Press.
———. 2001. *Problems of knowledge: A critical introduction to epistemology*. Oxford: Oxford University Press.
———. 2004. Is knowledge a natural phenomenon? In *The externalist challenge*, ed. Richard Schantz, 193–209. Berlin: de Gruyter.
Williamson, Timothy. 2007. *The philosophy of philosophy*. Oxford: Blackwell.
Wilson, Trevor. 2005. A continuous movement version of the Banach-Tarski paradox: A solution to De Groot's problem. *Journal of Symbolic Logic* 70: 946–952.
Wittgenstein, Ludwig. 1932–1933. *Muss sich denn nicht...* . In *Wittgenstein's Nachlass*, TS 219.
———. 1958. *Philosophical investigations*. Oxford: Blackwell.
———. 1969a. *The blue and brown books*. Oxford: Blackwell.
———. 1969b. *On certainty*. Oxford: Blackwell.
———. 1974. *Philosophical grammar*. Oxford: Blackwell.
———. 1975. *Philosophical remarks*. Oxford: Blackwell.
———. 1976. *Lectures on the foundations of mathematics, Cambridge 1939*. Brighton: Harvester Press.
———. 1978. *Remarks on the foundations of mathematics*. Oxford: Blackwell.
———. 1979. *Notebooks 1914–1916*. Oxford: Blackwell.
———. 1998. *Culture and value: A selection from the posthumous remains*. Oxford: Blackwell.
———. 2002. *Tractatus Logico-Philosophicus*. London: Routledge.
———. 2005. *The big typescript TS 213*. Oxford: Blackwell.
Wolpert, Lewis. 1992. *The unnatural nature of science*. Cambridge: Harvard University Press.
Worrall, John, and Elie Zahar. 1976. Editor's note. In Imre Lakatos, *Proofs and refutations: The logic of mathematical discovery*, 56–57. Cambridge: Cambridge University Press.

Zangwill, Nick. 1998. Aesthetic/sensory dependence. *British Journal of Aesthetics* 38: 66–81.

Zeki, Semir. 2014. Neurobiology and the humanities. *Neuron* 84: 12–14.

Zeki, Semir, John Romaya, Dionigi Benincasa, and Michael Atiyah. 2014. The experience of mathematical beauty and its neural correlates. *Frontiers in Human Neuroscience* 8: 1–12.

Zermelo, Ernst. 1967. A new proof of the possibility of a well-ordering. In *From Frege to Gödel: A source book in mathematical logic, 1879–1931*, ed. Jean van Heijenoort, 183–198. Cambridge: Harvard University Press.

Zilsel, Edgar. 1942. The genesis of the concept of physical law. *The Philosophical Review* 51: 245–279.

Zimecki, Michał. 2006. The lunar cycle: Effects on human and animal behavior and physiology. *Postępy Higieny i Medycyny Doświadczalnej* 60: 1–7.

Zukav, Gary. 1980. *The dancing Wu Li Masters: An overview of the new physics*. New York: Bantam Books.

Index

© Springer International Publishing AG 2017

C. Cellucci, *Rethinking Knowledge*, European Studies in Philosophy of Science 4,
DOI 10.1007/978-3-319-53237-0

FSC
www.fsc.org
MIX
Papier | Fördert
gute Waldnutzung
FSC® C083411

Kolibri 360 GmbH
Ferdinand-Jühlke-Straße 7
99095 Erfurt, Deutschland
produktsicherheit@kolibri360.de